Studies in Consciousness / Russell Targ Editions

Some of the twentieth century's best texts on the scientific study of consciousness are out of print, hard to find, and unknown to most readers; yet they are still of great importance. Their insights into human consciousness and its dynamics are still valuable and vital. Hampton Roads Publishing Company—in partnership with physicist and consciousness research pioneer Russell Targ—is proud to bring some of these texts back into print, introducing classics in the fields of science and consciousness studies to a new generation of readers. Upcoming titles in the *Studies in Consciousness* series will cover such perennially exciting topics as telepathy, astral projection, the after-death survival of consciousness, psychic abilities, long-distance hypnosis, and more.

BOOKS IN THIS SERIES

An Experiment with Time by J. W. Dunne

Mental Radio by Upton Sinclair

Human Personality and Its Survival of Bodily Death by F. W. H. Myers

Mind to Mind by René Warcollier

Experiments in Mental Suggestion by L. L. Vasiliev

Mind at Large edited by Charles T. Tart, Harold E. Puthoff, and Russell Targ

Dream Telepathy by Montague Ullman, M.D., and Stanley Krippner, Ph.D., with Alan Vaughan

Distant Mental Influence by William Braud, Ph.D.

Thoughts Through Space by Sir Hubert Wilkins and Harold M. Sherman

RUSSELL TARG EDITIONS

UFOs and the National Security State by Richard M. Dolan

The Heart of the Internet by Jacques Vallee, Ph.D.

STUDIES IN  CONSCIOUSNESS

R u s s e l l   T a r g   E d i t i o n s

Distant Mental Influence

Its Contributions to Science, Healing, and Human Interactions

William Braud, Ph.D.

HAMPTON ROADS
PUBLISHING COMPANY, INC.
for the evolving human spirit

Hampton Roads Publishing Company, Inc.
1125 Stoney Ridge Road
Charlottesville, VA 22902

434-296-2772
fax: 434-296-5096
e-mail: hrpc@hrpub.com
www.hrpub.com

If you are unable to order this book from your local
bookseller, you may order directly from the publisher.
Call 1-800-766-8009, toll-free.

Library of Congress Cataloging-in-Publication Data

Braud, William.
  Distant mental influence : its contributions to science, healing, and human
interactions / William Braud.
     p. ; cm.
Includes bibliographical references and index.
  ISBN 1-57174-354-5 (alk. paper)
  1. Parapsychology and science. 2. Parapsychology and medicine. 3.
Interpersonal relations--Psychic aspects.
  [DNLM: 1. Imagery (Psychotherapy) 2. Mental Healing. 3. Mind-Body
Relations (Metaphysics) WM 420.5.I3 B825d 2003] I. Title.
  BF1045.S33B74 2003
  133.8--dc22
                                    2003019817

ISBN 1-57174-354-5
10 9 8 7 6 5 4 3 2 1
Printed on acid-free paper in Canada

Permissions

The Rumi poem in the introduction is from *Open Secret: Versions of Rumi by John Moyne and Coleman Barks.* © 1984 by John Moyne and Coleman Barks. Reprinted by arrangement with Shambhala Publications, Inc., Boston, www.shambhala.com.

Contents of chapter 1 originally were published in Braud, W. G., and Schlitz, M. J. (1989). A methodology for the objective study of transpersonal imagery. *Journal of Scientific Exploration,* 3, 43–63. Copyright © 1989 by the *Journal of Scientific Exploration.* Used with permission.

Contents of chapter 2 originally were published in Braud, W. G. and Schlitz, M. J. (1983). Psychokinetic influence on electrodermal activity. *Journal of Parapsychology,* 47, 95–119. Copyright © 1983 by the Parapsychology Press. Used with permission.

Contents of chapter 3 originally were published in Braud, W. G. (1990). Distant mental influence of rate of hemolysis of human red blood cells. *Journal of the American Society for Psychical Research,* 84, 1–24. Copyright © 1990 by the *Journal of the American Society for Psychical Research.* Used with permission.

Contents of chapter 4 originally were published in Braud, W. G., and Schlitz, M. J. (1991). Consciousness interactions with remote biological systems: Anomalous intentionality effects. *Subtle Energies and Energy Medicine: An Interdisciplinary Journal of Energetic and Informational Interactions.* 2(1), 1–46. The contents of this article are Copyright © 1991 by the International Society for the Study of Subtle Energy and Energy Medicine (ISSSEEM), and the material is reprinted with permission. Contact information for ISSSEEM: 11005 Ralston Road, Suite 100D, Arvada, CO 80004; phone (303) 425-4625; fax (303) 425-4685; e-mail issseem@compuserve.com; website http://www.issseem.org

Contents of chapter 5 originally were published in Braud, W. G. (1992). Remote mental influence of electrodermal activity. *Journal of Indian Psychology,* 10(1), 1–10. Copyright © 1992 by the *Journal of Indian Psychology.* Used with permission.

Contents of chapter 6 originally were published in Braud, W. G. (1993). On the use of living target systems in distant mental influence research. In L. Coly and J. D. S. McMahon (Eds.), *Psi Research Methodology: A Re-examination* (pp. 149–181). New York: Parapsychology Foundation. The Parapsychology Foundation is the copyright holder of *Psi Research Methodology: A Re-examination,* Proceedings of an International Conference Held in Chapel Hill, North Carolina, October 29–30, 1988. Edited by Lisette Coly and Joanne D. S. McMahon, published in 1993. Used with permission.

This book is dedicated to the memories of
Frederic William Henry Myers (1843–1901)
William James (1842–1910)
Leonid Leonidovich Vasiliev (1891–1966)
and
Eric Douglas Dean (1916–2001)
whose thoughts, works, and writings
provided inspiration for the research
described in this book.

Table of Contents

FOREWORD (BY LARRY DOSSEY, M.D.) xi

ACKNOWLEDGMENTS xv

INTRODUCTION xvii

1: TRANSPERSONAL IMAGERY EFFECTS:
   INFLUENCING A DISTANT PERSON'S BODILY
   ACTIVITY USING MENTAL IMAGERY 1

2: CALMING OTHER PERSONS AT A DISTANCE 24

3: MENTALLY PROTECTING HUMAN
   RED BLOOD CELLS AT A DISTANCE 46

4: MENTAL INTERACTIONS WITH
   REMOTE BIOLOGICAL SYSTEMS 71

5: DISTANT MENTAL INFLUENCE OF PHYSIOLOGICAL
   ACTIVITY: NEW EXPERIMENTS AND
   THEIR HISTORICAL ANTECEDENTS 109

6: ON THE USE OF LIVING TARGET SYSTEMS IN
   DISTANT MENTAL INFLUENCE RESEARCH .118

7: REACTIONS TO AN UNSEEN GAZE (REMOTE
   ATTENTION): AUTONOMIC STARING DETECTION 150

8: ADDITIONAL STUDIES OF BODILY
DETECTION OF REMOTE STARING 166

9: EMPIRICAL STUDIES OF PRAYER, DISTANT
HEALING, AND REMOTE MENTAL INFLUENCE 183

10: HELPING OTHERS CONCENTRATE USING
DISTANT MENTAL INFLUENCE 196

11: DISTANT MENTAL INFLUENCE AND HEALING:
ASSESSING THE EVIDENCE 209

12: HEALTH IMPLICATIONS OF "BACKWARD-IN-TIME" DIRECT
MENTAL INFLUENCES 233

INDEX 257

Foreword

The most mysterious phenomenon in the universe is human consciousness. Despite our best efforts to understand it, its essential nature remains an enigma.

In recent decades, we have largely equated consciousness with the workings of the brain. Thus, on every hand, researchers assure us that in some yet-to-be-determined way the brain "makes" consciousness, much like the liver secretes bile or the pancreas produces insulin. Yet these crude images are appallingly inaccurate because in the whole of neuroscience there is no evidence whatever that neural tissue is capable of producing what we experience as conscious awareness.

The problems in equating consciousness with the material brain are becoming increasingly obvious to a growing number of researchers and philosophers. For example, John Searle, one of the most distinguished philosophers in the field of consciousness, has said: "At our present state of the investigation of consciousness, we *don't know* how it works and we need to try all kinds of different ideas."[1] Similarly, philosopher Jerry A. Fodor has observed: "Nobody has the slightest idea how anything material could be conscious. Nobody even knows what it would be like to have the slightest idea about how anything material could be conscious. So much for the philosophy of consciousness."[2] And Sir John Maddox, the former editor of *Nature* and one of science's elder statesmen, has soberly stated: "What consciousness consists of . . . is . . . a puzzle. Despite the marvelous success of neuroscience in the past century . . . , we seem as far away from understanding . . . as we were a century ago."[3]

Some scientists are convinced that consciousness must be considered fundamental in its own right. An example is mathematician and philosopher David J. Chalmers, who has proposed that consciousness be considered fundamental in the universe, reducible to nothing more basic, perhaps on a par with matter and energy.[4,5]

Although we cannot be sure which picture of consciousness will eventually emerge, one thing is certain. When the history of the science of consciousness is written, William Braud will occupy a place of high honor. For three decades his research has emphatically challenged the materialistic fashions that have dominated consciousness research. In terms of their conceptual importance and methodological rigor, his experiments are of Nobel caliber.

Braud's studies demonstrate compellingly that distant mental influence is an empirical fact and that these effects are relevant to humans. His experiments are anchored in the world of biology, employing subjects such as whole humans, human cells, and other living target systems. All researchers in the field of consciousness studies must eventually come to terms with his seminal work.

Braud has shown that conscious volition is correlated with changes in distant living systems that are remote not only in space, but also in time. His work therefore reveals that human consciousness is nonlocal or infinite in *both* space and time—consciousness as unconfinable to the brain, body, or the present moment.

Those who encounter Braud's work for the first time may consider it implausible. How can the mind cause changes to occur at a distance? Yet we should bear in mind that, for almost the entire history of the human race, this possibility was considered "the way things are." We should also recognize that certain developments within contemporary science permit the distant actions of consciousness. One such area is quantum physics, which describes nonlocal events that bear a striking resemblance to those demonstrated by Braud. Thus the eminent physicist Henry P. Stapp of University of California-Berkeley stated: "[T]he new physics presents prima facie evidence that our human thoughts are linked to nature by nonlocal connections: What a person chooses to do in one region seems immediately to effect what is true elsewhere in the universe. This nonlocal aspect can be understood by conceiving the universe to be not a collection of tiny bits of matter, but rather a growing compendium of 'bits of information.' . . . And, I believe that most quantum physicists will also agree that our conscious thoughts ought eventually to be understood within science and that when properly understood, our thoughts will be seen to DO something: They will be efficacious."[6]

Nonetheless, some individuals will criticize Braud's work as out of step with mainstream science, insisting that there are no hypotheses that are compatible with distant mental influence. Nothing could be further from the truth. Science is seething with hypotheses, in addition to Stapp's view, that fully permit the nonlocal, distant operations of consciousness, many of which have been elaborated in elegant mathematical detail.[7,8,9,10]

One of the hallmarks of Braud's work is, as mentioned, its relevance to human welfare. His experiments serve as the foundation or "bench science" for studies in distant healing and intercessory prayer that have recently emerged from clinical medicine.[11,12,13,14,15] Five positive systematic or meta-analytic reviews of these studies recently have been published.[16,17,18,19,20] Of the nation's 125 medical schools, 80 have developed courses exploring this work.[21]

Currently, the most glamorous areas of medical research are fields such as genome decoding, DNA manipulation, stem cell research, organ transplantation, and the development of new wonder drugs. While important, these areas pale in significance when compared to the research performed by Braud. His explorations touch the Big Questions that have occupied humans since prehistory—the origins and destiny of human consciousness. If an element of consciousness is indeed nonlocal, as Braud's work implies, this raises the possibility that some factor of the mind is eternal and immortal. It is difficult to conceive how any line of contemporary research could be more important than one that involves the question of survival of bodily death, as Braud's work does.

There is a tendency among contemporary scientists to select a field for research that is popular and commercially rewarding, and where one's reputation can be cultivated without excessive risk.[22] Yet many of the greatest breakthroughs in the history of science have occurred by risk takers working on the margins of science, in areas that have been considered outrageous. This requires great fortitude, which William Braud has in abundance. As one who has been inspired by his vision and courage, I bow deeply.

Larry Dossey, M.D.
Santa Fe, NM

[Larry Dossey is the Executive Editor of *Alternative Therapies in Health and Medicine* and the author of nine books on consciousness, spirituality, and health, most recently *Healing Beyond the Body*.]

References

1. Searle, J. (1995). Quotation on front cover. *Journal of Consciousness Studies,* 2(1).
2. Fodor, J. (1992, July). The big idea. *The [New York] Times Literary Supplement,* July 3, 20.
3. Maddox, J. (1999). The unexpected science to come. *Scientific American,* 281(6), 62–67.
4. Chalmers, D. J. (1995). The puzzle of conscious experience. *Scientific American,* 273(6), 80–86.
5. Chalmers, D. J. (1996), *The conscious mind: In search of a fundamental theory.* New York: Oxford University Press.
6. Stapp, H. (2001, February). Harnessing science and religion: Implications of the new scientific conception of human beings. *Research News and Opportunities in Science and Religion,* 1(6), 8.
7. Clarke, C. J. S. (1995). The nonlocality of mind. *Journal of Consciousness Studies,* 2(3), 231–240.
8. Jahn, R. G., and Dunne, B. J. (2001). A modular model of mind/matter manifestations (M5). *Journal of Scientific Exploration,* 15(3), 299–329.
9. Radin, D. (1997). Toward an adequate theory. In D. Radin, *The conscious universe* (pp. 278–287). San Francisco: HarperSanFrancisco.

10. Rauscher, E. A, and Targ, R. (2001). The speed of thought: Investigation of a complex space-time metric to describe psychic phenomena. *Journal of Scientific Exploration,* 15(3), 331–354.

11. Sicher, F., Targ, E., Moore, D., and Smith, H. S. (1998). A randomized double-blind study of the effect of distant healing in a population with advanced AIDS: A report of a small-scale study. *Western Journal of Medicine,* 169(6), 356–363.

12. Cha, K. Y., Wirth, D. P., and Lobo, R. (2001). Does prayer influence the success of in vitro fertilization-embryo transfer? Report of a masked, randomized trial. *Journal of Reproductive Medicine,* 46(9), 781–787.

13. Harris, W., Gowda, M., Kolb J. W., Strychacz, C. P., Vacek, J. L, Jones, P. G., Forker, A., O'Keefe, J. H., and McCallister, B. D. (1999). A randomized, controlled trial of the effects of remote, intercessory prayer on outcomes in patients admitted to the coronary care unit. *Archives of Internal Medicine,* 159(19), 2273–2278.

14. Byrd, R. (1988). Positive therapeutic effects of intercessory prayer in a coronary care unit population. *Southern Medical Journal,* 81(7), 826–829.

15. Krucoff, M. W., Crater, S. W., Green, C. L., Maas, A. C., Seskevich, J. E., Lane, J. D., Loeffler, K. A., Morris, K., Bashore, T. M., and Koenig H. G. (2001). Integrative noetic therapies as adjuncts to percutaneous intervention during unstable coronary syndromes: Monitoring and actualization of noetic training (MANTRA) feasibility pilot. *American Heart Journal,* 142(5), 760–767.

16. Astin, J. E., Harkness, E., and Ernst, E. (2000). The efficacy of "distant healing": A systematic review of randomized trials. *Annals of Internal Medicine,* 132, 903–910.

17. Abbot, N. C. (2000). Healing as a therapy for human disease: A systematic review. *Journal of Alternative and Complementary Medicine,* 6(2), 159–169.

18. Braud, W., and Schlitz, M. (1989). A methodology for the objective study of transpersonal imagery, *Journal of Scientific Exploration,* 3(1), 43–63.

19. Jonas, W. B. (2001). The middle way: Realistic randomized controlled trials for the evaluation of spiritual healing. *The Journal of Alternative and Complementary Medicine,* 7(1), 5–7.

20. Schlitz, M., and Braud, W. (1997). Distant intentionality and healing: Assessing the evidence. *Alternative Therapies,* 3(6), 62–73.

21. Better times for spirituality and healing in medicine. (2001, February). *Research News and Opportunities in Science and Religion,* 1(6), 12. [no author cited]

22. Ahuja, A. (2001, February 26). Where is the next Einstein? *The Times of London.* http://www.thetimes.co.uk/article/0..74–90352.00.html

Acknowledgments

The research reported in this book was carried out over a period of approximately 25 years. Various forms of support for this research were provided by Esalen Institute, Institute of Transpersonal Psychology, Mind Science Foundation, Parapsychology Foundation, SRI International, and University of Edinburgh. Colleagues and others who helped the author at different stages of this work, and in various ways, include Jeanne Achterberg, Sperry Andrews, Rick Berger, Edward Brame, Robert Brier, Gary Davis, Steve Dennis, Dennis DiBart, Larry Dossey, Alyce Green, Valerie Guerra, George Hansen, Charles Honorton, Scott Hubbard, David Hurt, Julian Isaacs, Jan Jackson, Stanley Krippner, Kay Mangus, Edwin May, Katherine McNeill, Michael Murphy, Bruce Pomeranz, Dean Radin, Robert Rosenthal, Marilyn Schlitz, Helmut Schmidt, Donna Shafer, Jerry Solfvin, Russell Targ, Charles Tart, Lynn Trainor, Jessica Utts, and Robert Wood. I am thankful to Russell Targ for suggesting that this book be prepared and included in the Studies in Consciousness series. I am grateful to my wife, Winona Schroeter, for her assistance and support during all stages of these endeavors.

Introduction

In an article entitled, "Final Impressions of a Psychical Researcher," published in October, 1909, less than a year before his death, the American philosopher and psychologist William James included the following:

> We with our lives are like islands in the sea, or like trees in the forest. The maple and the pine may whisper to each other with their leaves, and Conanicut and Newport hear each other's foghorns. But the trees also commingle their roots in the darkness underground, and the islands also hang together through the ocean's bottom. Just so there is a continuum of cosmic consciousness, against which our individuality builds but accidental fences, and into which our several minds plunge as into a mother-sea or reservoir. Our "normal" consciousness is circumscribed for adaptation to our external earthly environment, but the fence is weak in spots, and fitful influences from beyond leak in, showing the otherwise unverifiable common connection.[1]

A compelling poetical and metaphorical expression of this same notion is found, much earlier, in the writings of the thirteenth-century mystical poet, Jelaluddin Rumi, who wrote:

> I've heard it said there's a window that opens
> from one mind to another,
> But if there's no wall, there's no need
> for fitting the window, or the latch.[2]

The papers included in this book provide current empirical confirmations and elaborations of this widespread and enduring idea of nonlocal interconnectedness. Our Western scientific worldview maintains that we are isolated individuals who communicate and interact only locally, by exchanging information through our recognized senses and technological tools. However, carefully collected, but typically ignored, evidence suggests that James and Rumi were right—that beneath surface appearances, we are intimately and profoundly interconnected in our consciousness and our identity. This evidence comes from areas of psychology, parapsychology, consciousness research, and transpersonal studies. The purpose of this book is to present, interpret, and elaborate some of this evidence; to explore its implications for science; and to address its possible practical applications in our everyday lives and relationships.

Concisely stated, the evidence compiled in this volume indicates that, under certain conditions, it is possible to know and to influence the thoughts, images, feelings, behaviors, and physiological and physical activities of other persons and living organisms—even when the influencer and the influenced are separated by great distances in space and time, beyond the reach of the conventional senses. Because the usual modes of knowing and influence are eliminated in these studies, their success reveals modes of human interaction and interconnection beyond those currently recognized in the conventional natural, behavioral, and social sciences. Besides indicating areas of incompleteness and misapprehensions about such phenomena that exist in current scientific theories, these distant mental influence findings have important implications for our fuller understanding of consciousness, health and wellness, our typically untapped human potentials, and the spiritual aspects of our lives.

This book includes 12 previously published papers—each presented as a chapter—that provide empirical evidence for these distant mental influences. Besides providing evidence for the existence of these nonlocal influence effects, the papers describe what we have learned about the nature of these influences and the factors that make them more or less likely to occur. In this introduction, I provide background and supportive information that will help contextualize and elaborate the content of the 12 chapters. My intention, as well, is to bring the research up to date, indicate how the reported studies have been replicated and extended (by myself and by others), and discuss the meaning, significance, and implications of this work.

The Background of This Research Program: "The Context of Discovery"

In 1938, the philosopher of science, Hans Reichenbach, distinguished two kinds of activities of scientists, which he called the contexts of "discovery" and "justification."[3] The latter, in which the scientist formally publishes results for peers—presenting, justifying, and defending them—is typically the only form of scientific activity that we hear about. The former—in which the work actually is inspired, conceived, planned, and conducted, and in which discoveries actually are made—is rarely, if ever, mentioned by scientists themselves or by those who discuss their work. In neglecting the context of discovery, one loses sight of the most interesting, creative, and human aspects of the scientific enterprise. The 12 chapters that follow originally were presented in the mode of the context of justification. In order to keep the two modes in balance, in this preliminary section, I'll share aspects of the context of discovery in which this distant mental influence work originated and was carried out.

This research can be situated within the stream of psychical and parapsychological research. Such research concerns itself with ways in which we are able, under certain conditions, to obtain accurate knowledge about and influence events beyond the reach of our conventionally recognized senses, even at great distances through space and time. These ways of knowing and influencing events at a distance take the forms of telepathy (mind-to-mind communication), clairvoyance (accurate knowledge of remote events), precognition (accurate knowledge of future events), and psychokinesis (mind-over-matter effects or, better, direct intentional influences upon the physical world). These psychic experiences and events are collectively known as *psi*. Although these and similar experiences have occurred throughout history, in virtually all eras and cultures, they remain poorly understood even by those who experience them, and they have been greatly neglected—and their very existence often denied—by science.

My own interest in psychical events did not arise from any early or dramatic experiences of these things. Rather, this interest arose gradually as a form of intellectual curiosity. Temperamentally, I have always been drawn to unusual events, to unexpected exceptions—and psychical or paranormal events constitute one of the largest classes of exceptions.

I was not always favorably disposed toward these phenomena. There was even a period, during my graduate training in a very behavioristic university psychology department, when I was hostile to such experiences and lectured to my students about the problematical status and probable nonexistence of such things as telepathy, precognition, and so forth. The systems and theories of physics and psychology in which I had been trained had no room for such processes or experiences. Once I completed my graduate studies and began my own university teaching career, however, my horizons began to expand, and I found myself wondering about the nonordinary forms of consciousness, out-of-body experiences, mystical experiences, and paranormal experiences that some of my students were mentioning to me and that I was reading about. To learn more about such experiences, I started reading the professional journals devoted to psychical research and parapsychology, and I began attending professional conferences in these areas. The more I learned, the more I became intrigued by these phenomena.

A graduate student and I designed a study to learn whether we could find evidence for telepathy during experimental sessions conducted while another person—the "percipient" in these experiments—was hypnotized. This early study, in which I played the role of "agent" or "sender," yielded very accurate and dramatic results—even when the percipient and I were stationed on separate floors of the psychology building, in separate buildings, or in different parts of the city. For example, when I as sender was thinking about slow, happy music and about a piano, the distant, hypnotized percipient mentioned becoming aware of sheet music and a piano. When I put a glass of ice water to my neck, the percipient mentioned "hands around a glass . . . shaking because it's cold." When I watched a small spinning merry-go-round device, the percipient mentioned "merry-go-round, spinning." When I held my left hand over a candle flame, the percipient—many miles away—said, at that very moment, "flames . . . little flame which spread out . . . felt heat." These are only a few examples of the kinds of accurate, time-locked correspondences that we observed, and these are representative of what happened in many of the sessions. The extremely accurate results of this study convinced me that it was possible for me to conduct fruitful studies in this area, and the findings led, ultimately, to a long series of research projects designed to uncover some of the factors that might facilitate or impede psychic functioning in the laboratory.

Because I happened to have a keen interest, at the time, in progressive muscular relaxation (as a technique for fostering physical and psychological health and well-being), I immediately wondered whether the strong relaxation component of hypnosis might have been an important ingredient in the success of these distant telepathy sessions. A series of formal experiments indicated that this indeed appeared to be the case. Percipients who relaxed their bodies deeply through the use of progressive relaxation procedures performed more accurately in telepathy sessions than did nonrelaxed percipients. We also found that degree of psychic accuracy was related to degree of muscular relaxation, as measured though electromyographic recordings (in which muscle tension was monitored electronically).

We later expanded these studies to include a variety of techniques—other than or in addition to muscular relaxation—for helping percipients "relax" and quiet themselves at as many levels as possible. These techniques included using autogenic training exercises (self-suggestions for calmness and quietude, using phrases for physiological self-regulation) to reduce emotional and autonomic nervous system activity, sensory restriction techniques to reduce sensory distractions and induce perceptual quietude, exercises to reduce excessive logical and analytical thinking, meditationlike exercises to reduce excessive cognitive and mental activity in general, and exercises that helped reduce excessive effortful striving for success in these psychic tasks. The use of these various quieting techniques was indeed associated with increased accuracy in telepathy and clairvoyance tasks. These findings favored a view that, ordinarily, our psychic functioning might be inhibited by excessive bodily, emotional, and mental distractions—just as weak signals might be masked by too much noise. Reducing the interfering, distracting noise can allow us to detect formerly obscured signals. This view became known as the *noise-reduction* hypothesis or model of psychic optimization.[4]

As my colleagues and I were conducting these noise-reduction experiments, it became clear to us that, in addition to reducing noise, the techniques we were using were also freeing our percipients from various structuring or constraining forces and habits that ordinarily made it difficult for them to freely change or shift their minds and bodies in ways that could allow psychic functioning to occur. Lots of organized activities in our various bodily and mental channels can structure our thoughts, feelings, images, and bodily states in ways that tend to maintain or "freeze" them in their current forms and prevent them from changing easily and efficiently.

For example, at this moment your mental activity is being held and occupied (structured and constrained) by the words on this page. It is difficult for your thoughts and other subjective activities to shift when they are being structured so well. If you were to close your eyes right now, and not think about what you were reading or about anything else in particular, it would be much easier for your thoughts, images, and feelings to be swerved or influenced by subtle psychic inputs that ordinarily might not be effective or even noticed.

All of the techniques mentioned that reduce noise in their various "channels" also reduce distractions, structures, and constraints in those same channels. When those structures are dismantled or melt away, the raw materials in those channels—ingredients, if you will, of thoughts, feelings, sensations, and images—have greater opportunities to be reorganized or restructured in ways that might allow them to carry useful psychic information.

The ideas in the previous paragraph led to a *lability/inertia* model of psychic functioning. A system possesses *lability* when it can freely change; lability is another name for *free variability*. A thin stream of smoke, curling from the tip of a stick of incense, is quite labile—easily swayed or influenced by the slightest breeze. The mind of someone sitting before a fireplace, idling, watching the dancing flames, and thinking of nothing in particular, is a labile mind—one that is readily swayed by the subtlest wisp of thought or the subtlest suggestion. On the other hand, a computer marching its way through a predetermined program, and a mind that is deeply engaged in some attention-consuming mental task are not easily swerved from their very structured activities. Rather than being labile, such highly occupied computers and minds are characterized by *inertia*—they have strong tendencies to continue what they presently are doing and not doing; they are more resistant to change.

Another set of experiments indicated that systems rich in lability tended to yield greater psychic effects than those that were more inert. These findings applied both for percipients in receptive forms of psi—such as telepathy, clairvoyance, and precognition—and for the target systems for active forms of psi—i.e., the systems influenced through psychokinesis. A pattern was beginning to emerge: For telepathy, clairvoyance, and precognition, results seemed most accurate and successful if the percipient was characterized by a high degree of lability and if the agent (or target event) was characterized by a high degree of inertia. The opposite conditions—high inertia in the human influencer but high lability in the physical target system that one was attempting to influence—seemed to hold for the success of psychokinesis.[5]

While we were conducting these noise-reduction and lability studies, Helmut Schmidt, a theoretical physicist in our laboratory, was conducting a series of fascinating experiments involving psychokinesis effects using electronic random event generators as his to-be-influenced target systems. Psychokinesis work began in the 1940s, and the early psychokinesis (PK) studies employed mechanically random systems—such as tossed coins and bouncing dice—as the target systems that persons attempted to influence, mentally and at a distance. In the early 1970s, Schmidt provided the important technical and methodological innovation of introducing the electronic random event generator (REG) as the physical target system of choice for PK experiments.

These REGs are similar to electronic coin tossers or dice throwers. Rather than mechanical randomness, however, the devices make use of truly random physical processes such as the decay of radioactive materials and thermal (heat) noise inside electronic components. Radioactive emission is one of the most random processes known to contemporary science: The emission of individual particles from radioactive sources can be neither predicted nor controlled by any process known to science. Therefore, any devices driven by radioactive decay possess a high degree of indeterminateness or randomness. These devices—Schmidt's REGs—are excellent, sensitive targets for PK experimentation, and hundreds of REG-PK experiments have yielded significant and replicable PK results.[6]

As Schmidt was conducting his PK experiments with these inanimate (REG) target systems, I began to wonder if PK experiments with *living* systems might also be especially effective targets for PK research. Like the REGs, many animate systems also appeared to be characterized by a great amount of lability or free variability. Consider the many thoughts, images, and feelings that flit and course freely through our heads and hearts. The variability of our mental contents and processes are so well known that the restless activity of the mind has been likened to that of a drunken monkey (in yogic traditions) or to the flowing of a stream (in William James's famous metaphor of the "stream of consciousness"). Many of our physiological processes and movements seem to have indeterminate, freely variable aspects; and it has been suggested that random, quantum-like events take place in the tiny synapses (spaces) between the neurons of our nervous systems.

It seemed that the spontaneous activities of living systems might provide ideal targets for psychokinetic influences. In fact, there already were indications in the parapsychological literature that this might indeed be the case. Successful PK influences had already been noted in a variety of biological systems including bacteria, fungus colonies, cells, microscopic organisms, plants, animals, and certain physiological reactions of humans. It seemed a natural next step to repeat Schmidt's PK experiments with labile living target systems.

Early Distant Mental Influence Experiments with Living Target Systems

A number of considerations converged to determine my choices of living target systems and the methods used to explore their susceptibility to distant mental influence. The animate target systems would have to be convenient and relatively easy to measure. Because of my growing interest in the possible relevance of psychokinesis to some forms of unorthodox healing, it would be useful if the chosen targets could have some relationship to biological processes relevant to physical and psychological health and well-being. I was familiar with many of the bodily processes being explored in the contexts of biofeedback and psychophysiological self-regulation research and practical applications, so the activities typically monitored in biofeedback and self-regulation studies and clinical work became possible target systems. These systems included brain waves, muscle activity, heart rate, blood pressure, and electrodermal activity (the changing electrical activity of the skin).

Years earlier, I had read and been greatly impressed by work on distant mental suggestion that had been carried out in the 1920s and 1930s, in Russia, by Leonid Vasiliev and his coworkers. Vasiliev had published an account of his work in 1962, in Russian, in a book entitled *Experiments in Mental Suggestion*, and an English translation of this book appeared in 1963.* In a series of careful experiments—quite sophisticated for that time, and even by today's standards—Vasiliev's research team was able to induce motor acts, visual images and sensations, sleeping and awakening, and physiological reactions (breathing changes, changes in electrodermal activity) in persons stationed at remote locations and shielded from all conventional interactions. In these studies, the influencers and influencees were separated by distances that varied from 20 meters to 1,700 kilometers. Often, iron-, lead-, and Faraday-chamber screenings were used to block out possible conventional sensory and electromagnetic mediators.

From the time I first read Vasiliev's (1963) monograph, I had been intrigued by his experiments and curious about whether it would be possible to replicate them. I was particularly interested in his experiments on remote mental influence of physiological activity. During this same early twentieth-century time frame, similar investigations were being carried out in other countries. There were French experiments on inducing hypnosis at a distance

---

*Republished by Hampton Roads Publishing, Charlottesville, VA (2002).

(by Joire, Gibert, Janet, and Richet), Dutch experiments on the remote influence of motor acts (by Brugmans at Groningen), hypnotic experiments on "community of sensation" (in which a sensory experience of the hypnotist appeared to be experienced by the hypnotized subject), and international studies of telepathy and clairvoyance. I thought that the methods used in these Russian, French, and Dutch experiments could be modified and updated, and could be applied in the PK-on-living-targets studies that I was planning.[7]

For my first experiments on distant mental influence of physiological reactions, I chose to use electrodermal activity as the target response. These reactions—which involve changes in the electrical activity of the skin, due to changes in neural and sweat gland activity—are usually "unconscious"; they can be measured fairly easily and inexpensively (throughout these studies, my research budget was always quite limited); they, or similar reactions, had already been influenced in earlier studies by other investigators; and, because they reflected the degree of activation of the sympathetic branch of the autonomic nervous system and hence were associated with bodily, emotional, and mental arousal, they had relevance to physical and psychological health and well-being. Electrodermal activity also had been studied extensively in contexts of biofeedback and self-regulation research.

In fact, biofeedback ideas played an important role in these first distant influence experiments. In all biofeedback experiments or clinical applications, one's biological activity (e.g., heart rate, blood pressure, muscle tension, brain wave activity) is monitored, and one receives sensory feedback (usually in the form of changing meter readings, light changes, or sound changes) that corresponds to typically unnoticed changes in the biological activities. Biofeedback equipment provides new or augmented "conscious" information about internal bodily events of which we usually are unaware and "unconscious." Once we become aware of these activities (through the mediation of instrumentation), we can try various ways to deploy our will, volition, attention, and intentions to bring these previously "involuntary" activities under voluntary control. Once the techniques are mastered, the self-control methods can be internalized, and the monitoring and feedback equipment is no longer needed and can be discarded.

This procedure can be accurately described as *autobiofeedback,* because one receives information about, and ultimately comes to influence, one's own biological activity. The procedure also might be considered *closed-loop biofeedback,* because the person who receives the feedback and exerts control of the activity is the same person who generates the activity in the first place.

The first electrodermal distant influence experiment was arranged as a form of *allobiofeedback* (from the Greek *allos,* meaning *other*) or *open-loop biofeedback.* The design was such that the electrodermal activity of one person (the influencee) was monitored and recorded by means of a polygraph tracing;

however, this feedback tracing was seen not by the person being monitored, *but by another person, who was stationed in a distant room.* The "other person" (the influencer) attempted to influence the first person's electrodermal activity—using techniques that will be described later—mentally and at a distance, using the feedback for trial and error purposes and as an indicator of the success of various influence strategies.

In other words, we wished to learn whether the loop or gap between the motor (output) generation of the biological activity in one person, and the sensory (input) information and intentional aim in another person might be closed or bridged psychically through some form of interaction between the two separated research participants.[8]

For methodological purposes, we divided a long session into a large number of shorter segments. During some of these segments, the influencer attempted to mentally influence the distant influencee—with either calming (decrease activity) or activation (increase activity) aims. Other segments served as control periods in which no influence attempts were made, and during which the influencer attempted to think not about the experiment but about other, unrelated things. The different types of segments (influence versus control) were randomly interspersed. Of course, the influencee was "blind" as to the nature and timings of the influencing periods, so conventional self-regulation or "placebo" effects could not account for any obtained results.

The results could be evaluated by comparing the patterns of electrodermal activity during influence periods with those of control periods and assessing the differences using conventional statistical techniques. In the early experiments, the scoring was done manually by measuring the polygraph tracings in a blind fashion (i.e., without knowledge of whether the tracings were for influence or control periods). In later experiments, the measurements were digitized and automatically evaluated by means of a computer.

Successful Outcomes

These remote influence experiments yielded positive results. In their initial form—as allobiofeedback experiments—the monitored persons' own electrodermal activity levels differed, according to the intentional aims of the distant influencers. The influencees showed greater electrodermal activity during increase periods and less activity during decrease (calming) periods. Because all conventional influence pathways had been eliminated by the study design, the most reasonable interpretation of the positive results was that effective psychic interactions had occurred between the influencer and the distant influencee. In subsequent experiments, we repeated this basic experiment with many variations. Overall, the results tended to be positive and reliable.

We tried different physiological and biological activities as the measurable indicators of these remote influence effects, worked with many different people and types of people as research participants, tried different experimental designs, and explored several physical, physiological, and psychological factors that might influence the results. These investigations and their outcomes are detailed in the 12 chapters of this book. I will attempt to summarize the main things we learned from these studies a bit later in this introduction.

The bottom line conclusion of this work, however, is that one person's mental processes (chiefly, one's attention and intentions) are able to interact effectively with the bodily, emotional, and mental activities of another person— even when that second person is located at a distance in space (and, we will see later, even at a distance in time) and is beyond the reach of conventional informational and energetic influences. These findings have important implications for both science and for daily living, which we will explore shortly.

What's in a Name?

Over the years, we considered various names for the form of psychic influence that we were observing and studying. In a sense, the name one chooses for a phenomenon might be viewed as not very important. However, names can carry or suggest various meanings or associations which can be important, in that the names can reveal intentional or unintentional theoretical or practical interpretations—some desirable and useful, some undesirable and potentially misleading. I hope the reader will pardon a brief excursion into our naming adventures.

Allobiofeedback: This name fit the original design and findings. However, we eventually learned that for these influences to occur, feedback was not necessary. Feedback helped some influencers, but others did not use it, and still others found it disruptive. Most important, we found that here and in other contexts, successful outcomes could occur without the provision of the trial-by-trial continuous feedback that was provided in these early studies.

Transpersonal imagery effect: We once used this term to emphasize that one's imagery (of the sort that often is used to influence biological activity in biofeedback and self-regulation experiments and applications) not only can influence one's own bodily and mental conditions and activities (personal imagery effects) but also can act beyond *(trans)* oneself to influence the bodily and mental conditions and activities of other persons *(trans*personally).

Bio-PK: This term, short for *biological psychokinesis,* has much to recommend it, and it captures well our original aims in beginning this research. However, the term implies that the effects are truly influential ones, that a form of active psi (psychokinesis) is chiefly responsible, and that the participant who contributes most to the effect is the agent or influencer.

Distant mental influence: This is perhaps the most straightforward name for the process that we are exploring, and it is the name chosen for the title of this book. However, the name continues to imply that an active influencer plays the greatest role in bringing about the effects. Additionally, although the effects do typically occur at a distance, it also is possible for these effects to occur close at hand, although still in ways that our conventional understandings of physics and psychology would deem impossible—e.g., the effects might occur in nearby targets that are nonetheless well-shielded or protected in other ways from conventional influences.

Remote mental influence: Remote was once suggested as a better term than *distant,* in that it covers outcomes in which the target is remote and inaccessible not only because of its distance but also because it might be near but adequately screened from conventional access. Here, *remote* carries the meaning of *inaccessible through ordinary means.*

Direct mental influence: Both *distant* and *remote* suggest a true spatial or other effective separation between the human influencer and the target system. Although this seems to be true, at first glance, the two quotes (from William James and Rumi) at the beginning of this introduction remind us that in a deep, profound sense, such separateness and isolation may be illusory, and that we may really be in much more direct and intimate contact with others and with all aspects of the world around us. *Direct* suggests an immediate, unmediated connection with the influenced target system—whether or not that target is at a distance or remotely situated. I am indebted to the late parapsychologist, Charles Honorton, for suggesting this simple name. Even this name, however, continues to imply an active, influential process, in which the agent or influencer plays the greatest role.

Direct mental interactions with living systems (DMILS): Although criticized for being inelegant and cumbersome, this is perhaps the most accurate designation for the processes being studied. Note that *direct* has been retained, but *interactions* has been substituted for the earlier *influence.* This substitution is made to remove the presumption or conclusion that the process is essentially an active, psychokinetic, influential one, with the influencer, again, playing the major role. *Interactions* suggests that other psi processes—such as telepathy, clairvoyance, and precognition—may be involved as much as, or even to a greater extent than, psychokinesis; that the influencee might play a much more important—and cooperative—role than is immediately obvious; and that in all of these experiments, we are left, ultimately, with correlations between influencer intentions and influencee activities. For such correlations or mutual co-arisings, the term *interactions* seems more appropriate than *influences.* As these types of studies are being increasingly replicated and addressed by other investigators, this final name and its curious abbreviation (DMILS) are being increasingly used to designate this growing area of research.

Choosing Other Target Systems: Chance and Necessity

In addition to electrodermal activity, we explored various other human target systems in these distant influence studies. Other physiological activities included muscular tension, muscular tremor, unconscious muscular movements (ideomotor reactions), blood pressure, heart rate, and breathing rate. We also found that cognitive activities, such as the intensity of mental imagery and the ability to focus and concentrate one's attention, could be enhanced through distant intentional influence. In effect, these latter activities could be facilitated through a form of remote "psychic helping" by persons in distant locations.

Some have suggested that direct mental influences may be occurring at all times and as a matter of course within our own bodies, but these influences are so intertwined with other internal body/mind activities and are so common and familiar that they rarely are recognized as paranormal or unusual. Stated differently, these direct mental influence effects that seem so rare and unusual when observed outside of our bodies may also occur commonly and ordinarily within our bodies, and these may be part of, or essentially, our regular volitional activities. Crudely stated, if my mind moves your finger, we are filled with surprise and awe and call this *psychokinesis.* However, if, through the same process, my mind moves my own finger, this is nothing other than business as usual, and we call this *voluntary action* or *intentional movement.* The physiologist and Nobel laureate Sir John Eccles went so far as to suggest that mind and brain might interact through the will's direct influence upon (and its "cognitive caresses" of) certain dynamic patterns of cerebral activity in the brain, and that critically poised neurons or synapses might be the loci of interaction between will and body in all normal volitional acts.[9]

In pondering these ideas, I began to wonder about the following: If my mind can influence my own brain, perhaps it can influence other neural or biological material as well. Further, the degree to which my mind might continue to exert its influence on a distant target system could depend upon the similarity of that system to my own brain. Speculations such as these suggested two types of research programs: The first would focus on external living targets that might closely resemble brain tissue or neurons, and the second would focus on the possibility of continuing to influence cells, tissues, or biochemicals that had been freshly removed from one's body.

In considering the first strategy, it became clear that any telepathic interactions between two persons could be viewed as direct mental influences of one person upon the brain of the second person. Therefore, all instances of telepathically shared thoughts, images, or feelings might be special cases of distant mental influences of the type explored in the studies I am relating here. To test the idea that the most effective target systems for these influences might be other brains, other neurons, or materials similar to these, it would be useful to conduct experiments in which one might attempt to directly influence brain tissue, neurons, or other similar preparations that were maintained outside the body. If one immediately puts aside the obvious difficulties of working with cultured human neurons, there still remain various biological preparations that possess similar characteristics.

For example, there are giant neurons in simpler organisms (such as the large neural cells of the giant sea slug, *Aplysia,* and the giant axon of the squid) that could be studied. There are even plant cells and giant algae that exhibit electrical activities very much like those that occur in neural transmission. I spent some time investigating the possibility of using such preparations in my studies, but eventually dismissed these due to technical and logistical complexities.

In the course of this research, however, I came upon another interesting electrical preparation that I was able to investigate successfully. This was the electric knife fish, *Gymnotus carapo*—a species of fish that inhabits the murky water of the Amazon river and emits weak electrical signals, apparently for navigational and food-finding purposes. By placing an electric knife fish in a small tank with metal plates on its sides, and properly instrumenting this system, we were able to use the amplified fluctuating electrical field of the fish as a feedback signal to human influencers in a distant room, and these persons were able to intentionally influence the fish and its electrical activity.

We finally used the electrical characteristics of this fish simply as a convenient means of monitoring and quantifying the fish's spontaneous movements and orientation changes within its small tank. This, in turn, suggested that we might make use of the spontaneous movements of other intact animals as distance influence targets. We did conduct several experiments in which distant influencers were able to successfully influence the locomotor activities of Mongolian gerbils *(Meriones unguiculatus)* running freely in activity wheels.

The second type of research program involved considering various biological materials that might be removed from the body, so that possible distant influences of these freshly-removed materials could be explored. The first choice of such materials would be nervous system tissue that had been removed and cultured externally (or cloned). At first, it might seem that such an experiment would not be feasible. However, a study of this type could be performed with the cooperation of a neurosurgeon and a team of neuroscientists. Brain tissue removed for medical reasons, which ordinarily would be discarded, might be artificially cultured and maintained, and the chemical or electrical activities of these cells could be used as indicators of the efficacy of distant mental influence.

A problem with such a design is that the very tissue that might become available (e.g., samples from brain tumors) would likely consist of recalcitrant tissue that may not have been "cooperative" and appropriately responsive to the adaptive needs of the body, even in the tissue's usual environment. Whatever may have led to abnormal growth of these particular cells might continue to resist intentional/volitional influences when externalized in the hypothetical design just described.

Apart from this seemingly science-fiction approach, the next best biological target materials might be lymphocytes (white blood cells with important immune system roles, such as B-cells, T cells, and natural killer cells). Recent research has indicated similarities of neural cells and immune system cells, in terms of their neurochemical responsivity and signaling properties. However, the equipment and skills for measuring such cells or their biological activities are complex and were not available in our laboratory facility. Therefore, we considered a third best, logistically feasible material with which to work— osmotically stressed red blood cells (i.e., red blood cells placed in a solution in which the salt content was too low compared to normal conditions). Red blood cells could be removed from volunteer participants, the blood could be placed in test tubes in a distant room, and the rate of destruction of these blood cells could be monitored by means of a spectrophotometer-computer system. The donors of these cells, at the distant location, could attempt to protect certain samples of these cells (compared with other, control samples) through their distant mental influence.

Experiments of this type were conducted and were successful (see chapter 3). In these experiments, the remote influencers attempted to mentally "protect" the red blood cells by visualizing the cells with intact, resilient membranes that resisted the osmotic stress, rather than bursting; color slides of healthy, intact red blood cells were available to the influencers should they choose to use this sensory aid to enhance their protective mental imagery.

Eventually, we extended our distant mental influence work to include additional living target systems. Some of these were physiological processes (e.g., pulse rate, peripheral skin temperature, electromyographic activity [muscle tension of the frontalis—forehead—muscle group], breathing rate, blood pressure, muscular tremor, ideomotor reactions [subtle, "unconscious" movements]). Other processes were psychological—a distant person's intensity of mental imagery, a distant person's ability to concentrate and control attention. Our work with these systems of human activity is described in the various chapters of this book.

Reaching Out Mentally to Touch Someone: Research on Remote Staring Detection

During the course of our experiments on intentional (directional) influences of the electrodermal activity of distant persons, we realized that remote *attention* was always confounded with these remote intention effects. Our influencers not only were contacting the influencees mentally and at a distance to encourage specific, directional changes in their bodily activities, but they also were always reaching out to the distant influencees with their attention. It was as though two types of mental "signals" were involved.

The first was getting the influencee's attention (simply by the influencer focusing attention on the distant person), and the second was a kind of mental instruction, suggestion, or request to perform a particular bodily activity. We wondered if reaching out to someone mentally and at a distance simply by thinking about or observing that person might have its own direct influence on the distant person's physiological activity.

This reminded us of the familiar popular belief that it might be possible to gain someone's attention by simply staring at them, unobserved, from a remote location. Experiments had been conducted by others to test this staring detection idea—with various degrees of scientific rigor and various degrees of success. We thought the measurement of electrodermal activity might provide an ideal way to measure the possible subtle influences of remote staring. An electrodermal indicator of the detection of remote staring provided a bonus: It was an unconscious activity, and we suspected that unconscious bodily activities might provide more sensitive indicators of psychic awareness than could the usual conscious, verbal, or imagery-based indicators that are so popular in parapsychological research.

Therefore, we designed a study in which we would record possible changes in electrodermal activity of distant persons while they were being stared at from a distant room via a closed-circuit television system. During some periods (unknown to the staree), the distant influencer would stare intently at the staree's image on a television monitor and attempt to mentally reach out to the staree and get that person's attention. The staree's electrodermal activity during sets of remote staring periods could be compared directly with other sets of interspersed periods during which such remote staring did not occur. These experiments were successful, and they are described in chapters 7 and 8.

Major Findings

The major findings of our research on distant mental influence of the biological and psychological activities of other persons and of other living organisms may be summarized as follows:

1. Under certain conditions, it is possible for one person to effectively influence the bodily and mental activities of another person who is situated at a distance and shielded from all *conventional* sensory, informational, and energetic influences.

2. These distant influences appear to be *direct mental influences* because they cannot be accounted for in any obvious way by conventional explanations. The experimental designs used in these studies effectively ruled out the following alternative, conventional explanations of these effects: chance or coincidence, sensory cues, uncontrolled external stimuli, common internal rhythms, recording errors, motivated misreadings of records, conventional placebo effects, rational inference on the part of influencees of when influence periods were occurring, or errors due to progressive or systematic changes in the measured activities.

3. The effects occur when the influencer and the influencee (or influenced target system) are separated from one another by *distance* (as great as 20–25 meters) and by the walls and the closed and locked doors of their own and

intervening rooms. Distances greater than 25 meters have not yet been explored in my own research (although much greater distances have been explored in other research projects; see below).

4. The effects may occur when the influenced system and the influence attempt are separated in *time*. In some studies, the to-be-influenced bodily activities occurred 35–40 minutes before the influence attempts were made (in a study conducted in 1993) and 1–7 days before the influence attempts (in a study reported in 1979).

5. A *wide range* of bodily and mental activities have been successfully influenced mentally and at a distance. These activities include the spatial orientation of fish, the locomotor activity of small mammals, the autonomic nervous system activity of another person, the muscular tremor and ideomotor reactions of another person, the mental imagery of another person, a remote person's ability to concentrate and focus attention, and the rate of hemolysis of human red blood cells *in vitro*.

6. The ability to manifest these effects is apparently *widely distributed* in the population. Sensitivity to the effect appears to be normally distributed in the volunteers who have participated in our various experiments. Many persons are able to produce the effect with varying degrees of success, including unselected volunteers attempting it for the first time. More practiced individuals seem able to produce the effect more consistently. There are indications of improvement with practice in some influencers.

7. Based upon overall quantitative and statistical results, the distant mental influence effects are *relatively reliable and robust.*

8. The magnitudes of the effects are *not trivial* and under certain conditions may compare favorably with the magnitudes of self-regulation effects. In some cases, results can be dramatic and are comparable to effects caused by physical stimuli. For example, an influencer's *imagining* of going into a room and shaking the influencee's chair has been accompanied by this identical experience in the influencee along with appropriate, dramatic physiological reactions.

9. Persons with a *greater need* to be influenced (i.e., those for whom the influence is more beneficial) seem more susceptible to the effect.

10. Immediate, trial-by-trial analog sensory *feedback is not essential* to the occurrence of the effect; intention and visualization of the desired outcome is effective. Successful distant influence strategies have included: (a) attending

fully to the person or system one wishes to influence; (b) filling oneself as influencer with strong images and intentions of the desired outcome; (c) producing the desired outcome conditions in oneself, using self-regulation procedures, and intending for similar conditions to occur in the to-be-influenced person; and (d) vividly imagining that the influencee is in a physical or psychological setting that would naturally produce the desired reactions. Sometimes, *gentle wishing* (a more effortless form of intending) seems more effective than effortfully willing the desired outcome to come about. In addition to, or instead of, the more specific, process-oriented images and intentions just mentioned, the influencer may engage in *goal-oriented imagery* of a more general and overarching sort—i.e., images, visualizations, and intentions associated with a successful experiment outcome. These could include imagining the joy of the research personnel as they celebrate a positive outcome for a session or for the entire experiment, imagining a computer printing out significant findings, imagining reading a published report of positive findings of this session or this experiment, imagining how the outcome of the present session may contribute to the realization of some useful, health-related practical application of these principles, and so on.

11. The effect can occur *without the influencee's conventional knowledge* that such an influence is being attempted.

12. It may be possible for a person to *block or prevent an unwanted influence* upon his or her own physiological activity; psychological shielding strategies in which one visualizes protective surrounding shields, screens, or barriers may be effective.

13. Generally, our volunteer participants *have not evidenced concern* over the idea of influencing or being influenced by another person.

14. The effect can be *intentionally focused* or restricted to one of a number of physiological measures; it may also take the form of a generalized influence of several measures, if that is the intent of the influencer.

15. The living target systems can be influenced *bidirectionally;* i.e., their activity levels can be either increased or decreased.

16. The activity levels of at least some of the target systems (e.g., electrodermal activity, rate of hemolysis) and their susceptibility to distant mental influence appear to be associated with *geomagnetic field (GMF) activity;* i.e., the systems are more active and more susceptible to influence when the Earth's geomagnetic field activity is more "stormy" than during more "quiet" GMF periods.

17. Distant mental influence, in the expected direction, seems more successful when the intentions and images of the influencer are focused *specifically on the desired target activity*, rather than directed toward the target in a more general or global manner.

18. *Attention alone* (fully focusing attention on the person or system in question) can influence the distant person or other living system, even in the absence of an intention for a directional change; this is best evidenced in the studies of physiological detection of remote staring.

19. Successful distant influence episodes sometimes are *accompanied by distant knowing*, as well; the latter can take the form of telepathic awareness of specific images that the influencer happens to be using.

20. The degree to which one is able to distantly influence others, or be influenced by distant others, has been found to be *related to various psychological characteristics*, such as one's ability to concentrate or become absorbed in what one is attending to, one's degree of introversion, and one's degree of social avoidance and distress, as measured by standardized assessment instruments.

21. These distant mental influence effects recently have been *successfully replicated* in other laboratories by independent investigators (see below).

22. The effect *does not always occur.* The reasons for the absence of a significant effect in some experiments of a series that otherwise is successful are not clear. We suspect that the likelihood of a successful distant mental influence effect may depend upon the presence of certain psychological conditions in both influencer and influencee (and perhaps even in the experimenter) that are not always present. Possible success-enhancing factors may include belief, confidence, positive expectation, and appropriate motivation. Possible success-hindering factors may include boredom, absence of spontaneity, poor mood of influencer or influencee, poor interactions or poor rapport between influencer and influencee, and excessive egocentric effort (excessive pressure or striving to succeed) on the part of participants. We suspect that the effect occurs most readily in influencees whose nervous systems are relatively labile (i.e., characterized by free variability) and are momentarily free from external and internal constraints. Perhaps fullness of intention and intensity or vividness of visualization in the influencer facilitate the effect.

Replications, Extensions, and Related Research by Others

Not all our work in the area of distant mental influence is described in detail in this book. Some of our additional projects—e.g., remote influences on other persons' intensity of mental imagery—are cited in some of the review articles in this volume. In addition to our own expansion projects, our work has been replicated and extended by others. Some of these related projects are described in the reviews presented as chapters 4 and 11. Still others have suggested and implemented refinements and improvements of this work.[10]

Recent developments in three related research areas deserve special notice. These are projects in the areas of presentiment, focused group consciousness, and distant healing and intercessory prayer.

The presentiment (pre-feeling) studies are closely related to the retroactive intentional influence work described in chapter 12. In these studies—conducted chiefly by Dean Radin and Dick Bierman—participants are monitored for typically "unconscious" physiological reactions (heart rate, electrodermal activity, finger blood volume) before, during, and after sensory exposure to slides that have emotional or nonemotional content. Radin and Bierman have been finding that persons show different autonomic nervous system reactions to emotional and nonemotional slides, even *before* the slides are shown, and when the nature of the upcoming slide still is unknown (in conventional ways).[11] Although this presentiment effect usually is taken to reflect precognition (future-knowing) operating at an unconscious, bodily level, these interesting findings can just as well be interpreted as instances in which objective events (the presentation of the slide itself or the person's future *reaction* to the slide) may be acting backward in time to influence the person's physiological activity. Similar interpretations can be placed on Klintman's curious "time-reversed interference" effects described in chapter 12.

Indeed, all instances of precognition might be interpreted as the retroactive influence of future objective events upon the present or past mental activities of the persons who report these instances of future-knowing. Even if one takes the usual view of precognition—that one somehow reaches out into the future to access information about some yet-to-occur event (typically, some very meaningful, important, or disastrous event such as a railroad accident, the sinking of the *Titanic*, or the Aberfan mining disaster[12])—there remains the task of bringing this encountered information *back into present knowing*, so a process of backward-going-in-time still must be involved even in the usual interpretation of precognition.

The research on focused group consciousness is chiefly the work of Roger Nelson and his colleagues.[13] A number of electronic random event generators have been placed at many locations throughout the world—at the time of this writing, about 38 of these devices have been distributed—and their random activities are continually monitored. The remarkable finding is that at times during which there is an unusually strong degree of coherent or focused attention by large groups of people, the activities of these random generators deviate significantly from their usual baseline levels. The coherences of group attention to which the random generators are sensitive include such events as special moments during conferences or meetings, or moments of widespread viewings of televised events (such as the funeral ceremony of Princess Diana, the announcing of the jury verdict of the O. J. Simpson murder trial, various New Year's Eve ceremonies, special moments during Academy Awards telecasts, etc.). Some of the strongest and most consistent departures of random event generators from their normal patterns occurred just before and during the two crashes of terrorist-hijacked commercial airplanes into New York's World Trade Center twin towers on September 11, 2001.

In these and similar cases, there occur widespread (nonlocal) changes in the arrays of random generators and also indications of greater effects in the vicinities of the triggering events. One of the most straightforward interpretations of these effects is that events of great importance are associated with a greater alignment or coherence of attention of a large number of people, and the random event generators' activities shift in accordance with these enhanced instances of focused group consciousness.

Thus, the findings may be related to—but are much more amplified instances of—the staring detection findings mentioned above. In the studies reported in chapters 7 and 8, the focused, remote staring at one individual by another individual can be detected by a shift in the ongoing, freely-variable (quasi-random) physiological activities of the staree. In the group consciousness studies, the focused, remote attention of large numbers of individuals can be detected by shifts in the ongoing, freely-variable (random) activities of arrays of widely dispersed inanimate random generators. The most up-to-date descriptions and summaries of this group consciousness work are available on the Internet at http://noosphere.princeton.edu/.

Perhaps the most important related research and application is in the area of distant or nonlocal healing (including spiritual, psychic, mental, and "energetic" healings of others) and intercessory prayer (prayers said for the benefit or health of others). Although these practices have been carried out throughout history and in virtually all cultures, very little research attention has been devoted to these processes. Some of the earlier studies are reviewed in various chapters of this book. In recent years, there has been increased interest in studying the efficacy of distant healing and prayer. A number of reviews of such work recently have been published.[14]

Although the outcomes of the various studies are not always positive or consistent, the significant—and sometimes dramatic—effects reported in some of these studies provide promising indications of the potential effectiveness of these complementary and alternative healing techniques, and the overall results certainly suggest that it is important that additional work be carried out in these areas. Among the more interesting of these recent studies are those by Byrd, and Harris and co-workers (on efficacy of intercessory prayer for patients with coronary disease), and research by Sicher and co-workers (on efficacy of distant healing for patients with advanced AIDS).[15]

One of the most provocative recent reports is that of Leibovici on efficacy of remote, *retroactive* intercessory prayer for large groups of patients with blood stream infections; here, the prayer intervention was performed 4–10 years *after* the infections.[16] Distant healing and remote intercessory prayer findings are relevant to the topic addressed in this book in that distant mental influences, in the form of healing intentions, are involved in both nonlocal healing and intercessory prayer, and these healing intentions, especially if they are present in a large number of healers or prayers, may themselves play important influential roles in the positive results obtained in such studies.

What Does It All Mean? Explanations and Interpretations

The findings reported in this volume, along with other related findings, demonstrate that persons can remotely influence other persons and other living systems in ways that cannot be adequately accounted for or explained by our dominant scientific models and theories. At the very least, these findings suggest that the theories, models, and understandings of macroscopic physics, physiology, neuroscience, and cognitive psychology are incomplete and in need of modification and expansion if they are to adequately encompass and explain these distant mental influences and other similar, well-documented, nonlocal human interactions.

The most challenging aspects of these distant mental influence findings have been succinctly characterized by Larry Dossey in three words: these interactions appear to be *unmediated, unmitigated,* and *immediate.*[17] They are unmediated in that no presently known conventional sensorimotor, energetic, or informational processes seem capable of serving as vehicles for these effects. They are unmitigated in that the effects do not seem to be importantly diminished by increasing spatial (or temporal) distances between influencer and influencee. They are immediate in that the effects seem to occur instantaneously (although this latter claim is difficult to assert with full confidence).

In attempting to understand these distant influence effects, three classes of models have been proposed. In *transmission* models, it is suggested that remote influence is accomplished through some physical or quasi-physical force that carries information from one locus to another through some channel or medium in a manner analogous to mental radio: There is transmission and reception of information, intelligence, or "energy." Such models have many difficulties. The mediating force has not been identified, nor has the "channel," nor do we know of any mechanisms through which the conscious intention of the influencer might be coded into or modulated onto the "carrier" then decoded or demodulated from the carrier at the remote influence site. The distant mental influence process does not behave as other forms of transmission (i.e., the four conventional physical forces: electromagnetic, strong nuclear, weak nuclear, or gravitational) customarily behave with respect to physical factors such as distance, shields, screens, amplifiers, attenuators, the nature of the influenced system or of the conveyed information (message content), or (perhaps most problematically) time.

One of the leading contenders for possible physical carrier of distant mental influence (and paranormal events, in general)—extremely low frequency (ELF) radiation—does have the required great traveling and shield-penetrating characteristics; however, its low frequency would not seem sufficient to carry rich and detailed information sufficiently rapidly to account for complex, detailed, and quickly occurring paranormal events, nor are we aware of any conventional mechanism through which the body or brain of an influencer might be able to encode intentions or other biological information onto ELF carrier waves.[18]

In *reorganization* models, nothing is posited to be transmitted from point to point. Rather, the "noise," randomness, or disorder already present in the to-be-influenced system is reorganized in a manner that creates the desired goal outcome and *appears* forcelike. The process is one that is analogous to resonance, but without the typical mediators of familiar forms of resonance. The challenges facing such models are determinations of what precisely "feeds" the reorganization process at the target end and what precisely specifies the particular form the reorganization will take. Perhaps there are basic, axiomatic laws of the universe through which, under certain conditions, disorder in one area automatically becomes organized to match a strong, ordered pattern elsewhere, and perhaps distant mental influence is one particular manifestation of these more general laws. The latter may not yet have been "discovered" by conventional methods and theories of science, but they seem to have been known through other methods throughout history. The "nonscientific" knowledge of such laws can be glimpsed—sometimes, perhaps, only in distorted or partial forms—in various esoteric, magical, mystical, spiritual, and wisdom traditions.

In the third class of models, which could be called *holonomic* or *correspondence* models, nothing is either transmitted or reorganized. All information is already present throughout all parts of all systems, in some implicate or potential form, in a manner not unlike the complex interference patterns in which information is represented in a hologram. The problem then becomes one of accessing or manifesting a particular, desired outcome event as opposed to other potential events that might be latent (implicate) in the target system and in many alternative systems that *don't* exhibit the influence, specifying the grounds or fields that make all of this possible, and accounting for the creation of novelty within such target systems.

How do the *intended* outcomes occur at some particular time and in some particular place or person, as opposed to a vast number of alternative possibilities? This puzzle is analogous to the quantum-mechanical *measurement problem* of how one particular event is manifested or realized, as opposed to others, when the wave function (spread throughout space and throughout time) is collapsed to yield a particular observation.

The second and third classes of models call to mind similar statements found in Jung's concept of synchronicity, in Leibnitz's monadology in which "monads have no windows" but nonetheless perfectly mirror one another, in the ancient Hermetic maxim, "As it is above, so it is below," in the Buddhist notion of mutual arising or dependent origination, and in contemporary parallels such as David Bohm's implicate and explicate realities, Rupert Sheldrake's morphogenetic fields, and similar constructs.[19] Intimations of these principles may be found in the same esoteric, magical, and mystical systems and traditions alluded to previously.

There are two relatively unfamiliar areas of scientific thought that may hold promise for increasing our understanding of distant mental influence and related processes. These are the zero point field[20] and the eight-space metric.[21] At the very least, they—along with the more familiar notions of nonlocality in quantum physics—can provide useful analogies or metaphors that may aid our understanding of these distant influence phenomena.

Distant mental influences can occur with great specificity and rapidity, and without the influencer's knowledge or use of conventional causal processes or mechanisms of action in bringing these about. This *goal-directed* feature of these findings suggests that, in addition to the well-known causal laws of nature, there may also exist a complementary set of teleological or teleonomic principles that can govern outcomes and events, and such goal-achieving principles may be active not only in psychological, social, and cultural realms, but in biological, chemical, and physical realms as well.

Three Caveats

The reader will recognize that we have been discussing primarily *physical* explanations and possible *physical* mediators. It may be unwise, however, for us to look to physics for explanatory tools. There is danger in hitching one's wagon to a particular star in physics: When that star blinks out of favor or popularity, so does one's explanatory wagon. Ultimately, it may be best for psychology, parapsychology, and consciousness studies to consider and develop their own explanations and posited "mechanisms" through which various psychological, parapsychological, and consciousness-mediated events and experiences come about.

As human beings, we seem predisposed to become very uncomfortable when confronting what appear to be *gaps* of any kind, and we experience a strong need to fill these gaps, even in ways that are less than adequate. This gap-aversion applies also to scientists, and it helps account for the great scientific resistance to early observations and concepts of action at a distance (exemplified by electromagnetic field phenomena and gravitational field phenomena).

The quotes by William James and Rumi at the beginning of this introduction paradoxically free us from problematical gaps—by suggesting that what appear to be separate, isolated systems may be only illusorily or superficially so—and also indulge in their own gap-filling or gap-banishing by suggesting a deeper interconnectivity or intimacy among apparently unconnected events.

I think it is important for us to keep these issues in mind whenever we invoke notions of deep interconnectedness in order to "explain" interactions of seemingly separated events. To use interconnectedness in a *descriptive* sense certainly seems appropriate for describing the covariation of distantly separated events. To reify interconnectedness as some deeply interpenetrating and joining medium, however, may be but another, more subtle, form of gap-avoidance.

There is another area of potential reification of which we should remain aware. This is the exquisitely seductive tendency to reify the constructs of mind, consciousness, intentions, and attention. An alternative to this tendency is to treat these not as nouns, but as gerunds. In doing this, we would be heeding William James's early reminder that consciousness is a function, an ongoing *process,* a *stream,* and we would be following Alan Watts's more recent, and picturesque, lead in speaking of mind*ing,* conscious*ing,* matter*ing,* think*ing,* intend*ing,* and so on.[22]

Throughout this volume, the reader will encounter language that suggests that mind, consciousness, thoughts, images, intentions, and prayers are doing things. It may be more accurate to use gerundive forms and to say that minding, consciousing, thinking, imaging, intending, and praying are associated with other gerunds: moving, growing, increasing, decreasing, healing, and so on. This would remind us that we are discussing complex, ongoing processes—dynamic systems having particular patterns.

To the extent that brain and body (braining and bodying?) are facets of these processes, their conditions would be expected to play important roles in these events and their possible outcomes. Indeed, apprehending "mental" processes in such a way—minding?—could allow us to better appreciate how physical factors such as changes in information, geomagnetic field conditions, and local sidereal time might become associated with differences in accuracy of psychic functioning, as some recent studies have suggested.[23]

Implications and Possible Applications: A Cornucopia of Questions

In the chapters that follow, the reader will find several suggested implications and potential practical applications of these distant mental influence findings. Here, I depart from the usual presentational format and address these implications in the form of a series of questions.

- I intend to wiggle my finger, and the finger wiggles. This set of events typically is explained in terms of neuromuscular structures, chemicals, and electrical processes. But what if I intend to wiggle your finger, and your finger wiggles (which, in principle, is what the findings of this volume suggest is possible)? In which ways might neuromuscular structures, chemicals, and electrical process enter into the production of that intention-wiggle relationship?

- If conventional anatomical, physiological, and physicochemical processes cannot completely account for distant mental influence effects, what does this mean for a worldview that insists upon these kinds of processes as the only possible causal agencies?

- If direct mental influence can occur at a distance, could similar processes be occurring—perhaps ubiquitously—within our own bodies? If so, what might be their implications for processes such as hypnosis, biofeedback, physiological self-regulation, placebo effects, psychoneuroimmunology effects, volitional actions, and memory?

- If direct and distant mental influences occur in the laboratory, might these also be occurring in everyday life?

- If your intentions are able to influence my reactions (or feelings, or thoughts, or images), what does this imply about where you end and I begin?

- What does your answer to the previous questions imply about the nature of our identities as human beings?

- What do the findings of direct mental influence imply about our potentials as human beings?

- What might direct mental influence findings imply about some of the unusual events and experiences reported in spiritual and wisdom traditions (unusual feats, "miracles," etc.)?

- If our intentions can play active roles in influencing, or even creating, various physical, physiological, and psychological changes and conditions, might our intentions also be able to play active roles in influencing or creating various social, cultural, or planetary conditions?

- Might our intentions play active roles in influencing or even creating spiritual experiences or spiritual realms or realities?

- What might be the possible ranges and limits of events or experiences that may be susceptible to distant mental influence?

- What are the implications of distant mental influence research for our understanding of the nature of consciousness and the role of consciousness in the physical world?

• If distant mental influence effects are attributable to the direct action of consciousness, and if these effects occur nonlocally, does this imply that consciousness itself is nonlocal?

• If consciousness is nonlocal, does this imply that consciousness is omnipresent and eternal (immortal)?

• If consciousness is nonlocal, does this imply that consciousness is one—i.e., as the physicist Erwin Schrödinger suggested, "consciousness is a singular of which the plural is unknown; that there is only one thing and that, what seems to be a plurality, is merely a series of different aspects of this one thing, produced by a deception (the Indian MAYA); the same illusion [that] is produced in a gallery of mirrors"?[24]

• In any dyadic (two-person) situation—e.g., health practitioner/patient, therapist/client, counselor/counselee, spiritual guide/spiritually guided, teacher/student, trainer/trainee—could the mental practice or realization, in the first member of each pair, of an intended outcome directly facilitate or help realize the same desired outcome in the second member of the pair?

• How might direct or distant mental influence best be used for purposes of facilitating health, well-being, and healing?

• If beneficial changes or conditions can be fostered through distant mental influence, could harmful changes or conditions also be fostered in similar ways?

• If our intentions can act retroactively—i.e., if our present intentions can influence the likelihood of occurrence of past events—what does this imply about our understanding of the nature of time?

• If our intentions can act backward in time, might they act even transgenerationally?

• If distant mental influence effects are real, why do they seem to occur so infrequently? *Do* they occur infrequently?

• If such effects are real, why do we know so little about them?

• If such effects are real, why has so little work been directed toward their exploration and study?

• Because these findings have implications for the nature of our identities and capabilities while we are alive, what implications do these same findings have for the possibility of human survival of physical death and the possibility of an afterlife existence in some form?

Interconnectedness

For many, one of the major lessons of distant mental influence and related findings is that these findings suggest a deep and profound inter-connectedness among people and also between people and all of animate and inanimate nature. Such a view can provide a corrective for the more apparent and more familiar view of persons and things as having separate, isolated essences and existences. Thomas Wolfe's "Every man is an island" is complemented by John Donne's "No man is an island." This interconnected view of things is, indeed, the view suggested by the James and Rumi quotes.

It also is compatible with Henri Bergson's image of "our large body"—co-extensive with our consciousness, comprising all we perceive, reaching even to the stars—which, Bergson maintained, we commonly ignore in limiting our consciousness to our more familiar small body.[25] This view corresponds also with the Iroquois metaphor or reality of the "long body"—the greater, more extensive, and inclusive "body" or self that we share not only with other persons, but also with all sentient beings and all inanimate nature.[26]

Direct mental influences, and other paranormal events and experiences, may be our way of informing ourselves—through impressive and sometimes elaborate indicators—about the deeply interconnected and interrelated conditions in which we always participate, but which we too often forget.

> What better way to dramatize to ourselves that we are truly one than to share—especially at great distances and in defiance of powerful conventional barriers—each other's thoughts, feelings, images, sensations, and reactions? And what better way to demonstrate that we are in intimate contact with all of reality than to touch and move things with our minds? Perhaps the apparent transfer of information and the apparent forces that we seem to see in [paranormal] functioning are not really what they appear to be. Rather, they may be quick yet effective and convincing indicators that are readily at hand (paths of least resistance, so to speak) when we wish to remind ourselves of our forgotten interconnections.
>
> In general, [these] experiences may be self-created metaphors and dramatizations—extremely real and concrete teaching stories—that hold important latent meanings, lessons, and reminders that may have little to

do with the more obvious literal and "informational" content [and "causal influence" aspects] of the experiences. It could be fruitful to ask ourselves: What is the real message, and what is merely the medium? Experimental parapsychology reveals interconnectedness only indirectly. Its impact is primarily upon the intellect and is but a shadow of the fuller, more direct impact of oneness felt and known in . . . mystical [and related] experiences.[27]

The value of realizing the interconnectedness suggested or revealed by distant knowing and distant influence findings is not only its contribution to our better apprehension of the way things are, but also—some maintain—in the ethical and moral understandings and actions that may flow out of this apprehension. This is part of the message embodied in the Sanskrit motto *tat tvam asi—that thou art*. If I am deeply and profoundly interconnected with other persons or with nature, then in some important way, I am one with others and with nature. To harm others and nature is to harm myself; to be loving, compassionate, understanding, protective, and caring for others and nature is to treat myself in these benevolent ways.

A greater appreciation of our connectedness with others can foster greater feelings of compassion than otherwise might be possible. Additionally, such interconnectedness can serve to remind us that our own thoughts, feelings, and actions can directly affect others and the environment, thereby increasing the likelihood of more responsible actions toward others and toward the world at large.

No doubt, we can treat one another with kindness, understanding, and compassion even if we were not profoundly and intimately interconnected in nontrivial ways. However, having direct experiences and knowledge of this interconnectedness—as revealed by distant mental influences and related work—can provide, for some of us, additional rationales and direct personal experiences that can support our love for other persons and all of nature and can, thus, enhance our ethical actions toward others and nature. So, it is with these possible beneficial consequences, as well as increased knowledge and understanding, in mind, that I offer the remaining chapters of this book.

References

1. James, W. (1977). Final impressions of a psychical researcher. In J. McDermott (Ed.), *The writings of William James: A comprehensive edition* (pp. 787–799). Chicago: University of Chicago Press. (Original work published in 1909)
2. Rumi, J. (1984). *Open secret* (J. Moyne and C. Barks, Trans.). Putney, VT: Threshold Books.
3. Reichenbach, H. (1938). *Experience and prediction.* Chicago: University of Chicago Press.
4. Braud, W. G. (1978). Psi conducive conditions: Explorations and interpretations. In B. Shapin and L. Coly (Eds.), *Psi and states of awareness.* New York: Parapsychology Foundation, pp. 1–41.
5. Braud, W. G. (1981). Lability and inertia in psychic functioning. In B. Shapin and L.

Coly (Eds.), *Concepts and theories of parapsychology.* New York: Parapsychology Foundation, pp. 1–36.

6. Radin, D. I. (1997). *The conscious universe: Truth of psychic phenomena.* San Francisco: HarperCollins.

7. Vasiliev, L. L. (1963). *Experiments in mental suggestion* (A. Gregory, Trans.). London: Institute for the Study of Mental Images. (Original work published in Russian in 1962.) Vasiliev, L. L. (2002). *Experiments in mental suggestion.* Charlottesville, VA: Hampton Roads.

8. Braud, W. G. (1978). Allobiofeedback: Immediate feedback for a psychokinetic influence upon another person's physiology. In W. Roll (Ed.), *Research in parapsychology 1977.* Metuchen, NJ: Scarecrow Press, pp. 123–134.

9. Eccles. J. C. (1953). *The neurophysological basis of mind.* Oxford: Clarendon Press.

10. Brady, C., and Morris, R. (1997). Attention focusing facilitated through remote mental interaction: A replication and exploration of parameters. *Proceedings of Presented Papers: Parapsychological Association 40th Annual Convention,* pp. 73–91. Schmidt, S., and Walach, H. (2000). Electrodermal activity (EDA): State-of-the-art measurement and techniques for parapsychological purposes. *Journal of Parapsychology,* 64, 139–163. Schneider, R., Binder, M., and Walach, H. (2000). Examining the role of neutral versus personal experimenter-participant interactions: An EDA-DMILS experiment. *Journal of Parapsychology,* 64, 181–194. Schmidt, S., Schneider, R., Binder, M., Buerkle, D., and Walach, H. (2001). Investigating methodological issues in EDA-DMILS: Results from a pilot study. *Journal of Parapsychology,* 65, 59–82. Schneider, R., Binder, M., and Walach, H. (2001). A two-person effort: On the role of the agent in EDA-DMILS experiments. *Journal of Parapsychology,* 65, 273–290. Stevens, P. (2000). Human electrodermal response to remote human monitoring: Classification and analysis of response characteristics. *Journal of Parapsychology,* 64, 391–409. Watt, C., and Brady, C. (2002). Experimenter effects and the remote facilitation of attention focusing: Two studies and the discovery of an artifact. *Journal of Parapsychology,* 66, 49–71.

11. Radin, D. I. (1997). *The conscious universe: Truth of psychic phenomena.* San Francisco: HarperCollins.

12. Dean, E. D. (1974). Precognition and retrocognition. In J. White (Ed.) *Psychic exploration: A challenge for science* (pp. 153–177). New York: G. P. Putnam's Sons.

13. Nelson, R. (2001). Correlation of global events with REG data: An Internet-based, nonlocal anomalies experiment. *Journal of Parapsychology,* 65, 247–271.

14. Abbot, N. C. (2000). Healing as a therapy for human disease: A systematic review. *The Journal of Alternative and Complementary Medicine,* 6(2), 159–169. Astin, J. E., Harkness, E., and Ernst, E. (2000). The efficacy of "distant healing": A systematic review of randomized trials. *Annals of Internal Medicine,* 132, 903–910. Benor, D. (1990). Survey of spiritual healing research. *Complementary Medical Research,* 4, 9–33. Benor, D. J. (1993). *Healing research,* Vols. 1–2. Munich, Germany: Helix Verlag. Jonas, W. G. (2001). The middle way: Realistic randomized controlled trials for the evaluation of spiritual healing. *The Journal of Alternative and Complementary Medicine,* 7(1), 5–7. Krucoff, M. W., Crater, S. W., Green, C. L., Maas, A. C., Seskevich, J. E., Lane, J. D., Loeffler, K. A., Morris, K., Bashore, T. M., and Koenig, H. G. (2001). Integrative noetic therapies as adjuncts to percutaneous intervention during unstable coronary syndromes: Monitoring and actualization of noetic training (MANTRA) feasibility pilot. *American Heart Journal,* 142(5), 760–767. Walach, H. (2001). The efficacy paradox in randomized control trials of CAM and elsewhere: Beware of the placebo trap. *The Journal of*

Alternative and Complementary Medicine, 7(3), 213–218. Wiesendanger, H., Werthmueller, L., Reuter, K., and Walach, H. (2001). Chronically ill patients treated by spiritual healing improve in quality of life: Results of a randomized waiting-list controlled study. *The Journal of Alternative and Complementary Medicine,* 7(1), 45–51.

15. Byrd, R. (1988). Positive therapeutic effects of intercessory prayer in a coronary care unit population. *Southern Medical Journal,* 18(7), 826–829. Cha, K. Y., Wirth, D. P., and Lobo, R. (2001). Does prayer influence the success of in vitro fertilization-embryo transfer? Report of a masked, randomized trial. *Journal of Reproductive Medicine,* 46(9), 781–787. Harris, W. G., Gowda, M., Kolb, J. W., Strychacz, C. P., Vacek, J. L., Jones, P. G., Forker, A., O'Keefe, J. H., and McCallister, B. D. (1999). A randomized, controlled trial of the effects of remote, intercessory prayer on outcomes in patients admitted to the coronary care unit. *Archives of Internal Medicine,* 159(19), 2273–2278. Matthews, W. J., Conti, J. M., and Sireci, S. G. (2001). The effects of intercessory prayer, positive visualization, and expectancy on the well-being of kidney dialysis patients. *Alternative Therapies in Health and Medicine,* 7(5), 42–52. O'Laoire, S. (1997). An experimental study of the effects of distant, intercessory prayer on self-esteem, anxiety, and depression. *Alternative Therapies in Health and Medicine,* 3(6), 38–53. Sicher, F., Targ, E., Moore, D., and Smith, H. S. (1998). A randomized double-blind study of the effect of distant healing in a population with advanced AIDS: Report of a small-scale study. *Western Journal of Medicine,* 169(6), 356–363. Walker, S. R., Tonigan, J. S., Miller, W. R., Comer, S., and Kahlich, L. (1997). Intercessory prayer in the treatment of alcohol abuse and dependence: A pilot investigation. *Alternative Therapies in Health and Medicine,* 3(6), 79–86.

16. Leibovici, L. (2001). Beyond science? Effects of remote, retroactive intercessory prayer on outcomes in patients with bloodstream infection: Randomized controlled trial. *British Medical Journal,* 323, 1450–1451.

17. Dossey, L. (2002). How healing happens: Exploring the nonlocal gap. *Alternative Therapies in Health and Medicine,* 8(2), 12–16, 103–110.

18. Persinger, M. A. (1979). ELF field mediation in spontaneous psi events: Direct information transfer or conditioned elicitation? In C. Tart, H. Puthoff, and R. Targ (Eds.), *Mind at large* (pp. 191–204). New York: Praeger.

19. Bohm, D. (1980). *Wholeness and the implicate order.* London: Routledge and Kegan Paul. Sheldrake, R. (1981). *A new science of life.* Los Angeles, CA: J. P. Tarcher.

20. McTaggart, L. (2001). *The field: The quest for the secret force of the universe.* London: HarperCollins.

21. Rauscher, E. A., and Targ, R. (2001). The speed of thought: Investigation of a complex space-time metric to describe psychic phenomena. *Journal of Scientific Exploration,* 15(3), 331–354.

22. Watts, A. W. (1967). The world is your body. In *The book: On the taboo against knowing who you are* (pp. 80–100). New York: Collier.

23. May, E. C., Spottiswoode, S. J. P., and James, C. L. (1994). Shannon entropy: A possible intrinsic target property. *Journal of Parapsychology,* 58, 384–401. Spottiswoode, J. (1997a). Apparent association between effect size in free response anomalous cognition experiments and local sidereal time. *Journal of Scientific Exploration,* 11, 109–122. Spottiswoode, J. (1997b). Geomagnetic fluctuations and free response anomalous cognition: A new understanding. *Journal of Parapsychology,* 61, 3–12.

24. Schrödinger, E. (1993). *What is life?* with *Mind and matter* and *Autobiographical*

sketches (p. 89). Cambridge: Cambridge University Press. (Original work published 1944)

25. Bergson, H. (1935). *The two sources of morality and religion.* New York: Henry Holt and Company.

26. Aanstoos, C. (1986). Psi and the phenomenology of the long body. *Theta,* 13(14), 49–51. Roll, W. (1989) Memory and the long body. In L. Henkel and R. Berger (Eds.), *Research in parapsychology 1988* (pp. 67–72). Metuchen, NJ: Scarecrow Press.

27. Braud, W. G. (1997). Parapsychology and spirituality: Implications and applications. In C. T. Tart (Ed.), *Body, mind, and spirit: Exploring the parapsychology of spirituality* (135–152). Charlottesville, VA: Hampton Roads.

William Braud
Mountain View, CA
June 21, 2002

1

Transpersonal Imagery Effects: Influencing a Distant Person's Bodily Activity Using Mental Imagery

William Braud and Marilyn Schlitz

The imagination of man can act not only on his own body, but even on others and very distant bodies. It can fascinate and modify them; make them ill, or restore them to health.

—Ibn Sînâ

This chapter describes experiments in which persons were able to influence the autonomic nervous system activity of other, distant, persons using mental techniques of imagery and intention.

This information originally was presented at the Second World Conference on Imagery, Toronto, Canada, 1987. Later, the work was published in Braud, W. G., and Schlitz, M. J. (1989). A methodology for the objective study of transpersonal imagery. Journal of Scientific Exploration, 3, *43–63. Copyright © 1989 by the* Journal of Scientific Exploration. *Used with permission.* —William Braud

Abstract—Abundant methodologies already exist for the study of *preverbal imagery*, in which one's imagery acts upon one's own cellular, biochemical, and physiological activity. This paper reports a new methodology for the objective study of *transpersonal*

A condensed version of this paper was presented at the Second World Conference on Imagery, Toronto, Canada, June 25–28, 1987.

Acknowledgement. We are grateful to Dr. Dean Radin for his helpful suggestions and comments on this paper.

imagery, in which one person's imagery may influence the physical reactions of *another* person. The method involves the instructed generation of specific imagery by one person and the concurrent measurement of psychophysiological changes in another person who is isolated in a distant room to eliminate all conventional sensorimotor communication. Thirteen experiments were conducted using this methodology. A significant relationship was found between the calming or activating imagery of one person and the electrodermal activity of another person who was isolated at a distance (overall $z = 4.08$, $p = .000023$, mean effect size = 0.29). Potential artifacts which might account for the results are considered and discounted. The findings demonstrate reliable and relatively robust anomalous interactions between living systems at a distance. The effects may be interpreted as instances of an anomalous *"causal"* influence by one person directly upon the physiological activity of another person. An alternative interpretation is one of an anomalous *informational* process, combined with unconscious physiological self-regulation on the part of the influenced person. Additional research is being conducted in an attempt to increase our understanding of the processes involved, as well as to learn the various physical, physiological and psychological factors that may increase or decrease the likelihood of occurrence of the effect.

Introduction

In her book *Imagery in Healing* (Achterberg, 1985), psychologist Jeanne Achterberg distinguished two types of imagery which may have positive impacts upon health. In *preverbal imagery,* the imagination acts upon one's own physical being to alter cellular, biochemical, and physiological activity. The study of such imagery has a long history, and there exist a variety of successful methodologies for its objective evaluation. The second type of imagery that Achterberg identified is *transpersonal imagery,* which "embodies the assumption that information can be transmitted from the consciousness of one person to the physical substrate of others" (p. 5). She suggested that the validation of transpersonal imagery must be sought in the more qualitative types of observational data gathered by anthropologists, theologians, and medical historians, and in intuitive philosophical speculation.

Indeed, the power of preverbal imagery in influencing one's own chemical, cellular, physiological and behavioral reactions has been well documented. We find extensive evidence for such psychosomatic influences in the areas of dreaming, hypnosis, relaxation, autogenic training, biofeedback, meditation, therapeutic imagery, mental rehearsal, and placebo effects. Some of the most exciting (and potentially useful) findings regarding the influence of imagery on somatic functioning are now being reported by researchers within the new interdisciplinary field of psychoneuroimmunology, in which it is being discovered that individuals, through use of relaxation, hypnosis, and imagery techniques, may be able to exert rapid and quite specific influences upon certain subpopulations of their white blood cells (see, for example, Hall, 1984a, 1984b, 1987; Peavey, 1982; Schneider, Smith, and Whitcher, 1984).

Less well known are the various observations which tend to support the reality and effectiveness of transpersonal imagery effects. There are, of course, abundant anecdotes and field observations that the sensations, thoughts, feelings and images of one person may, under certain conditions, directly affect the bodily reactions of another person, even when the two persons are separated by great distance, and when the influenced person is not aware that an influence attempt is being made. Observations of ostensible distant mental influence in the context of anthropology have been reviewed by Angoff and Barth (1974), Long (1977) and Van de Castle (1977). The late Eric Dingwall, in his four-volume work, *Abnormal Hypnotic Phenomena* (1968), surveyed many cases of putative distant mental influence which occurred in 19th-century practices of hypnosis (or "mesmerism," as it was then called). Two of the more interesting of these "higher phenomena of hypnosis" were (a) *community of sensation,* in which hypnotized subjects were reported to have responded appropriately to sensory stimuli presented to a distantly located hypnotist, and (b) *mental suggestion,* in which the hypnotist was alleged to have exerted an influence upon a distant subject's behavior (while the latter was in a hypnotic "trance") or even to have induced hypnosis itself at a distance. These phenomena, as well as the results of more modern hypnotic investigations, have been examined by Honorton (1974, 1977). Finally, possible distant mental influence effects occurring within the context of mental healing have been reviewed by Ehrenwald (1977) and by Solfvin (1984).

The possibility of distant somatic effects of imagery is also suggested by anecdotal reports of various investigators involved in clinical biofeedback applications who sometimes observed unusual correlations between the changes in electrophysiological activity of one client and those of another client (in group biofeedback training sessions) or between the client's activity and that of the investigator himself or herself. If such coincident physiological patterns are reliable and applicable, they might be explained most parsimoniously by assuming that they result from either (a) gross or subtle external stimuli that influence both persons in the same manner, or (b) internal rhythms that happen to be in phase in the two persons and interact with the monitored activities in identical ways. A third possibility, however, is that at least some proportion of these physiological congruences may be attributable to transpersonal imagery effects. Such a possibility would be highly speculative were it not for several reports of experimental findings of similar interactions between, for example, the electroencephalic (Duane and Behrendt, 1965; Putoff and Targ, 1976; Targ and Putoff, 1974) or autonomic (Dean, 1966) activity of one person and that of another person, when those persons were remotely situated, shielded, and the possibility of conventional energetic and informational exchanges between them had been eliminated. Indeed, the entire body of research findings in the areas of psychical research and of parapsychology is relevant to and supportive of the notion that the mental activity of one person may influence the bodily activity of another person at a distance. Quite complete and useful reviews of the concepts, methods, findings, and theories of modern parapsychology may be found in Edge, Morris, Palmer and Rush (1986); Krippner (1977, 1978, 1982, 1984); Nash (1986); and Wolman (1977).

The Present Research Program: Purpose and Overview

In this paper, we describe an objective, quantitative methodology for the study of transpersonal imagery which allows the investigation of the latter within the framework of experimental psychology. In addition to the methodology itself, we shall present the promising results of 13 experiments that we already have conducted in order to test the usefulness of the procedure.

The method involves the instructed generation of specific imagery by one person, and the concurrent measurement of psychophysiological changes in *another* person. Throughout the experiment, the two persons occupy separate, isolated rooms, and all conventional sensorimotor communication between the two persons is eliminated in order to insure that any obtained effects are truly transpersonal. In a typical experiment, Person A is instructed to use specific mental imagery in order to induce a specific physiological change in Person B, who is isolated in a distant room. The expected psychophysiological effect is assessed by measuring the spontaneous electrodermal activity (skin resistance responses, SRR) of Person B during 20 30-second recording epochs. During 10 of these epochs, interspersed randomly throughout the sequence of 20 epochs, Person A generates imagery designed to produce a specific somatic effect (decreased sympathetic nervous system activity in some cases, increased sympathetic activation in other cases); the remaining 10 epochs serve as Control periods during which Person A does not generate the relevant imagery. Person B is, of course, unaware of the sequence of the two types of epochs (the sequence is randomly determined) and is also "blind" to the exact starting time of the experiment, the number and timing of the various periods, etc. Electrodermal activity is objectively assessed by an electrodermal amplifier interfaced with an analog-to-digital converter and a microcomputer. The amount of electrodermal activity during the Imagery epochs is compared with that of the Control epochs using conventional parametric statistical techniques.

If the experimental protocol just described is not violated, and yet it is found that significantly greater somatic activity of an appropriate, imagery-relevant type is found to occur during the Imagery periods than during the Control periods, we can conclude with confidence that a transpersonal imagery effect (TIE) has occurred, and that the results cannot be attributed to

(a) conventional communication channels or cues (since the two parties are isolated from contact with each other through the use of distant, isolated rooms), (b) common external signals, common internal rhythms, or rational inference of the imagery/nonimagery schedule and resultant appropriate self-regulation (since the imagery/nonimagery schedule is truly randomly determined and is unknown to Person B), or (c) "chance coincidence" (since the level of responding to be expected on the basis of chance alone may actually be determined and compared statistically with the obtained response levels).

Method

Subjects

The experiments involved the participation of unpaid male and female volunteer subjects, ranging in age from 16 to 65 years. Participants were selected from a pool of volunteers from the San Antonio community who had learned about the Foundation's experiments through local newspaper advertisements and articles, notices posted throughout the city, lectures given by Foundation staff at local colleges and universities, and comments from other participants, and whose interest in the experiments and time schedules permitted participation. Approximately equal numbers of males and females participated in the various studies. In most cases, participants were not selected on the basis of any special physical, physiological, or psychological characteristics, and could best be described as "self-selected" on the basis of their interest in the topics being researched. In only one experiment were "special" subjects recruited and selected. This was an experiment in which we were interested in whether persons having a greater "need" for a possible *calming* influence would evidence stronger results than persons without such a need. Therefore, for that experiment, we selected individuals who self-reported symptoms of greater than usual sympathetic autonomic activation—i.e., stress-related complaints, excessive emotionality, excessive anxiety, tension headaches, high blood pressure, ulcers, or mental or physical hyperactivity. The subjects for this experiment were also screened in an initial electrodermal activity recording session to guarantee that they did in fact exhibit greater than average sympathetic autonomic activity.

The persons who served as "influencers" in these experiments (i.e., those who regulated their own images and intentions in order to influence the subjects at a distance) were selected from a similar pool of volunteers. In some experiments, the experimenters themselves served as influencers. In still other experiments, the influencers were individuals who were interested in unorthodox healing and who themselves practiced certain mental healing techniques, such as "therapeutic touch" (see Borelli and Heidt, 1982; Krieger, 1979; Kunz, 1985) or "Reiki healing" (see Schlitz and Braud, 1985). Many of the influencers were practitioners of various forms of meditation and self-exploration.

In most cases, however, the influencers were simply interested persons from the local community who wished to give the experiments a try.

The authors served as the experimenters for the series of studies, assisted in some experiments by two other experimenters, J. C. and H. K. The first author had extensive research experience in the areas of experimental psychology, physiological psychology, and parapsychology. The second author had extensive experience in parapsychological and anthropological research. The third experimenter, J. C., had research experience in the area of nursing. The fourth experimenter, H. K., was a student at a local college who was participating in a research practicum at the Foundation.

In all, 337 persons participated in these experiments. Of these participants, 271 served as subjects, 62 as influencers, and 4 as experimenters.

Procedure

Physical Layout. During the experimental sessions, it was essential to guarantee that the influencer and the experimenter would not be able to communicate with the subject via conventional sensorimotor channels. This was accomplished by situating the experimenter and the influencer in one closed room, while the subject occupied a distant second room, which was also closed. Figure 1 illustrates the floor plan of the rooms used in Experiments 1 through 10. The rooms used in Experiments 11 through 13 are shown in Figure 2. The distance (20 meters or more) between the two rooms used in an experiment, and the presence of several intervening closed doors and corridors, isolated the participants from possible sensory interaction. Additionally, verbalization of any information regarding the imagery/non-imagery schedule (see below) by the influencer or the experimenter was not allowed during the experimental sessions. There were no active microphones in either room, through which participants could communicate. The headphones through which the participants in the two rooms received required auditory information were attached to independent electrical circuits so that possible "crosstalk" between two sets of headphones was eliminated (i.e., it was impossible for one person's headphone to function as a microphone for the other person's headset).

Subject's Instructions and Activities. Throughout an experimental session, the subject sat in a comfortable armchair in a dimly illuminated, closed room. In Experiments 1 and 3, the subject was exposed to visual and acoustic ganzfeld stimulation throughout the session (see Bertini, Lewis, and Witkin, 1964; Schacter, 1976); this was accomplished by having the subject view a uniform red light field through translucent, hemispherical acetate eye covers while listening to moderately loud white noise through headphones. In Experiments 2 and 4, ganzfeld stimulation was not employed; rather, the subject simply sat quietly in the dim room, with freedom to open or close the eyes as desired. In Experiments 5 through 13, the subject watched randomly changing patterns of colored lights on a 12-inch display screen 2 meters away, while listening to computer-generated random sounds through headphones. The subject was instructed to make no deliberate effort to relax or to become more active, but rather to remain in as ordinary a condition as possible and to be open to and accepting of a possible influence from the distant influencer whom he or she had already met. The subject remained unaware of the number, timing or scheduling of the various influence attempts, and was instructed not to try to guess consciously when influence attempts might be made. The subject was asked to allow his or her thought processes to be as variable or random as possible and to simply observe the various thoughts, images, sensations, and feelings that came to mind without attempting to control, force, or cling to any of them.

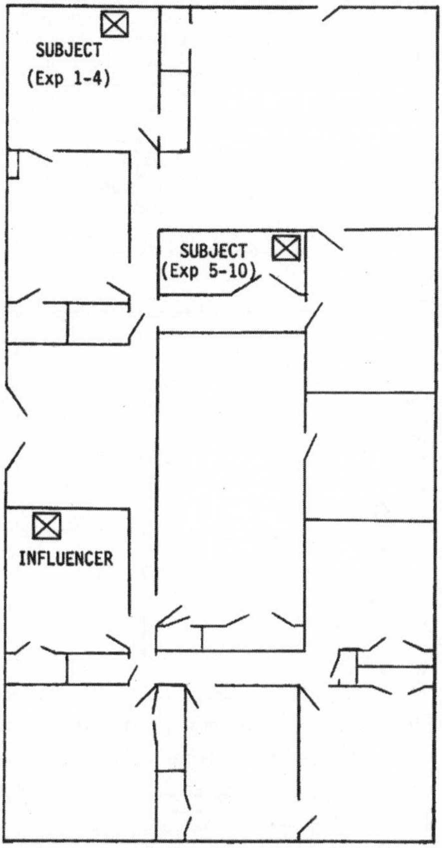

Fig. 1. Laboratory floor plan showing locations of subject and influencer for Experiments 1 through 10.

Influencer's Instructions and Activities. The influencer sat in a comfortable chair in front of a polygraph in another closed room. The polygraph provided a graphic analog readout of the concurrent electrodermal activity of the distant subject. For half of each session for Experiment 6, this polygraph was turned off and no feedback was allowed. For all other sessions of all other experiments, polygraph feedback information about the momentary physiological activity of the subject was available to the influencer. The influencer had the option of attending to this polygraph feedback or ignoring it. In most cases, the influencer watched the polygraph tracing throughout a session. In some cases, the influencer closed his or her eyes and ignored the polygraph tracing during the actual 30-second imagery or non-imagery periods (see below), but looked at the tracings following those periods in order to learn of the success or failure of the influence attempts.

An experimental session contained 20 30-second recording periods or epochs. Each epoch was signaled to the experimenter and to the influencer by an

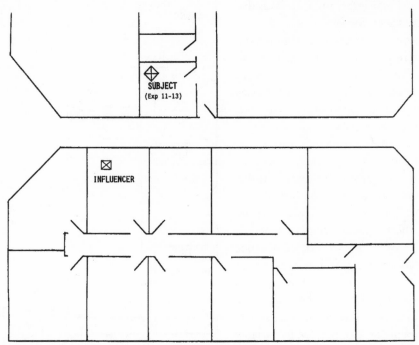

Fig. 2. Laboratory floor plan showing locations of subject and influencer for Experiments 11 through 13; subject and influencer rooms are in separate suites of the same building, separated by an outside corridor and several closed doors.

auditory signal that could not be heard by the distant subject. Immediately before each signal, the experimenter exposed a card to the influencer. This card contained an instruction for the upcoming epoch. The word "influence" indicated that the next 30-second period was to be an imagery epoch during which the influencer would attempt to influence the distant subject; the word "control" indicated a nonimagery or noninfluence period. The influencer had been instructed beforehand that during each influence period, he or she was to attempt to influence the electrodermal activity of the distant subject through the use of self-generated imagery. In some experiments (Experiments 5, 6, 8, 10 and 11), the goal of the imagery influence attempts was the *calming* of the distant subject—the reduction of the subject's sympathetic autonomic nervous system activity and hence the reduction of the frequency and magnitude of spontaneous skin resistance responses. In other experiments (Experiments 2, 4, and 7), the goal of the imagery influence attempts was the *activation* of the distant subject— an increase in the subject's sympathetic autonomic nervous system activity and hence an increase in the frequency and magnitude of spontaneous skin resistance responses. In still other experiments (Experiments 1, 3, 9, 12 and 13), both calming and activation strategies were used within a single session; in those experiments, there were 10 calm-aim periods and 10 activate-aim periods.

During control periods, the influencer attempted not to think about the subject or about the experiment, and to think of other matters. During influence periods, the influencer used the following strategies (either alone or in combination) in an attempt to influence the somatic activity of the distant subject.

1. The influencer used imagery and self-regulation techniques in order to induce the intended condition (either relaxation or activation, as demanded by the experimental protocol) in *himself or herself,* and imagined (and intended for) a corresponding change in the distant subject.

2. The influencer imagined the *other person* in appropriate relaxing or activating settings.

3. The influencer imagined the desired outcomes of the polygraph pen tracings—i.e., imagined few and small pen deflections for calming periods and many and large pen deflections for activation periods.

There were rest periods, ranging in duration from 15 seconds to 2 minutes in the various experiments, between the 30-second recording epochs. During those periods, the influencer was able to rest and to prepare for the upcoming epoch.

Scheduling of Influence Attempts. In order to eliminate the possible influence of common internal rhythms and to remove the possibility that the influencer and the subject just happened to respond at whim in the same manner and at the same times, it was necessary to *formally assign* to the influencer specific times for engaging in imagery; such assignments had to be truly random and, of course, could not be known to the subject (lest the subject self-regulate his or her own physiology on the basis of such knowledge, in order to confirm the expectations of the experimenter). The subject's blindness with respect to the imagery/nonimagery sequence was maintained by keeping *all* participants (including the experimenter) blind regarding the sequence until preparatory interactions with the subject had been completed and the session was about to begin. Only then, when the subject and the influencer/experimenter team were stationed in their separate rooms, did the experimenter become aware of the proper epoch sequence for that session. In Experiments 1 and 3, the epochs were scheduled in a truly random manner by means of an electronic binary random event generator (see Schmidt, 1970). In Experiments 2 and 4, the epochs were randomly scheduled by means of a set of 20 cards (10 influence and 10 control cards) which were shuffled by the experimenter 20 times before each session. In the remainder of the experiments, the epochs were scheduled in an ABBA or BAAB sequence; the experimenter learned whether a particular session's sequence was to be ABBA or BAAB by consulting a sealed envelope immediately before the beginning of each session. The envelopes had been prepared beforehand by someone who had no further role in the experiments. The "preparer" had prepared each session's sequence envelope through the use of a table of random numbers, with the only restriction for its use being the occurrence of equal numbers of ABBA and BAAB sequences in an experiment. [We used an ABBA design in order to minimize possible progressive error in the experiments; such a design allows any progressive error (i.e., the contribution of any extraneous variable which varies systematically with time) which may have autonomic concomitants to contribute equally to the A and B periods, thus avoiding a biasing contribution to any one condition alone.]

Monitoring of Electrophysiological Activity. The subject's sympathetic autonomic nervous system activity was assessed by monitoring his or her spontaneous skin resistance responses (SRR) on a continuous basis throughout the 20 minutes of an experimental session. In Experiments 1 and 3, SRR activity was recorded by means of silver/silver chloride electrodes (7.0 mm in diameter) with partially conductive electrode gel, attached by adhesive collars to the subject's right palm. Phasic electrodermal activity was recorded by means of a Stoelting Model SA 1473 GSR amplifier and a Stoelting Model 22656 Multigraphic Recorder. Sensitivity was adjusted so that an internal calibrating signal of 1.0 kilohm resulted in a 10.0 mm recording pen deflection. In Experiments 2 and 4, a Lafayette Model 76405 multiplex GSR amplifier was used, along with the Stoelting chart-mover/penwriter described above; chrome-plated stainless steel finger electrodes (each with a surface area of 585 mm²) without electrode paste were attached to the first and third fingers of the left hand by means of Velcro bands. In Experiments 5, 6, and 7, the Lafayette amplifier was used along with a Harvard Apparatus chart mover and pen writers; the steel/pasteless finger electrodes were attached to the subject's right hand. In Experiments 8 through 13, the Lafayette amplifier and Harvard chart recorder were used, but with silver/silver chloride electrodes and partially conductive gel; electrodes were attached to the subject's right palm. For Experiments 1 through 4, electrodermal activity was evaluated by blind-scoring of pen tracings by someone who had no other role in the experiments.[1] For Experiments 5 through 13, scoring was automated through the addition of an analog-to-digital converter interfaced with a microcomputer. This equipment sampled the subject's SRR activity 10 times each second for the 30 seconds of a recording epoch and averaged these measures, providing what is virtually a measure of the area under the curve described by the fluctuation of electrodermal activity over time (i.e., the mathematically integrated activity). The computer provided a paper printout of the results at the end of the session. For all experiments, with the exception of Experiment 13, a 5-minute adaptation/habituation period for the subject preceded the actual experimental session. For Experiment 8, other physiological measures were recorded in addition to electrodermal activity (*viz.,* pulse rate, hand temperature, breathing rate, and electromyographic activity of the frontalis muscle group); those measures, however, will not be described in this paper.

Assessment of Physiological Responses. Each session of each experiment yielded 10 assessments of electrodermal (SRR) activity recorded during an influencer's attempts to influence that activity in a specific direction using specific imagery, and 10 assessments of activity recorded in the absence of such attempts. (The sole exception to this occurred in Experiment 13, in which there was a total of only 12 recording epochs for each session, rather than the usual 20.) Our evaluation of whether the influencer's imagery influenced the subject's somatic activity was carried out on a session-by-session basis, and involved a determination of the proportion of somatic activity in the prescribed direction which occurred during the influence periods, relative to its occurrence during control periods. For each session, we calculated the total activity for that session by summing the SRR scores for all 20 30-second recording epochs (or for all 12 epochs, in the case of Experiment 13). Next, we calculated the activity that occurred during the 10 30-second influence or imagery epochs of a session by summing those 10 scores; separately, we calculated the activity occurring during the 10 30-second control (i.e., noninfluence or nonimagery) epochs of the session by summing those 10 scores. Dividing the influence and control sums, respectively, by the total activity yielded two activity proportions. In the absence of a transpersonal imagery effect (TIE), each of these two proportions would be expected to approximate 0.50; i.e., on the basis of chance alone, half of a subject's total electrodermal activity would be expected to occur during the influence periods and half during the control periods. A significant departure of these proportions from 0.50, in the appropriate predicted direction, would constitute evidence for the presence of a transpersonal imagery effect.

Results

We have completed 13 experiments using the methodology described above. Experiments 1, 2, 3, 4, and 11 were "demonstration studies" conducted to test the effectiveness of the method with different samples of subjects and influencers. In the remaining eight experiments, we sought to determine how the transpersonal imagery effect might be influenced by various psychological factors. Since our purpose in this paper is to describe the method itself, we shall not present the rationales, details, or specific outcomes of the individual experiments, but will limit our remarks to the common features of the studies and to their overall results.

In each experiment, the primary method of analysis involved a comparison of the proportion of electrodermal activity which occurred during the imagery influence epochs of a session with the proportion expected on the basis of chance alone, i.e., 0.50. Chi-square goodness of fit tests indicated that the distribution of obtained session scores did not differ significantly from a normal distribution; therefore, parametric statistical tests were used for their

Table 1. Quantitative Summary of Transpersonal Imagery Experiments

Experiment	Influencer(s)	Number of Sessions	Hit Sessions	Mean % Influence	t	p	z	d	Type of Study
1	Experimenter	10	9	9%	3.07	.0065	2.73	.97	Demonstration
2	M. M.	10	8	9%	2.04	.035	1.81	.64	Demonstration
3	10 unselected volunteers	10	8	8%	2.96	.0077	2.42	.94	Demonstration
4	10 unselected volunteers	10	5	-3%	-0.76	.736	-0.63	-.24	Demonstration
5	Experimenters	16	12	10%	2.40	.014	2.20	.60	Need (greater)[4]
5	Experimenters	16	6	0%	-0.09	.537	-0.04	-.02	Need (lesser)[4]
6	24 unselected volunteers	24	17	7%	1.77	.043	1.72	.36	Feedback (within)[5]
7	Experimenters	32	16	3%	1.15	.13	1.13	.20	Blocking[5]
8	Experimenters	30	15	2%	0.45	.33	0.44	.08	Specificity[5]
9	Experimenters	30	18	1%	0.44	.33	0.43	.08	Direction[6]
10	Experimenters	16	7	3%	1.31	.10	1.28	.33	Magnitude (within)[6]
11	3 healing practitioners	15	9	1%	0.62	.28	0.58	.16	Demonstration (Reiki method)[7]
12	5 selected volunteers	40	19	1%	0.21	.41	0.23	.03	IDS pilot (within)[8]
13	8 selected volunteers	32	20	7%	2.14	.02	2.08	.38	IDS confirmation single seed (within)[8]
13	8 selected volunteers	32	14	-2%	-0.53	.70	-0.52	-.09	IDS confirmation multiple seeds (within)[8]

evaluation. Single-mean t tests were used to compare the obtained session scores with an expected mean of 0.50.

Summary statistics for the 13 experiments are presented in Table 1.

For experiments (such as Experiments 5 and 13) in which significant differences obtained between different subconditions and/or in cases in which *a priori* decisions had been made to evaluate certain groups separately, scores are presented for each subcondition; otherwise, scores of subconditions are combined and presented for the experiment as a whole. The number of sessions contributing to each experiment varied from 10 to 40. The single-mean t tests produced independently significant evidence for the transpersonal imagery effect (i.e., an associated p of 0.05 or less) in 6 of the possible 15 cases, yielding an experimental success rate of 40%. The experimental success rate expected on the basis of chance alone is, of course, 5%.

Results for the 13 experiments are presented in another form in Figure 3. For this presentation, we calculated z scores and effect size scores for the overall results of each experiment. The z scores were calculated according to the Stouffer method [see Rosenthal (1984)] which involves converting the studies' obtained p values into z scores, summing these z scores, and dividing by the square root of the number of studies being combined; the result is itself a z score that can be evaluated by means of an associated p value. For Figure 3, this method was used to provide an overall or combined z score for each of the 13 experiments, for ease of graphical portrayal. The effect sizes shown in Figure 3 are "Cohen d" measures which are recommended by those interested in meta-analyses of scientific experiments [see Cohen (1969); Glass, McGaw, and Smith (1981); Rosenthal (1984)]; the effect sizes were calculated according to the formula $d = t\sqrt{1/n}$. These effect sizes varied from −0.24 to 0.97, with a

Fig. 3. Overall z scores and effect sizes (Cohen's d measures) for the 13 successive transpersonal imagery experiments.

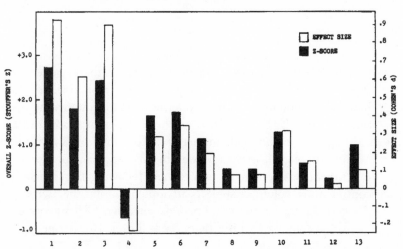

mean $d = 0.29$, and compare favorably with effect sizes typically found in traditional behavioral research.

A global analysis of the 13 experiments is presented in Table 2. There were 15 assessments of the transpersonal imagery effect. Contributing to those assessments were 323 sessions conducted with 271 different subjects, 62 influencers, and 4 experimenters. Six of the 15 assessments (40%) were independently significant statistically ($p < .05$); this is to be compared with the 5% experimental success rate expected by chance. Fifty-seven percent of the *sessions* were successful (i.e., these were sessions in which the influence imagery epochs accounted for more than 50% of the subject's electrodermal activity during activation attempts and less than 50% of the total activity during calming attempts); this is to be compared with the 50% session success rate to be expected on the basis of chance. The overall mean magnitude of the TIE for all experiments differed from chance expectation by 3.73%; when only the six independently significant experiments are considered, the obtained mean TIE had a magnitude of 8.33%. The two most important entries of Table 2 are the combined z score (for the experimental series as a whole, calculated according to the Stouffer method) and the mean effect size (Cohen's d, for the entire series). The overall z is 4.08 and has an associated $p = .000023$; the average effect size for all 13 experiments is 0.29.

Table 2. Summary Statistics for Transpersonal Imagery Experiments

Total Experiments	Psi Assessments	Number of Sessions	Number of Subjects	Number of Influencers
13	15	323	271	62

Successful Studies	Successful Sessions	Mean Percent Influence		Combined z	Overall p	Mean Effect Size (d)
		All	Successful			
6/15 (40%) (MCE = 5%)	183/321 (57%) (MCE = 50%)	3.73%	8.33%	4.08	.000023	.29

Inspection of Table 1, Figure 3, and Table 2 indicates that the effect occurring in these 13 experiments is a relatively consistent, replicable, and robust one. It should also be pointed out that, in terms of its magnitude, the effect is not a negligible one. Under certain conditions, the transpersonal imagery effect can compare favorably with an imagery effect upon one's own physiological activity. Although it is not reviewed in this paper, an autonomic self-control experiment was conducted immediately following Experiment 5. In the self-control study, volunteer subjects attempted to calm themselves using relaxing imagery during 10 30-second periods, and their SRR activity during those periods was compared with activity levels during 10 interspersed nonimagery control periods. The strength of the self-control imagery effect in that study (an 18.67% deviation) did not differ significantly from the strongest transpersonal imagery effect of Experiment 5 (a 10% deviation).

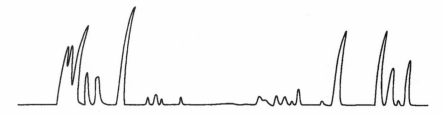

Fig. 4. Sample of polygraph tracing of electrodermal activity.

Discussion

The results of this series of 13 experiments indicate that the present methodology is effective for the objective assessment of transpersonal imagery effects. It was demonstrated that the psychophysiological activity of one person varied, to a significant degree, with the imagery content of another person. The experimental design guaranteed that the effect could not be attributed to conventional sensorimotor cues, common external stimuli, common internal rhythms, or chance coincidence. A number of additional potential artifacts can be mentioned here and can be effectively dismissed.

1. *The findings are the result of recording errors and motivated misreadings of polygraph records.* This explanation is rejected on the basis of blind-scoring of polygraph records (see Figure 4 for sample record) and, later, by the use of completely automated assessment techniques and computer-scoring of response activity.

2. *The subjects knew beforehand when influence attempts were to be made and "cooperated" by changing their own autonomic activity when appropriate.* This explanation may be rejected because the subjects were not told when or how many influence attempts would be made, nor was the experimenter aware of the influence/control epoch schedule until all preliminary interactions with the subject had been completed. Subjects did not know of the existence of, or have access to, the envelopes containing schedule information.

3. *Subjects could have become aware during the experimental sessions themselves of when influence epochs were in progress and could have altered their own physiological reactions during those periods.* This possibility was eliminated by isolating the subject from any such cues from the influencer. Subject and

influencer were in separate, closed rooms at least 20 meters apart. No auditory cues could have impinged upon the subject to indicate when recording epochs were in progress or whether such epochs were influence or control periods. Neither the influencer nor the experimenter made any vocalizations that could have informed the subject about whether influence or control periods were in progress. The epoch-indicating tones heard by the influencer and experimenter, and the random tones heard by the subject, were provided by independent audio systems which eliminated the possibility of electrical crosstalk and also the possibility of headphones functioning as microphones and inadvertantly cueing the subject.

4. *Differences in autonomic activity between influence and control periods are due to systematic error—i.e., some progressive change in electrodermal activity over time.* This objection may be rejected. Progressive (time-based) errors could have been contributed by (a) changes in equipment sensitivity as the equipment warmed up, (b) changes in electrodermal activity due to adaptation or habituation to the experimental environment, or (c) changes in electrodermal activity due to polarization of the recording electrodes. Equipment was allowed to warm up for 15 to 20 minutes prior to the beginning of a session and therefore had become thermally stable before the experiment began. The use of electrodes with large surface areas, and the use of a constant-current electrodermal recording device reduced the possibility of polarization problems. The use of silver/silver chloride electrodes and partially conductive paste in other experiments further minimized a polarization problem. A special analysis of the data from Experiment 5 is relevant to the habituation question. Statistical evaluation of total electrodermal activity for the first halves *versus* the second halves of the sessions indicated no evidence of an habituation effect. This absence of habituation could be attributed to the use of an adaptation period before the actual recording session began, and to the use of constantly changing auditory and visual stimulation of the subject (i.e., the use of the random tones and colored lights display). Thus, there was no progressive change in electrodermal activity due to any of the three possible processes mentioned above. However, even if a progressive change had occurred, the use of the ABBA counterbalanced design and the use of truly random influence/control sequencing in other experiments would have prevented this error from contributing differentially to influence *versus* control epochs.

5. *The findings are due to arbitrary selection of data.* This explanation may be rejected since total numbers of subjects and trials were prespecified, and the analyses reported include all recorded data.

6. *The results are due to fraud on the part of the subjects.* This explanation may

be rejected. The subjects were unselected volunteers; it may be assumed that such subjects had no motive for trickery. However, even if a subject were motivated to cheat, such an opportunity was not present. Cheating would have required knowledge of a session's influence/control epoch sequence and of the precise starting time for the session, or the assistance of an accomplice. Both of these requirements were eliminated.

7. *The results are due to fraud on the part of the experimenters.* No experiment, however sophisticated, can ever be absolutely safe from experimenter fraud. Even if an experiment were controlled by an outside panel of disinterested persons, a hostile critic could still argue that collusion was involved. The imagined extent of such a conspiracy would be limited only by the imagination and degree of paranoia of the critic. We can only state that we used multiple-experimenter designs so that one experimenter's portion of the experiment served as a kind of control for another experimenter's portion. Only the successful replication of these findings by investigators in other laboratories would reduce experimenter fraud to a non-issue. We hope that this report will stimulate such replication attempts.

We conclude that our results cannot be attributed to any of the various potential artifacts or confounds mentioned above, and therefore are not spurious. Rather, the results reflect an anomalous psychophysical interaction between two individuals separated from one another in space.

A Range of Reactions

In addition to responding physiologically in a manner consistent with the imagery of the distant influencer, subjects often reported subjective responses which corresponded to the influencers' images. Sometimes these reports were of relatively vague feelings of relaxation or activation. However, there were also reports of extremely *specific* thoughts, feelings, and sensations which strikingly matched the imagery employed by the influencer. For example, a subject reported spontaneously that during the session he had a very vivid impression of the influencer coming into his room, walking behind his chair, and vigorously shaking the chair; the impression was so strong that he found it difficult to believe that the event had not happened in reality. This session was one in which the influencer had employed just such an image in order to activate the subject from afar.

Subjects sometimes spontaneously reported mentation which corresponded closely to that of the influencer or the experimenter, even when that mentation was incidental and not employed consciously as part of an influence strategy. For example, at the beginning of one session, the experimenter remarked to an influencer that the electrodermal tracings of the subject were very precise and regimented and that they reminded him of the German techno-pop instrumental musical group, Kraftwerk. When the experimenter went to the subject's room at the end of the session, the subject's first comment was that early in the session, for some unknown reason, thoughts of the group Kraftwerk had come into her mind. The subject could not have overheard the experimenter's earlier comment to the influencer. Such correspondences were not rare.

There appeared to be a continuum of reactions possible for the subject. At one extreme, there would be no resemblance whatsoever between the imagery of the influencer and the imagery and physiological reactions of the subject. Next on the continuum would be cases in which autonomic reactions occurred which were appropriate to the influencer's imagery, but the subject was completely unaware of those reactions. Next were appropriate physiological reactions of which the subject was only vaguely aware, and next would be reactions accompanied by very definite subjective experiences. Closer to the "resemblance" end of this continuum would be cases of reactions accompanied by images in the subject which were virtually identical to those of the influencer. Even closer to the resemblance end would be cases of appropriate electrodermal activity, quite similar imagery, *plus* behavioral and/or gross physical changes consistent with the influencer's imagery. An example of the latter occurred in a subject who experienced a dramatic reddening of the face and neck during a session. Other subjects experienced muscle tremors, tingling of body parts, awareness of a pounding heart and rushing blood, a felt need to take deep breaths, decreased awareness of body parts, etc. Although our overall statistical findings leave no doubt that the subjects' recorded autonomic reactions were in fact related to the imagery of the influencers, such certainty is not possible in the case of these subjective or physical reactions, since no time-correlated records of those latter reactions were kept. Some of the physical symptoms observed may simply have been bodily conditions that were present all along, but which were brought to the subjects' awareness more forcefully during the experimental sessions due to the demand characteristics of the study. However, some reactions may have been directly influenced or even brought about by the influencers' imagery. We intend to pursue this issue more analytically in future studies in which the temporal distribution of such reactions will be monitored by having subjects verbalize their reactions as they are occurring, or indicate unusual or noteworthy feelings by pressing a button that will mark an event channel of the polygraph. This will allow a determination of whether particular experiences or reactions of subjects are "time-locked" to specific images used by influencers during the sessions. A similar monitoring of the details of the influencers' imagery would permit a determination of the most and least effective forms of imagery, and could teach us a great deal about the varieties and manifestations of transpersonal imagery.

Some Preliminary Findings

We indicated earlier that we did not intend to describe specific details of the various 13 experiments in this paper. However, it does seem appropriate to mention some of our preliminary findings and tentative conclusions at this point.

1. The transpersonal imagery effect (TIE) is a relatively reliable and robust phenomenon; this conclusion is based upon overall statistical results.

2. The magnitude of the effect is not trivial, and under certain conditions it compares favorably with the magnitude of a self-regulation effect.

3. The ability to manifest the effect is apparently widely distributed in the population. Sensitivity to the effects appears to be normally distributed among the 271 volunteer subjects tested in these experiments. Altogether, 62 different influencers were able to produce the effect, with varying degrees of success. Many persons were able to produce the effect, including unselected volunteers attempting it for the first time. More practiced individuals seem able to produce the effect more consistently. There are indications of improvements with practice for some influencers.

4. The TIE can occur at a distance, typically 20 meters; greater distances have not yet been explored.

5. Subjects with a greater need to be influenced (i.e., those for whom the influence is more beneficial) seem more susceptible to the effect.

6. Immediate, trial-by-trial analog sensory feedback is not essential to the occurrence of the effect; intention/visualization of the desired outcome is effective.

7. The TIE can occur without the subject's knowledge that such an influence is being attempted.

8. It may be possible for the subject to block or prevent an unwanted influence upon his or her own physiological activity; psychological shielding strategies in which one visualizes protective surrounding shields, screens, or barriers may be effective.[2]

9. Generally, our volunteer participants have not evidenced concern over the idea of influencing or being influenced by another person.

10. The TIE may generalize to other physiological measures (such as heart rate), but the effect may also be intentionally focused or restricted to one of the number of physiological measures.[3]

11. The TIE does not always occur. The reasons for the absence of a significant effect in some experiments of a series which is otherwise successful are not clear. We suspect that the likelihood of a successful TIE may depend upon the presence of certain psychological conditions, in both influencer and subject (and perhaps even in the experimenter), which are not always present. Possible success-enhancing factors may include belief, confidence, positive expectation, and appropriate motivation. Possible success-hindering factors may include boredom, absence of spontaneity, poor mood of influencer or subject, poor interactions or poor rapport between influencer and subject, and excessive egocentric effort (excessive pressure or striving to succeed) on the part of participants. We suspect that the effect occurs most readily in subjects whose nervous systems are relatively "labile" (i.e., characterized by free variability) and are momentarily free from external and internal constraints. Perhaps fullness of intention and intensity or vividness of visualization in the influencer facilitate the effect.

Additional research is needed to determine the validity of these conclusions, and to explore more thoroughly the various physiological and psychological factors which are favorable or antagonistic to the occurrence of the TIE.

Implications and Applications

The methodology employed in these experiments reveals that, under certain conditions, mental imagery does indeed have a transpersonal aspect. The results suggest a fundamental inter-connectedness among people through which the transpersonal imagery effect may be mediated. The findings provide an additional illustration of the power of the imagination. The method extends research possibilities for the further laboratory study of imagery, transpersonal functioning, psychic functioning, emotional contagion, and other related processes.

If the effects of transpersonal imagery prove to be sufficiently strong and robust, it is not inconceivable that the phenomenon could be practically applied. Possible applications include the use of transpersonal imagery as an adjunct in medical and psychological healing practices; as an aid in therapy, counseling, and training for biofeedback, hypnosis, and meditation; and as an additional educational tool. Each of the processes just mentioned could conceivably be facilitated in one person (the learner or client) if appropriate and powerful images are held concurrently in the mind of another person (the teacher or therapist).

We hope this presentation of our methodology and preliminary findings will prompt other researchers and practitioners to conduct further experimental, theoretical, and applied investigations of the important but relatively ignored phenomenon of transpersonal imagery.

Endnotes

1. The scorer measured, to the nearest millimeter, the amplitudes of all skin resistance responses greater than 2 mm. The amplitudes of all reactions during a 30-second epoch were summed, yielding a total SRR activity score for that period. This was done for each of the 20 30-second trial epochs. The trial sequence was then decoded and the SRR activity was summed for the 10 control and for the 10 influence periods. The scorer, of course, had been blind during the measurement phase.

2. This tentative conclusion derives from certain segments of Experiment 7; the reader should consult Braud, Schlitz, Collins, and Klitch (1985) for details.

3. This tentative conclusion derives from certain segments of Experiment 8; see Braud, Schlitz, Collins, and Klitch (1985) for further details.

4. In Experiment 5, we studied the influence of the strength of "need" or motivation; for further details see Braud and Schlitz (1983).

5. For additional information about Experiments 6, 7, and 8, see Braud, Schlitz, Collins, and Klitch (1985).

6. In Experiment 9, we sought to determine whether increments or decrements in SRR activity might be easier to produce via distant mental influence; in Experiment 10, we sought to determine whether the magnitude of a distant mental influence effect could be self-modulated by the influencer. Detailed results of these experiments will be published at a later date.

7. The influencers for Experiment 11 were practitioners of a Reiki healing method; see Schlitz and Braud (1985) for details.

8. Experiments 12 and 13 were conducted to examine the possible role of "intuitive data sorting" (IDS) in these experiments; details may be found in Braud and Schlitz (1988).

References

Achterberg, J. (1985). *Imagery in healing.* Boston: New Science Library.

Angoff, A., and Barth, D. (1974). *Parapsychology and anthropology.* New York: Parapsychology Foundation.

Bertini, M., Lewis, H., and Witkin, H. (1964). Some preliminary observations with an experimental procedure for the study of hypnagogic and related phenomena. *Archivo di Psicologia Neurologia e Psychiatria, 6,* 493–534.

Borelli, M., and Heidt, P. (Eds.). (1982). *Therapeutic touch: A book of readings.* New York: Springer Publishing Company.

Braud, W., and Schlitz, M. (1983). Psychokinetic influence on electrodermal activity. *Journal of Parapsychology, 47,* 95–119.

Braud, W., and Schlitz, M. (1988). Possible role of intuitive data sorting in electrodermal biological psychokinesis (bio-PK). In D. Weiner and R. Morris (Eds.), *Research in parapsychology 1987* (pp. 5–9). Metuchen, NJ: Scarecrow Press.

Braud, W., Schlitz, M., Collins, J., and Klitch, H. (1985). Further studies of the bio-PK effect: Feedback, blocking, specificity/generality. In R. White and J. Solfvin (Eds.), *Research in parapsychology 1984* (pp. 45–48). Metuchen, NJ: Scarecrow Press.

Cohen, J. (1969). *Statistical power analysis for the behavioral sciences.* New York: Academic Press.

Dean, E. (1966). Plethysmograph recordings as ESP responses. *International Journal of Neuropsychiatry, 2,* 439.

Dingwall, E. (Ed.). (1968). *Abnormal hypnotic phenomena.* London: Churchill. 4 vols.

Duane, T., and Behrendt, T. (1965). Extrasensory electroencephalographic induction between identical twins. *Science,* 150, 367.

Edge, H., Morris, R., Palmer, J., and Rush, J. (1986). *Foundations of parapsychology.* Boston: Routledge and Kegan Paul.

Ehrenwald, J. (1977). Parapsychology and the healing arts. In B. Wolman (Ed.), *Handbook of parapsychology* (pp. 541–556). New York: Van Nostrand Reinhold.

Glass, G., McGaw, B., and Smith, M. (1981). *Meta-analysis in social research.* Beverly Hills, CA: Sage Publications.

Hall, H. (1984a). Imagery and cancer. In A. Sheikh (Ed.), *Imagination and healing* (pp. 159–170). Farmingdale, NY: Baywood Publishing Company.

Hall, H. (1984b). Hypnosis, imagery and the immune system: A progress report three years later. Paper presented at the 36th Annual Convention of the Society for Clinical and Experimental Hypnosis, San Antonio, Texas.

Hall, H. (1987). Imagery in the treatment of life-threatening illness. Paper presented at the 2nd World Conference on Imagery, Toronto, Ontario, Canada.

Honorton, C. (1974). Psi-conducive states of awareness. In E. Mitchell (J. White, Ed.), *Psychic exploration: A challenge for science* (pp. 616–638). New York: Putnam.

Honorton, C. (1977). Psi and internal attention states. In B. Wolman (Ed.), *Handbook of parapsychology* (pp. 435-472). New York: Van Nostrand Reinhold.

Krieger, D. (1979). *The therapeutic touch: How to use your hands to help or to heal.* Englewood Cliffs, NJ: Prentice-Hall.

Krippner, S. (Ed.). (1977). *Advances in parapsychological research. Volume 1: Psychokinesis.* New York: Plenum.

Krippner, S. (Ed.). (1978). *Advances in parapsychological research, Volume 2: Extrasensory perception.* New York: Plenum.

Krippner, S. (Ed.). (1982). *Advances in parapsychological research, Volume 3.* New York: Plenum.

Krippner, S. (Ed.). (1984). *Advances in parapsychological research, Volume 4.* Jefferson, NC: McFarland and Company.

Kunz, D. (Ed.). (1985). *Spiritual aspects of the healing arts.* Wheaton, IL: Quest.

Long, J. (Ed.). (1977). *Extrasensory ecology: Parapsychology and anthropology.* Metuchen, NJ: Scarecrow Press.

Nash, C. (1986). *Parapsychology: The science of psiology.* Springfield, IL: Charles C Thomas.

Peavey, B. (1982). Biofeedback-assisted relaxation: Effects on phagocytic immune functioning. Unpublished doctoral dissertation, North Texas State University, Denton, Texas.

Puthoff, H., and Targ, R. (1976). A perceptual channel for information transfer over kilometer distances: Historical perspective and recent research. *Proceedings of the IEEE,* 64, 329–354.

Rosenthal, R. (1984). *Meta-analytic procedures for social research.* Beverly Hills, CA: Sage Publications.

Schacter, D. (1976). The hypnagogic state: A critical review of the literature. *Psychological Bulletin,* 83, 452–481.

Schlitz, M., and Braud, W. (1985). Reiki-plus natural healing: An ethnographic/experimental study. *Psi Research,* 4, 100–121.

Schmidt, H. (1970). Quantum-mechanical random-number generator. *Journal of Applied Physics,* 41, 462–468.

Schneider, J., Smith, C., and Whitcher, S. (1984). The relationship of mental imagery to white blood cell (neutrophil) function: Experimental studies of normal subjects. Paper presented at the 36th Annual Convention of the Society for Clinical and Experimental Hypnosis, San Antonio, Texas.

Solfvin, J. (1984). Mental healing. In S. Krippner (Ed.), *Advances in parapsychological research, Volume 4* (pp. 31–63). Jefferson, NC: McFarland and Company.

Targ, R., and Puthoff, H. (1974). Information transfer under conditions of sensory shielding. *Nature, 252,* 602–607.

Van de Castle, R. (1977). Parapsychology and anthropology. In B. Wolman (Ed.), *Handbook of parapsychology* (pp. 667–686). New York: Van Nostrand Reinhold.

Wolman, B. (Ed.). (1977). *Handbook of parapsychology.* New York: Van Nostrand Reinhold.

| 2

Calming Other Persons at a Distance
William Braud and Marilyn Schlitz

*Under experimental conditions, persons were successful in influencing
other persons to reduce (calm and quiet) the activity of their sympathetic,
autonomic nervous systems (as indicated by changes in the electrical activity of the skin)—mentally and at a distance. The effect was strongest in persons with greatest needs to be calmed. The effect compares favorably with
physiological self-regulation effects.*

*The material in this chapter originally was published in Braud, W. G.
and Schlitz, M. J. (1983). Psychokinetic influence on electrodermal activity.*
Journal of Parapsychology, 47, 95–119. *The contents of this article are
Copyright © 1983 by the Parapsychology Press, and the material is used
with permission.* —William Braud

Abstract: We conducted a "bio-PK" experiment to determine whether target persons
with a relatively strong need to be influenced (calmed) would evidence a greater psi
effect than would persons without such a need. Serving as the influencers, we attempted
to psychokinetically decrease the electrodermal activity of distant target persons during
certain prespecified periods as compared to an equal number of control epochs in which
PK attempts were not made. Sixteen target persons had relatively high sympathetic
nervous system activity and thus had a need to be calmed. Sixteen other target persons

A preliminary account of these experiments was presented at the 1983 Parapsychological
Association Convention in Madison, N.J. These experiments were conceived by W. B., with
M. S. contributing importantly to their design. The volunteers were recruited primarily by
M. S. and were tested by W. B. and M. S. Statistical analyses were performed independently
by W. B. and M. S. The final paper was written by W. B.

had moderate or low activity and no particular need to be calmed. A significant PK-calming effect occurred for the active (needy) persons, but not for the inactive persons. The PK-calming effect was significantly greater for active than for inactive persons. Various nonpsychic and psychic explanations for these results are discussed.

For comparison, we conducted an experiment on self-control of autonomic activity in sixteen active subjects (a nonpsi experiment). It indicated that the magnitude of self-control did not greatly exceed the magnitude of psychic hetero-influence of autonomic activity.

Since 1976, the first author has been developing a series of experiments involving psychokinetic influences upon living systems. Such "bio-PK" experiments are important for both theoretical and practical reasons. Theoretically, there is reason to suspect that biological systems may be unusually susceptible to psychokinetic influence because these systems are quite complex and possess considerable plasticity and potential for change ("lability," "randomness," "noise")—characteristics which have been hypothesized by several recent theorists (Braud, 1981; Puthoff and Targ, 1975; Stanford, 1978; Walker, 1975) as increasing the likelihood and magnitude of observed PK effects. Practically, experiments involving PK influence on organic systems can be conceptualized as prototypes or analogs of at least some forms of psychic healing. By means of such experiments, it may be possible to determine the magnitude, permanence, limitations, and potential applicability of biologically meaningful PK effects.

The protocol for these bio-PK experiments is as follows. Some arbitrarily selected behavior or physiological activity of a freely responding target organism is monitored for a period of time. This period is divided into an equal number of influence and control epochs. During the influence epochs, an influencer (beyond the sensory range of the target organism) attempts to psychically influence the ongoing activity of the organism in a predetermined direction. In all experiments conducted thus far, the influencer has received instantaneous and continuous analog feedback concerning the state of the target system. During the control epochs, no such psychokinetic attempts are made. The design is such that, in the absence of a psi influence, equal amounts of activity in the prescribed direction are expected to occur during control and influence epochs. A statistically significant excess of prescribed activity during the influence epochs is evidence for a psi effect.

Using this protocol, we have observed significant PK influences on the locomotor behavior of gerbils, the swimming orientation of electric knife fish, and the electrodermal activity (EDA) of human volunteers (Braud, 1978, 1979; Braud, Davis, and Wood, 1979). A very preliminary study has also suggested a possible PK influence on the rate of hemolysis of human red blood cells (Braud, et al., 1979). Successful influencers in these experiments have included the experimenter (W. B.), a gifted psychic (Matthew Manning), and many unselected volunteers. A summary of the results of 13 bio-PK experiments is presented in Table 1.

Table 1. Statistical Summary of 13 Previous Bio-PK Experiments

Influencer	Target activity[a]	No. of sessions	Magnitude of effect[a]	p[b]	Reference
1. Experimenter (W. B.)	Spatial swimming orientation of an electric knife fish	10	2%	.01*	Braud, 1979
2. Matthew Manning		10	3%	.004	Braud et al., 1979
3. Ten unselected volunteers		10	4%	.03	Braud, 1979
4. Ten unselected volunteers		10	1%	n.s.	Braud, 1979
5. Experimenter (W. B.)	Increased locomotor activity of a gerbil	10	5%	n.s.*	Braud, 1979
6. Matthew Manning		10	3%	.03	Braud et al., 1979
7. Ten unselected volunteers		10	5%	.02	Braud, 1979
8. Ten unselected volunteers		10	2%	.009	Braud, 1979
9. Matthew Manning	Decreased rate of hemolysis of human red blood cells	10	7%	.00006*	Braud et al., 1979
10. Experimenter (W. B.)	Increased electrodermal activity of another person	10	9%	.01*	Braud, 1978
11. Matthew Manning		10	9%	.03	Braud et al., 1979
12. Ten unselected volunteers		10	8%	.008	Braud, 1979
13. Ten unselected volunteers		10	-3%	n.s.	Braud, 1979

[a]The "magnitude of effect" percentage score represents the deviation of the psi-indicating response measure from chance expectation. In these experiments, a 50% score is expected by chance; thus, a 57% score would yield a magnitude of effect score of 7%.

[b]An asterisk denotes a two-tailed test; all other p's are one-tailed.

In these 13 experiments, four different animate systems were found to be susceptible to PK influence. The levels of responsivity of the target systems included: gross behavioral activity of whole organisms (Experiments 1 through 8), peripheral autonomic activity of another person (Experiments 10 through 13), and in vitro cellular activity (specifically, membrane permeability in Experiment 9).

The results summarized in Table 1 suggest a number of interesting conclusions. First, the psi effect is a relatively reliable and robust one. Let us exclude the hemolysis system, because only one preliminary experiment (Experiment 9) was conducted. We are left with three groups of four experiments each. Each group involves a different target system, and each group includes results for 40 sessions, contributed by 22 different influencers. The success rate for each group is approximately 75%. Three out of each four experiments tend to yield significant outcomes. Although not indicated in the table, this same 75% success rate appears when scores of individual sessions are used as the units of analysis; approximately three out of four sessions yield successful outcomes.[1]

A second conclusion is that the results do not seem to vary in any systematic way as a function of who serves as the influencer. Results for the experimenter (W. B.), for a gifted psychic (Matthew Manning), and for unselected volunteer subjects are relatively comparable. Perhaps there is a slight tendency for the gifted psychic to perform consistently well across a variety of experiments; and perhaps there exists a slight tendency for unselected subjects to perform more poorly. However, these effects could be attributed to differences in strength of motivation and to differences in feelings of familiarity and ease in the experimental setting.

A third conclusion—actually not more than a suggestion at this point—is that the magnitude of the PK effect appears to increase as the target system becomes more meaningful to the influencer. The magnitude is smallest in the case of the fish targets, greater for the warm, fuzzy gerbil targets, better still for the human red blood cell targets, and greatest when the target system is another person. It is almost as if the magnitude of the effect increases as the similarity of the target to the influencer increases.[2]

We have decided to focus our research efforts on the human electrodermal target system because this system seems to yield the strongest psi effect and because work with this system has the greatest likelihood of yielding findings relevant to a possible practical application in psychic healing. If some arbitrarily selected aspect of human physiology can be psychokinetically influenced in an arbitrary direction, will the application of a psychokinetic influence to a judiciously selected response system prove medically and psychologically beneficial? The possibility of an affirmative answer to this question suggests the usefulness of further research in this area—research that might explore the conditions which make bio-PK more or less likely, as well as investigations into the possible magnitude, degree of permanence, degree of specificity or generality of the effect, and possible limitations or undesirable side effects of the procedure. It is in this context that we designed the present (14th) bio-PK experiment.

The Influence of Need

In everyday life, psychic interactions occur in contexts characterized by *need*. Information may be acquired or effects produced which are meaningful, relevant, and important to those involved. Psi may serve an adaptive function, providing information or outcomes which are biologically or psychologically significant. On the other hand, the information and outcomes contingent on psi manifestations in the laboratory are usually relatively trivial and typically lack this feature of significance or meaning. From the point of view of the psychology of motivation and learning, motivation, incentive, and reinforcement are greatly degraded in the laboratory situation.[3] Ehrenwald (1978) has speculated about these differences in terms of "need-determined" and "flaw-determined" psi.

The use of experiments designed as psychic-healing analogs may provide a way of upgrading the meaningfulness of laboratory experimentation and optimizing the motivation of the participants. In psychic-healing experiments, there exists a definite *need* in the target organism, and a successful psi influence aids in the satisfaction of that need. Examples of such need satisfaction may be found in Grad's work on psychic facilitation of growth in plants and wound-healing in mice (Grad, 1967), the Watkinses' work on psychic resuscitation of anesthetized mice (Watkins and Watkins, 1971; Watkins, Watkins, and Wells, 1973), and the recent work of Rauscher and Rubik (1980) on the psychic protection of microorganisms from the effects of toxic chemicals.

In a psychic-healing analog experiment, a strong need is also present in the influencer. There exists a strong wish to *help* the ill or injured target organism; feelings of empathy with the target organism may also be present. The need of the influencer may even conflict with that of the target organism, as in cases in which the goal of the experiment is an *inhibitory* influence upon what is viewed as negative, human life-endangering, target organisms or growths such as cancer cells (see Braud, et al., 1979), tumors (Onetto and Elguin, 1966), fungus (Barry, 1968; Tedder and Monty, 1981), or harmful bacteria (Nash, 1982).

In the four human EDA bio-PK experiments listed in Table 1 (Experiments 10 through 13), the need factor was minimal. The target persons were unselected volunteers who had no particular need to be influenced, and the direction of the influence (an *increase* in EDA) was not particularly beneficial to the target persons. In the present experiment, we sought to optimize the need factor by choosing as target persons individuals with excessive sympathetic nervous system activity who would benefit from a reduction in such activity, and letting the goal for the influencers' bio-PK efforts be a *decrease* in sympathetic nervous system activity (as indexed by a decrease in EDA). Thus, target persons were well motivated to cooperate with the desired PK influence, and the influencers were well motivated to improve the conditions of their target persons through PK influence. The role of the need factor was assessed experimentally by comparing the magnitude of the bio-PK effect in needy or active target persons with the magnitude of the effect in control persons in whom such a need was not present.

Experiment 1—The Main Study

Overview of the experimental design. Both of us (W. B. and M. S.) functioned as influencers in the study. Our goal was to psychically decrease the EDA of the target persons during only the influence periods, which were intermixed with control periods of no PK attempts. There were 32 sessions, one target person for each. Half of the target persons (*n* = 16) were persons with excessive EDA (active subjects); half of the target persons *(n* = 16) were individuals with normal or low levels of EDA (inactive subjects). For each of the 32 sessions, the target person's EDA (skin conductance reactions) was sampled and scored by computer during ten 30-second influence epochs and ten 30-second control epochs. Influencers watched a polygraph tracing of the target person's EDA throughout each session and therefore received instantaneous and continuous feedback about the effectiveness of their bio-PK attempts. The magnitude of the bio-PK effect was expressed as a percentage and was derived by dividing the EDA obtained during influence (decrease) epochs by the total EDA of both epochs (the influence plus the control epochs). Mean chance expectation (MCE) in the absence of a psi effect was 50%.

There were three a priori predictions about the outcome of the study. The primary prediction was that the percent EDA during influence epochs would be significantly lower for active subjects than for inactive subjects; that is, the bio-PK effect would be greater for subjects with greater need. Two secondary a priori predictions were that (a) for the inactive subjects, the percentage of EDA occurring during influence epochs would be significantly below MCE, and that (b) for the active subjects, the percentage of EDA occurring during influence epochs would be significantly below MCE. Alpha levels for each of these three predictions were set at p = .05, and one-tailed tests were to be employed.

Method

Subjects. The Foundation advertised in newsletters, posters, and by word of mouth for volunteers for a psychic-healing analog experiment. We indicated an interest in persons with excessive sympathetic autonomic activity and suggested that such excessive activity might be present in persons who were very emotional, anxious, or under stress, or who might have tension headaches, high blood pressure, ulcers, or hyperactivity. The study itself was described in general terms. From among the volunteers who expressed interest in the study, we selected the first 32 persons who met the specific criteria of autonomic responsiveness described under Procedure—16 active and 16 inactive subjects. It is important to note that all subjects were chosen from the same pool of persons who considered themselves autonomically active and were interested in the study; thus, there were no systematic differences between the active and inactive subjects other than their levels of autonomic activity.[4]

As influencers, both of us had participated previously as successful bio-PK influencers—W. B. in the experiments described earlier, and M. S. in mouse-resuscitation experiments at the Foundation for Research on the Nature of Man.[5] We each worked with 16 target persons, 8 of whom were autonomically active and 8 autonomically inactive.

Apparatus. Electrodermal reactions were monitored through the use of a Lafayette Instruments Model 76405 GSR/BSR amplifier[6] and Harvard Apparatus pen writers and chart mover. The skin resistance amplifier is a constant current device with a subject current of 24 μA. Chrome-plated stainless steel finger electrodes were used without electrode paste. The surface area of each electrode was 584.2 mm[2]. An analog-to-digital converter interfaced to a microprocessor allowed automated sampling and averaging of the ongoing EDA.[7] The computer provided a paper printout of the results at the end of the session.

Procedure. Each subject made two visits to the laboratory, approximately one week apart; the two visits were at approximately the same time of day. The purpose of the first visit was to familiarize the subject with the lab, the experiment, and the experimenter, to establish rapport between subject and experimenter, and to conduct an EDA evaluation. Upon arriving at the lab, the subject was met by the experimenter who was later to serve as his or her influencer. After discussing the experiment, the subject's expectations, and the subject's reasons for considering himself or herself autonomically active, the experimenter escorted the subject to a small, comfortable, dimly illuminated room and seated the subject in a reclining chair. Electrodes were attached by means of Velcro bands to the volar surfaces of the first and second fingers of the right hand of the subject. The subject was asked to make no deliberate effort to relax, but rather to remain in as ordinary a condition as possible while listening to computer-generated random sounds through headphones and watching randomly changing patterns of colored squares on a 12-inch display screen about 2 m away. These sounds and colored lights were used to help occupy the subject's attention during the 25-minute session, to produce some degree of activation or arousal in the subject, and to encourage in the subject a more random, flowing mode of cognition (see Braud, 1980, 1981). The experimenter then dimmed the lighting even further, closed the door of the subject's room, and went to a control room 20 m away. After a 5-minute delay, the experimenter recorded the subject's initial basal skin resistance (BSR) level, then activated the computer program and the cassette deck which played the random-sounds tape to the subject. The computer sampled the subject's EDA at a rate of 10 samples per second for each of twenty 30-second recording epochs, then averaged and stored each of these epoch scores. The computer similarly assessed the subject's EDA during twenty 30-second intertrial intervals, and during five initial pre-session baseline periods. In addition to computer-monitoring, a permanent polygraph chart record of the EDA was generated. During this first recording session, the experimenter busied himself or herself with reading or other activities and made no deliberate attempt to influence the subject in any way. At the end of the recording period, the printer printed the mean EDA scores for each of the 20 epochs, and the experimenter recorded the subject's final BSR level. After the subject had written a brief description of his or her experiences during the session and had indicated his or her feeling of degree of rapport or connectedness with the experimenter (on a 10-point scale), the experimenter reentered the subject's room and discussed the session in general terms. An appointment was then made for the subject's second visit to the laboratory, approximately one week later and at the same time of day. After the subject had left the lab, the experimenter scored the polygraph record by counting the number of electrodermal reactions greater than 2 mm in amplitude during the session as a whole. Subjects who produced fewer than 3 reactions were excluded from the study as being too inactive. Subjects producing more than 2 but fewer than 100 reactions were assigned to the inactive group. Subjects producing more than 100 reactions were assigned to the active group.

From the subject's point of view, the procedure for the second session was identical to that for the first session. However, this time the subject was told that during certain parts of the session, the experimenter would attempt to psychically *calm* the subject, and that the subject should wish for and allow such calming effects to occur. The subject, of course, was not told how many calming attempts were to be made, or when they would occur. Indeed, at this point the experimenter was still unaware of the predetermined sequence of the calming versus control periods for that subject. The experimenter asked the subject to "make a wish that you will become calm at the appropriate times, and that such effects will occur effortlessly and automatically," then went to the control room as before.

The EDA recording procedure and the experimenter's activities in the distant control room were identical to those of the first session, with the following important exceptions. After recording the subject's initial BSR level, the experimenter opened an envelope which indicated the influence versus the control epoch sequence for that session. The envelopes had been prepared beforehand by another person, using a table of random numbers. Recording epochs were scheduled in an ABBA sequence to minimize progressive error.[8] The envelope revealed whether the "A" recording epochs were to be influence (decrease) or control periods. After ascertaining whether A or B epochs were to be devoted to influence attempts, the experimenter arranged a deck of 20 cards (10 marked

Decrease, 10 marked *Control*) in the proper order and prepared himself or herself for the first recording epoch. A low-frequency tone, heard only through the influencer's headphones, sounded during each epoch. Upon hearing the tone, the influencer consulted the topmost card for instructions. If the card indicated that a control epoch was in progress, the influencer attempted not to think about the subject or "allowed" the subject to be active during that 30-second period. During a decrease epoch, the influencer attempted to psychically calm the subject so that little EDA would be produced during that 30-second epoch. The influencer attempted such a psychic influence by relaxing himself or herself and imagining that the subject was doing the same, by sending mental "messages" or feelings of quietude to the subject, and by visualizing the polygraph pen producing a flat tracing, free of EDA.[9]

The computer printouts of the session results were filed away, along with coded polygraph tracings, and were not carefully examined until the completion of the study. This was done to prevent the experimenters' emotional reactions to early data returns from influencing the eventual outcome of the study.

Results

For each session, a "percent decrease" score (i.e., the percentage of total activity occurring *during* the decrease periods) was calculated by summing the mean EDA scores for each of the 20 recording epochs and dividing this total activity score into the sum of the mean EDA scores for the 10 decrease (i.e., PK influence) epochs. A chi-square goodness of fit test indicated that the scores of these 32 sessions did not differ significantly from normality; therefore, parametric statistics were used in their evaluation. The scores were next divided into the active and inactive subgroups. Within each subgroup, statistical comparisons indicated that the scores of the two experimenters did not differ significantly from each other; therefore, the scores of the two experimenters were pooled. The primary prediction was that the scores for the 16 active subjects would be significantly lower than those for the 16 inactive subjects. An independent samples t test comparing these two sets of scores indicated a significant effect in the predicted direction (active $\bar{X}$ = 40.39%, inactive $\bar{X}$ = 50.33%; $t[30]$ = 1.86; p = .035, one-tailed). Single mean t tests comparing the percent decrease scores with MCE (MCE = 50%) were used to test the two secondary predictions. These tests indicated a significant departure from chance in the predicted direction for the active subjects ($t[15]$ = 2.40; p = .014, one-tailed), but not for the inactive subjects ($t[15]$ = –0.09; p = .54, one-tailed). Thus, the primary prediction and one of the two secondary predictions were confirmed. Results are depicted graphically in Figure 1.

To determine the likelihood that similar effects might actually have occurred by chance, we conducted parallel analyses on the corresponding *first-session* scores of the two subgroups. During the first session, it will be recalled, no influence attempts were made. These first-session scores differed neither

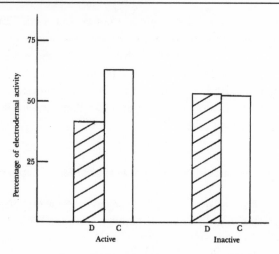

Figure 1. Percentage of mean electrodermal activity occurring during decrease (D) and control (C) recording epochs for the 16 active subjects and the 16 inactive subjects during the second (experimental) session of the experiment.

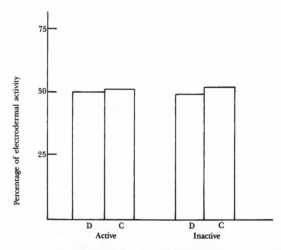

Figure 2. Percentage of mean electrodermal activity occurring during decrease (D) and control (C) recording epochs for the 16 active subjects and the 16 inactive subjects during the first (control) session of the experiment.

from one another nor from MCE (active $\bar{X}$ = 49.86%, inactive $\bar{X}$ = 48.77%; all ts nonsignificant). The comparison is depicted graphically in Figure 2.[10]

Discussion

Our major finding in the present study was that persons with a *need* to have their sympathetic autonomic nervous system activity reduced evidenced a greater bio-PK effect than did persons without such a need. This finding was not an artifact of higher absolute activity levels and hence greater room for change in the active condition, since relative percentage measures which control for initial absolute differences were used in the analyses. The magnitude of the bio-PK in this study, 10%, was the largest we have observed in such experiments. Table 2 allows a comparison of the present results with results of previous bio-PK experiments (Table 1).

Table 2. Statistical Summary of Present Bio-PK Experiment

Influencer	Target activity	No. of sessions	Magnitude of effect	p
Experimenters (W. B. and M. S.)	Decreased electrodermal activity of another person	16 Active subjects	10%	.01
		I6 Inactive subjects	0%	n.s.

Some rejected nonpsi explanations. In view of possible skepticism about the validity of psi results on the part of persons both within and without the parapsychological community, we present a consideration of various alternative explanations of the present finding. A number of nonpsi hypotheses can be dismissed:

1. *The variations observed are coincidental, chance variations.* This chance hypothesis is rejected on the basis of results, obtained with conventional statistical methods, which reached conventional significance levels for rejection of the null hypothesis.

2. *The findings are the result of recording errors and motivated misreadings of polygraph records.* This explanation is rejected on the basis of the use of completely automated assessment techniques and computer-scoring of response activity.[11]

3. *The target persons knew beforehand when influence attempts were to be made and "cooperated" by changing their own autonomic activity when appropriate.* This explanation may be rejected because the subjects were not told when or how many influence attempts would be made, nor was the experimenter aware of the influence/control epoch schedule until all preliminary interactions with the subject had been completed. Subjects did not know of the existence of, or have access to, the envelopes containing schedule information.

4. *Subjects could have become aware during the experimental sessions themselves of when influence epochs were in progress and could have altered their own physiological reactions during those periods.* This possibility was eliminated by isolating the subject from any such cues from the influencer. Subject and influencer

were in separate, closed rooms 20 m apart from one another. No auditory cues could have impinged upon the subject to indicate when recording epochs were in progress or whether such epochs were influence or control periods. The influencer made no vocalizations whatsoever while in the control room. The epoch-indicating tones heard by the influencer and the random tones heard by the subject were provided by independent audio systems which eliminated the possibility of electrical crosstalk and also of headphones functioning as microphones and inadvertently cueing the subject. It should be noted that even if the subject had heard these epoch-signalling tones, it still would not have been possible to know whether a decrease or a control epoch was in progress because identical tones signalled these two types of recording periods.

5. *Differences in autonomic activity between influence and control periods are due to systematic error—that is, some progressive change in electrodermal activity over time.* This objection may be rejected. Progressive (time-based) errors could have been contributed by (a) changes in equipment sensitivity as the equipment warmed up, (b) changes in EDA due to adaptation or habituation to the experimental environment, and (c) changes in EDA due to polarization of the recording electrodes. Equipment was allowed to warm up for 15 to 20 minutes prior to the beginning of a session and had therefore become thermally stable before the experiment began. The use of electrodes with large surface areas, and the use of a constant-current electrodermal monitoring device reduced the possibility of a polarization problem. Statistical evaluation of total EDA for the first halves ($\bar{X} = 47.58$)[12] vs. the second halves ($\bar{X} = 52.42$) of the sessions indicated no evidence of an habituation effect *(t = 0.76, p = n.s.)*. Therefore, there was no progressive change in EDA due to any of these three possible processes. However, even if a progressive change had occurred, the use of the ABBA counterbalanced design would have prevented this error from contributing differentially to influence versus control epochs.[13]

6. *The findings are due to arbitrary selection of the data.* This explanation may be rejected. Total numbers of subjects and trials were prespecified, and the analyses reported include all recorded data.

7. *The significant difference in scores for the active versus the inactive subjects is an artifact of the active scores being at an initially higher absolute level, and therefore having more room to decrease; the inactive scores are initially too low to decrease further and therefore suffer a ceiling (floor) effect.* This explanation can be rejected. Absolute scores were not used in the analyses, but rather, percentage-change scores which reflect the *relative* difference (within a given subject) between influence and control activity, regardless of what the initial absolute level of activity happens to be. These percentage scores were used

for this very reason: they control for differences in absolute activity levels and therefore allow valid comparisons of subjects or groups with initially unequal levels of activity.

8. *The results are due to fraud on the part of the subjects.* This explanation may be rejected. The subjects who participated in the experiment were unselected volunteers. It may be assumed that such subjects had no motive for trickery. However, even if a subject were motivated to cheat, such an opportunity was not present. To cheat, a subject would have to know the sequence of influence and control epochs for his or her session. Such knowledge could only be acquired by learning the contents of the envelope for that session, and this could only be done by peeking at the envelope's contents during the experimental session. (Peeking *beforehand* would not work because the subject would not have known which of the many numbered envelopes would be used for his or her session.) If the subject had attempted to cheat unaided by an accomplice, he or she would have had to leave the subject room and disconnect the finger electrodes, which would have been immediately obvious from its effect on the polygraph tracing. Thus, the subject would have required an accomplice. However, the subjects came unaccompanied to the laboratory. Additionally, there were no openings in the control room which could have allowed visual access to the instruction cards from outside the room. Thus, to determine the sequence in force for a given session, the accomplice would have had to enter the closed control room, and this would have been obvious to the experimenter. Finally, it should be mentioned that the present experiment involved a large number $(N = 32)$ of subjects. The statistical tests used to evaluate the results required not only a certain magnitude but also a certain *consistency* of results. A substantial number of subjects must exhibit an effect for the results to be significant. Thus, a critic suggesting that our results were attributable to subject fraud would have to suggest that many of the subjects were engaging in such fraud. This is exceedingly unlikely. Subject fraud is not a very useful explanation in studies employing a large number of subjects. When the results depend entirely on one or two special subjects, however, the subject-fraud hypothesis is a viable one, and special precautions should be taken to prevent such a possibility.

9. *The results are due to fraud on the part of the experimenters.* No experiment, however sophisticated, can ever be absolutely safe from experimenter fraud. Even if an experiment were controlled by an outside panel of disinterested persons, a hostile critic could still argue that collusion was involved. The imagined extent of such a conspiracy would be limited only by the imagination (and degree of paranoia) of the critic. We can only state that we used a two-experimenter design so that one experimenter's portion of the exper-

iment served as a kind of control for the other experimenter's portion. Only the successful replication of these findings by other investigators would reduce experimenter fraud to a nonissue. Some of the bio-PK experiments described earlier in this paper have already been conceptually replicated by independent researchers (e.g., Gruber, 1979, 1980; Kelly, Varvoglis, and Keane, 1979).[14] We encourage independent investigators to attempt to replicate the present experiment, and hope that many researchers will succeed in doing so.

Alternative psi hypotheses. Having established that the psi effect seen in the present study is a real one, we present a number of alternative psi hypotheses:

1. *The true psi effect is not exerted on the subject's physiological activity, but rather on the selection of the influence/control sequences; that is, sequences are psychically selected so as to best match the subject's ongoing activity to the intended outcome of the study.* This would appear to be a more likely possibility in cases in which more random sequence selection procedures were used. If card shuffles or random-event generators were used to select influence vs. control epochs, it is conceivable that an unconscious "psychic shuffle" or psychokinetic influence upon the random generator could occur, which would maximize the likelihood that a subject's ongoing activity in the prescribed direction was captured by influence rather than control epochs. However, the use of a fixed (ABBA) sequence of influence and control epochs in the present study would seem to make this possibility quite unlikely. Even the decision as to which type of epoch began the sequence (whether "A" was a decrease or control epoch) involved few degrees of freedom and was based on a table of random numbers that was used in a manner which would seem to minimize a psi influence at that locus (see Stanford, 1981).

2. *There was a PK influence, but it was exerted on the equipment rather than on the living target system.* Such a possibility cannot be ruled out in designs such as the one used here. The only way to eliminate this possibility would be to conduct bio-PK experiments in which instrumentation was not used at all, but in which naked eye measurements were made instead. The use of rate of growth of plants or rate of recovery of visible wounds in animals (as in Grad's 1967 research) would exemplify this strategy. We suspect, however, that in these bio-PK experiments, the PK influence is exerted on the most labile link of the chain of events participating in the recording, and this link would be the activity of the biological target system. Fortunately, this suspicion is not merely speculative. In these EDA experiments, subjects have spontaneously remarked that during certain periods within the session they actually *felt* that someone had entered the room to activate them (in the activation series of Experiments 10 through 13) or calm them (in the present study). They

reported definite physical feelings of activation or quietude during specific periods of the session that could be correlated with activity seen in the polygraph records and with psychic influence attempts. This suggests that *they* were responding appropriately, and not merely the recording equipment.

3. *An effect occurred, but it was a telepathy or clairvoyant effect localized in the subject rather than a PK effect localized in the influencer.* Such a possibility cannot be ruled out, and indeed we suspect that such an effect did occur in the present experiment, along with a PK effect. In fact, it was stressed (at least by W. B.) in initial discussions with the subject that the experiment would involve a joint effort of influencer and subject. The subject was told that the influencer would attempt to psychically influence the subject, while the subject was to remain open to such an influence and also attempt to reach out psychically to the influencer, psychically determine when to become calm, and cooperate by becoming calm. Thus, we suspect that the experiment involved a mixture of PK and GESP components.

It is interesting to conjecture about whether such a mixture of processes is maintained in bio-PK experiments involving nonhuman life forms as target systems. We can easily imagine human target persons utilizing GESP and cooperating by changing their physiological activities appropriately. But what about gerbil targets or electric fish or human red blood cells? It seems more plausible to exclude the possibility of GESP in such target systems and assume the presence solely of PK effects. However, such a view is based on an assumption of the emergence of a GESP-possessing form of mentality or consciousness at a particular phylogenetic level—the human level. This is an assumption which some of us may be unwilling to make.

Additional comments. A number of additional aspects of this experiment remain to be discussed. One such issue is the absence of a psi effect in the inactive condition. This may be an instance of a preferential or differential effect (Rao, 1966) or of a psi-mediated experimenter effect. If this is indeed an experimenter effect, it may have been encouraged by the experimenters' knowledge of the group membership of the subjects. This knowledge was unavoidable. We wished to provide the influencer with immediate analog feedback in this experiment, and receiving such feedback allowed the polygraph observer to quickly distinguish active and inactive subjects. Therefore, the influencer could not be kept blind regarding the activity condition of the subject.[15]

An alternative explanation for the absence of a psi effect in the inactive condition is that the psi effects in this study were weaker than usual except in the active condition, in which the weak effect may have been amplified by the strong need factor. It should be noted that the present experiment marked our first attempt to *decrease* autonomic activity via PK; previous experiments involved increases in such activity. It may well be the case that decrements in autonomic activity are more difficult to produce psychically than increments are. This was indeed W. B.'s impression in comparing his subjective feelings for incremental versus decremental psi attempts. It may be the case that the initial psychic contact with a target person produces some initial autonomic activation. The effect of such a contact or attention signal would be oriented in the same direction as the preferred outcome in incremental-aim studies and would be expected to facilitate that outcome. However, the same tendency would conflict with the preferred outcome in decremental-aim studies, possibly reducing the net psi effect. It is of interest to note that there exists a general trend in the literature on autonomic biofeedback for increments to be easier to produce than decrements, and some authors have suggested that the two effects may even involve different mechanisms (see Blizard, Cowings, and Miller, 1975; Engel, 1972).

Another topic requiring comment is the psi strategy used by the influencers in this experiment. Both influencers used essentially the same strategy to attempt to calm the subject during influence epochs. However, they differed somewhat in the strategies used for control epochs. M. S. actively attempted to increase electrodermal activity during control periods and subsequently reported experiencing some uneasiness in doing so because the purpose of influencing these autonomically active persons was to calm them. This conflict could have influenced her psi performance. W. B. did not actively attempt to increase activity during control periods, but rather "allowed" the subject to return to normal during such periods. His strategy was to imagine that the subject had a certain amount of activity to display during the session and his psychic instruction to the subject was to *schedule* this activity so that it would tend to occur less often during influence and more often during control epochs. He frequently used an image of the subject holding back on activity during influence epochs and then releasing that activity during the control epochs. Interestingly, some of W. B.'s subjects spontaneously reported just such a subjective feeling—one of holding, then being allowed to release activity during certain periods. Sometimes changes in breathing or urges to breathe in different ways were associated with these holding and release patterns. Additionally, W. B. emphasized, in initial discussions with his subjects, the possibility of the subject actively cooperating with the influence in producing a calming effect—that is, reaching out psychically to learn when to be calm, then cooperating by calming himself or herself. This active GESP aspect was not encouraged by M. S., who tended to instruct her subjects to take a more passive attitude and simply let the influence happen. Results for W. B.'s subjects were somewhat better than those for M. S.'s subjects, although not significantly so. The use of these different strategies may have been important in yielding this difference in effect. Alternatively, W. B.'s generally better results may be due simply to his greater experience with these kinds of bio-PK experiments.

Our final comment concerns the magnitude of effect in the present experiment. The 10% effect in the active subjects condition is a large one in the context of PK effects. We wondered how well an effect of this magnitude would compare with the magnitudes of EDA changes produced by more conventional procedures such as biofeedback, autogenic exercises, meditation, and so on. We considered reviewing the literature in an attempt to estimate the percent magnitude of a typical conventional effect. It soon became evident, however, that there existed far too much variability in procedures and outcomes of such studies to allow any meaningful comparisons. Our use of a number of short epochs, for example, made comparisons quite impossible. We therefore decided to conduct our own comparison study in which exactly the same epoch durations and distributions would be used, but in which subjects would attempt to control their own electrodermal activity. The percent magnitude of activity change produced in this conventional manner could be compared directly with the magnitude of a similar effect which was psychically produced.

Experiment 2—Autonomic Self-Control

To directly compare the magnitude of the psi effect obtained in the bio-PK/need experiment just described with the magnitude of a similar effect obtained through nonpsychic means, we tested 16 additional active subjects in an autonomic nervous system self-control experiment. The subjects were tested individually by W. B. Following a 5-minute adaptation period, electrodermal activity was monitored while the subject attempted to calm himself or herself as much as possible during ten 30-second epochs, each signalled by a low-frequency tone. During 10 interspersed 30-second control epochs signalled by higher frequency tones, the subject was asked to be normal, that is, not to attempt to increase autonomic activity. The experimenter made no attempt to psychically influence the subject.

Method

Subjects. The 16 subjects were volunteers, chosen from the Foundation's pool of persons interested in our experiments. Because we wished the conditions of this experiment to resemble closely those of the first experiment, we did not retest the original 16 active subjects, who were now somewhat habituated to these test conditions. Rather, we wished to test active subjects who were naïve regarding this test arrangement. Thus, we included the first 16 new volunteers who met the minimal criterion of autonomic activity described hereafter in Procedure.

Apparatus. The equipment used for this study was identical to that described in Experiment 1.

Procedure. W. B. greeted the subject and escorted him or her to the subject room, where the experiment was described and instructions were given. The volunteer was told that there was to be no deliberate psychic component in this study, but that we wished to learn how well a person can control his or her own autonomic activity using conventional techniques. No hints were given regarding specific techniques. The subject was told that, following a 5-minute period of silence, a number of tones would be presented through headphones. During the higher pitched tones, the subject was to remain in a normal state of consciousness. During the lower pitched tones, the subject was to attempt to become as calm, quiet, and relaxed as possible. Note that the tones were not used in a biofeedback manner, but simply as indicators for periods of self-control without feedback. The same random colored-lights display as was used in Experiment 1 was available for viewing in this experiment; however, the random sounds were eliminated since these would have interfered with hearing the two tones used in this study. The subject was told the entire experiment would last about 25 minutes. At the end of the session, the subject completed a brief questionnaire which asked about the subject's experiences during the session, the methods used to become calm during the low-tone periods, and the estimated degree of success at autonomic self-control (on a 10-point scale).

A new computer program was written which presented 10 high-pitched and 10 low-pitched 30-second tones to the subject in an ABBA or BAAB sequence. Intertone intervals, computer sampling rates, and all EDA recording and assessment procedures were identical to those of Experiment 1.

We included in the formal experiment the first 16 subjects who met a criterion of producing at least fifty 2-mm EDA pen deflections on the polygraph record for the entire 25-minute session.[16]

Results

As in Experiment 1, a percent decrease score was calculated for each subject by summing the mean EDA scores for each of the 20 recording epochs and dividing this total activity score into the sum of the mean EDA scores for the 10 decrease (low-tone) epochs. These percent decrease scores reflected the degree of autonomic self-control, with a score of 50% indicating no control, and a score of 0% indicating maximal control. In fact, the scores varied from 5.07% to 53.33%, with a mean of 31.33% and a standard deviation of 11.74. Comparison of these 16 percent-decrease scores with chance expectation (50%) yielded evidence of highly significant autonomic self-control (single mean $t[15] = 6.16$; $p = 3.8 \times 10^{-5}$, one-tailed). Comparison of the 16 self-control scores with the 16 psi-influenced scores for the active subjects of Experiment 1 ($\bar{X} = 40.39\%$) yielded an independent samples $t = 1.80$ which, with 30 df, did not quite reach significance ($p = .08$, two-tailed). Thus, although self-control yielded a stronger effect (an approximately 19% deviation from chance) than did psi influence (an approximately 10% deviation from chance), the effect was not significantly stronger.[17] Note, however, that the actual difference between the active (psi) and the inactive (psi) groups was

9.94%, with a *t* of 1.86, whereas the actual difference between the active (psi) and the self-control groups was 9.06%, with a *t* of 1.80. These scores are quite similar, and the conclusion of no group difference in the latter case depends on the use of a two-tailed test and a very exacting distinction between *p* values of slightly less than .05 and those that are slightly greater than .05.

Discussion

The magnitude of psi influence in Experiment 1 was approximately 10%; the magnitude of self-control influence in Experiment 2 was approximately 20%. Although the latter is 10% greater than the former, that difference yielded only a marginal significance level and was surprisingly less than we expected. This indicates that under certain conditions, a psi influence on another person's physiological activity may compare favorably with a person's more direct or more conventional influence on his or her own physiological activity. It should be realized, of course, that the measure of degree of conventional control in the present comparison was derived from subjects who were making their very first attempt at autonomic self-control. Undoubtedly, additional practice would have yielded greater self-control. However, additional practice of the psi procedure also may have yielded greater psi control.

Subjects used a variety of methods to quiet their own autonomic nervous systems. Most imagined themselves relaxing and visualized tranquil natural scenes. A comparison of the relative efficacy of different autonomic nervous system self-control strategies might be quite useful. If it were found that a particular method of self-control is most effective, such a method could be incorporated more fully into the psi-influence procedure, possibly increasing the effectiveness of the latter.[18]

Endnotes

1. Since W. B. was involved in all 12 of these experiments, either as experimenter or as influencer, his possible psi contribution to their success cannot be overlooked.

2. It was not the case that one type of experiment was begun only after another type had been completed. Rather, the various types of experiments were conducted concurrently. Thus, the suggested third conclusion cannot be attributed to an artifact of *when* the experiments happened to be conducted.

3. Although it is true that experimenters and special subjects may be highly motivated to have their laboratory experiments succeed, we feel such motivation is still less than the motivation present in more meaningful everyday life situations. It is perhaps this very motivation to succeed which accounts for the successful experiments we are able to carry out in the laboratory.

4. One potentially significant difference may have been that active subjects had a correct perception of their physiological state, whereas inactive subjects did not. This difference could be correlated with other variables (e.g., personality) which might influence psi performance. Alternatively, the heightened autonomic arousal of those subjects classified as inactive may have been more situation-dependent and may not have been adequately tapped during the first (screening) session.

5. M. S. performed significantly in a successful but unpublished FRNM study in which she had served as one of the subjects.

6. This amplifier, which had a time constant of 0.5 sec., could be set to record either phasic (AC) responses (i.e., galvanic skin responses or GSRs) or tonic (DC) skin resistance levels (i.e., basal skin resistance, or BSR). The device was set momentarily to the DC-BSR mode to record initial and final BSR levels. Throughout the 20 recording epochs, the device was set to the AC-GSR mode so that only *fluctuations* in skin conductance were recorded.

7. Electrodermal activity is the electrical activity of the sweat glands, which are sympathetically innervated. Just as the heart and brain emit electrical signals which may be recorded as EKG and EEG activity, respectively, so too do the fibers innervating the sweat glands. Increased sweat gland activity is "seen" as increased voltage or increased skin conductance or decreased skin resistance by devices monitoring electrodes placed over concentrations of sweat glands in the skin. Phasic changes in sweat gland electrical activity are commonly called "galvanic skin responses (GSRs)" or "electrodermal activity (EDA)," as in this paper. High frequencies and amplitudes of EDA indicate heightened activity of the sympathetic branch of the autonomic nervous system which underlies emotionality. Thus, high EDA levels indicate heightened arousal, activation, emotionality, stress, etc. Low levels of EDA indicate relaxation, the absence of stress, physical and mental calmness.

While the EDA measures used in this experiment indicate momentary fluctuations in sympathetic activity, the two basal skin resistance (BSR) measurements indicate the tonic or ongoing baseline aspects of this activity. High resting BSR level generally indicates a relatively relaxed and stress-free condition; low BSR values are generally correlated with tension or arousal. Note that GSR (frequency and amplitude) and BSR are "inversely related," so that stress is associated with high GSR activity and low BSR levels, while calmness is associated with low GSR activity and high BSR levels.

The subjects' ongoing, fluctuating EDA (GSR) responses were monitored during the recording epochs of the present study. The computer and analog-to-digital converter sampled the EDA 10 times each second for 30 seconds and averaged these measures, thus providing what is virtually a measure of the area under the curve described by the fluctuating EDA (i.e., the mathematically integrated activity).

8. *Progressive error* refers to the contribution of any irrelevant variable which varies systematically with time (e.g., hunger or fatigue). Our use of the ABBA design allows any progressive error which may have autonomic concomitants to contribute equally to the A and B periods, thus avoiding a biasing contribution to any one condition alone.

9. This design, which employs many rapidly alternating, brief influence periods, would not be appropriate in cases in which longlasting "linger effects" are observed, of the type reported by the Watkinses. We have not found evidence of such linger effects in our various bio-PK experiments.

10. Target persons' self-ratings of felt degree of rapport or connectedness with the influencers did not correlate significantly with psi scores.

11. W. B. and M. S. independently calculated the various scores and independently did the statistical tests, thus effectively double-checking these computations.

12. This first-half measure was based on the total mean EDA for the first 10 recording epochs, collapsing across control and influence periods; the second-half measure was based on the total mean EDA for the last 10 recording epochs for a given subject.

13. The nonsignificant results from the exactly parallel analyses of the pseudo scores for Session 1 (see last paragraph of Results section) also argue against this fifth nonpsi alternative.

14. An additional report of a similar bio-PK effect, this time on the electric fish *Gnathonemus petersii,* has recently been published. The reference for this experiment is: V. R. Protasov, V. D. Baron, L. A. Druzhkin, and O. Yu Chistyakova, "Nile elephant" *Gnathonemus petersii* as a detector of external influences. *Psi Research,* 1983, 2, 31–37. It should be mentioned that the Kelly et al. and the Protasov et al. studies are conceptual replications of the bio-PK design inasmuch as attempts were made to psychically influence biological systems. However, because immediate, trial-by-trial, analog feedback was not provided in these two studies, they do not qualify as "allobiofeedback" (see Braud, 1978) conceptual replications.

15. It should be pointed out that if the influencer were aware of the greater ostensible need of subjects in the active condition and therefore was more motivated to influence those subjects, this increased need or motivation *of the influencer,* if it were to facilitate psi-scoring, would also be evidence favoring the need hypothesis. The locus of the need (in the subjects or in the influencer or in both, as was probably the case in the present study) is not critical in this case.

16. Originally, it was planned to use a criterion of 100 responses per session. However, we were finding that two or three volunteers had to be screened to find one who met this stricter criterion of activity. Therefore, to complete this nonpsi experiment before the deadline for submission of this paper to the 1983 Parapsychological Association Convention, we relaxed the criterion to 50 responses per session. An examination of the results indicates that degree of autonomic self-control does not differ appreciably for the 50–100 and the <100 response subgroups.

17. Estimated ratings of degree of success at autonomic self-control correlated positively and rather highly ($r = +.41$) with actual autonomic control scores; this correlation, however, did not quite reach the .05 level of significance.

18. As an additional point of interest, it can be reported that an actual real-time bio-PK experiment was filmed in its entirety on January 31, 1983, for a special BBC Television "Horizon" parapsychology documentary to be aired later [that] year. Whereas the mean bio-PK effect in the experiment reported here was approximately 10% greater than chance expectation, the magnitude of the successful bio-PK effect during the BBC filming was 35% greater than chance expectation. This provides still another indication of the robustness of the effect and of the influence of *need* or motivation.

References

Barry, J. (1968). General and comparative study of psychokinetic effect on a fungus culture. *Journal of Parapsychology,* 32, 237–243.

Blizard, D. A., Cowings, P., and Miller, N. E. (1975). Visceral responses to opposite types of autogenic-training imagery. *Biological Psychology,* 3, 49–55.

Braud, W. (1978). Allobiofeedback: Immediate feedback for a psychokinetic influence upon another person's physiology. In W. G. Roll (Ed.), *Research in parapsychology, 1977.* Metuchen, N.J.: Scarecrow Press.

Braud, W. (1979). Conformance behavior involving living systems. In W. G. Roll (Ed.), *Research in parapsychology, 1978.* Metuchen, N.J.: Scarecrow Press.

Braud, W. (1980). Lability and inertia in conformance behavior. *Journal of the American Society for Psychical Research,* 74, 297–318.

Braud, W. (1981). Lability and inertia in psychic functioning. In B. Shapin and L. Coly (Eds.), *Concepts and theories of parapsychology.* New York: Parapsychology Foundation.

Braud, W., Davis, G., and Wood, R. (1979). Experiments with Matthew Manning. *Journal of the Society for Psychical Research,* 50, 199–223.

Ehrenwald, J. (1978). Psi phenomena, hemispheric dominance and the existential shift. In B. Shapin and L. Coly (Eds.), *Psi and states of awareness.* New York: Parapsychology Foundation.

Engel, B. T. (1972). Operant conditioning of cardiac function: A status report. *Psychophysiology,* 9, 161–177.

Grad, B. (1967). Some biological effects of the "laying on of hands": A review of experiments with animals and plants. *Journal of the American Society for Psychical Research,* 61, 286–305.

Gruber, E. R. (1979). Conformance behavior involving animal and human subjects. *European Journal of Parapsychology,* 3, 36–50.

Gruber, E. R. (1980). PK effects on pre-recorded group behavior of living systems. *European Journal of Parapsychology,* 3, 167–175.

Kelly, M. T, Varvoglis, M., and Keane, P. (1979). Physiological response during psi and sensory presentation of an arousing stimulus. In W. G. Roll (Ed.), *Research in parapsychology, 1978.* Metuchen, N.J.: Scarecrow Press.

Nash, C. B. (1982). Psychokinetic control of bacterial growth. *Journal of the Society for Psychical Research,* 51, 217–221.

Onetto, B., and Elguin, G. (1966). Psychokinesis in experimental tumorogenesis. *Journal of Parapsychology,* 30, 220.

Puthoff, H., and Targ, R. (1975). Physics, entropy, and psychokinesis. In L. Oteri (Ed.), *Quantum physics and parapsychology.* New York: Parapsychology Foundation.

Rao, K. R. (1966). *Experimental parapsychology: A review and interpretation.* Springfield, Illinois: Charles C. Thomas, pp. 123–129.

Rauscher, E. A., and Rubik, B. A. (1980). Effects on motility behavior and growth rate of *Salmonella typhimurium* in the presence of a psychic subject. In W. G. Roll (Ed.), *Research in parapsychology, 1979.* Metuchen, N.J.: Scarecrow Press.

Stanford, R. G. (1978). Toward reinterpreting psi events. *Journal of the American Society for Psychical Research,* 72, 197–214.

Stanford, R. G. (1981). Are we shamans or scientists? *Journal of the American Society for Psychical Research,* 75, 61–70.

Tedder, W. H., and Monty, M. L. (1981). Exploration of long-distance PK: A conceptual replication of the influence on a biological system. In W. G. Roll and J. Beloff (Eds.), *Research in parapsychology, 1980.* Metuchen, N.J.: Scarecrow Press.

Walker, E. H. (1975). Foundations of paraphysical and parapsychological phenomena. In L. Oteri (Ed.), *Quantum physics and parapsychology.* New York: Parapsychology Foundation.

Watkins, G., and Watkins, A. (1971). Possible PK influence on the resuscitation of anesthetized mice. *Journal of Parapsychology,* 35, 257–272.

Watkins, G., Watkins, A., and Wells, R. (1973). Further studies on the resuscitation of anesthetized mice. In W. G. Roll, R. L. Morris, and J. D. Morris (Eds.), *Research in parapsychology, 1972.* Metuchen, N.J.: Scarecrow Press.

Mind Science Foundation
102 W. Rector, Suite 215
San Antonio, TX 78216

3
Mentally Protecting Human Red Blood Cells at a Distance
William G. Braud[1]

Individual research participants were able to "protect" human red blood cells (from themselves and from others) from osmotic stress, mentally and at a distance, using techniques of intention and imagery.

This research originally was published in Braud, W. G. (1990). Distant mental influence of rate of hemolysis of human red blood cells. Journal of the American Society for Psychical Research, *84, 1–24. The contents of this article are Copyright © 1990 by, and reprinted by permission of, the* Journal of the American Society for Psychical Research.   *—William Braud*

Abstract: A formal investigation was conducted in order to determine whether a relatively large number of unselected subjects would be able to exert a distant mental influence upon the rate of hemolysis of human red blood cells. For each of 32 subjects, red blood cells in 20 tubes were submitted to osmotic stress (hypotonic saline). The subjects attempted to protect the cells in 10 of the tubes using visualization and intention strategies; the remaining 10 tubes served as noninfluence controls. For each tube, rate of hemolysis was measured photometrically over a 1-minute trial period. Subjects and experimenter were "blind" regarding critical aspects of the procedure, and subjects and tubes were located in separate rooms in order to eliminate conventional influences. Results indicated that a significantly greater number of subjects than would be expected on the basis of chance alone showed independently significant differences between their "protect" and "control" tubes ($p = 1.91 \times 10^{-5}$). Overall, blood source (i.e., whether the influenced cells were the subject's own cells or those of another person) did not significantly influence the outcome, although there was a trend toward stronger hitting in the "own blood" condition. Additional analyses of the results were performed by SRI International researchers to determine whether the data were better described by remote action (causal) or by intuitive data sorting (informational) predictions; the results of those mathematical analyses were inconclusive. This research is presented in the context of methodologies for investigating a possible role of psi in self-healing.

Laboratory research has indicated that under certain conditions persons are able to psychokinetically influence a variety of biological systems (see Solfvin, 1984). At the Mind Science Foundation (MSF), we have been able to observe successful distant mental influences upon the spatial orientation of fish, the locomotor activity of small rodents, and the physiological activity of another person (Braud, 1978a, 1979; Braud, Davis, and Wood, 1979; Braud and Schlitz, 1983; Schlitz and Braud, 1985). Such findings are consistent with an interpretation of psychic healing in which the healing might be contributed, at least in part, by the healer's conscious or unconscious psychokinetic influence upon the healee's bodily processes.

It is a relatively straightforward matter to design experiments to isolate a possible psi component in one person's healing of another. The healee may be kept blind regarding the timing and nature of a healing attempt, and various isolation and control procedures can be applied. However, it is ordinarily impossible to study the role of psychic functioning in *self*-healing *in vivo* using such techniques. The healer is necessarily aware of the timing and nature of self-healing attempts, and therefore the psychic process cannot be isolated from the many nonpsychic factors that can influence healing. We cannot distinguish a psychic effect from the more conventional influences of expectation (the "placebo" effect), changes in emotionality, self-regulation of neural, muscular, and hormonal activities, etc. Fortunately, it may be possible to eliminate the influence of these nonpsychic processes through the use of biological psychokinesis (bio-PK) protocols that have proved useful in the study of distant mental influence. The strategy would involve the selection of some specific biological activity to be targeted psychokinetically and then the *removal* of the measured biological activity, in space or in time, from the reach of all possible nonpsychic self-influences.

Spatial or temporal isolation could be accomplished in three ways:

1. The experimental participant could attempt to influence cells, tissue, or biochemicals that have been freshly *removed* from his or her body.

2. The participant could attempt to influence his or her own biological activity that had been *prerecorded* at an earlier time and that had remained unobserved until the subsequent influence attempt; this strategy involves the so-called "time-displaced" or "retro-psychokinesis" design explored by Schmidt (1976, 1981; Schmidt, Morris, and Rudolph, 1986). Another method of removing the target biological activity in time would be to ask participants to attempt to influence their own *future* activity. In this case, the influence epochs would have to be coded in a way that would eliminate

the possibility that the participant might remember the desired outcome and consciously or unconsciously self-regulate at the time of the future test in order to bring about the desired changes.

3. The participant could attempt to influence a particular physiological or biochemical process within his or her body while, at the same time, the activity of a similar, externalized process is monitored to determine whether the latter responds in "resonance" or in "sympathy" with the former.

If research participants succeed in influencing their biological materials through psychic means when those materials are isolated, then it is likely that the biological materials remain susceptible to psi influence under ordinary, nonisolated conditions as well. Psi influence thus becomes a viable component in self-regulation and hence in self-healing. We might even find that persons are psychically able to influence their own cells or activities more readily than they are able to influence the cells or activities of another person. If this is indeed the case, it would suggest that psi's contribution to self-healing may be greater than its contribution to the healing of another person. It would also suggest that more impressive results might be obtained in bio-PK studies if experimenters selected target systems that were more closely associated with the participant-influencers in their experiments.

We have conducted numerous bio-PK experiments in which both selected and unselected participants were able to alter significantly the activity of specific biological target systems, mentally and at a distance. In all of those studies, the amount of target system activity in the prescribed direction during a number of influence periods was compared statistically with its activity during an equal number of interspersed control (noninfluence) periods. The obtained bio-PK results have been reliable and relatively robust. For example, a recent review of all of the MSF's electrodermal bio-PK experiments conducted to date indicated a $z = 4.08$, $p = .000023$, and mean effect size $= 0.29$ for all 13 experiments combined (Braud and Schlitz, 1989). Experiments could be conducted, using an identical methodology, in which persons attempt to influence their *own* biological materials.

If Eccles (1977) and others are correct in maintaining that one's mind routinely exerts a true psychokinetic influence upon one's own brain through "cognitive caresses" of the synapses of cortical neurons, then perhaps the most ideal target material would be nervous tissue that had been removed and cultured externally (or cloned). At first, it might appear that such an experiment would not be feasible. However, a study of this type could actually be done with the cooperation of a neurosurgeon. Brain tissue that is removed for medical reasons and that is ordinarily discarded might be artificially cultured and maintained as target material for bio-PK attempts by the patient, following his or her recovery from the operation.

Outside of this seemingly science-fictional scenario, the next best target material might be lymphocytes (white blood cells with important roles in the immune process: B-cells, T cells, natural killer cells). Unfortunately, the procedures required for the measurement of these cells or of their biological activities are relatively complex and beyond the capabilities of most psi research facilities. It is noteworthy that several researchers recently have reported results of experiments in which their subjects were strikingly successful in self-regulating specific sub-populations of their leucocytes (e.g., neutrophils) using relaxation and imagery techniques (see Braud, 1986; Hall, 1984a, 1984b; Schneider, Smith, and Whitcher, 1984). The work of Schneider et al. is especially interesting in view of the *accuracy, rapidity,* and *specificity* of the influences that occurred without the provision of conventional sensory feedback. The process appears goal-directed and immediately reminds one of the similar goal-directed nature of psychokinesis.

The third best, and logistically the easiest, material with which to work would be osmotically stressed red blood cells. The goal of the experiment would be to protect one's own red blood cells and retard their rate of hemolysis. If red blood cells are maintained in fluids having a salinity similar to that of the blood plasma, the cells survive for long periods. However, if placed in a fluid having a salinity lower than that of the plasma (e.g., in distilled water or in very dilute saline), the corpuscles swell due to the movement of water through their semipermeable membranes. Eventually the cells rupture and release their hemoglobin content into the surrounding medium. The rate of hemolysis can be measured using a spectrophotometer with an output to a computer and pen recorder. Light of a given wavelength is passed through a tube of blood cells suspended in dilute saline. Intact cells are relatively opaque to the light. As the cells rupture, the solution becomes increasingly transparent. The spectrophotometer/computer/pen recorder system provides an objective readout of the change in light passage over time, and hence the time course of hemolysis. Rate of hemolysis can be measured during influence trials in which the research participant attempts to mentally retard the hemolytic process, that is, attempts to psychokinetically protect the red blood cells from osmotic injury. These results can be compared with hemolysis rates obtained during interspersed control trials in which no psychokinetic attempts are made.

We conducted a preliminary experiment of this sort in our laboratory several years ago. In that experiment, involving a single selected subject and a small number of trials, a significant influence of hemolysis rate was observed (Braud, Davis, and Wood, 1979). In the present experiment, we sought to repeat that experiment with many more trials, a large number of unselected subjects, and an improved methodology, in order to test the reproducibility and generality of the finding. Briefly, subjects attempted mentally to retard the rate of hemolysis of osmotically stressed human red blood cells that were isolated from all conventional influences. The subjects and the target system were kept in separate rooms. Rate of hemolysis was monitored accurately by a spectrophotometer interfaced by means of an analog-to-digital converter to a microcomputer. The experimenter operating the equipment was blind regarding the scheduling of the influence (protect) versus noninfluence (control) attempts. Half of the subjects worked with their own blood cells, and half worked with cells from another person; both experimenter and subject were blind regarding the blood source. The experimental design included features that we hoped would help us determine whether any obtained psi effect might be most parsimoniously interpreted as a true psychokinetic (remote action) effect or, alternatively, as an instance of "intuitive data sorting" (see May, Radin, Hubbard, Humphrey, and Utts, 1985). Prior to the formal experiment to be described in detail in this paper, two additional studies were conducted— a pilot study and an "intermediate phase" study. Space limitations permit only relatively brief summaries of those two studies.

Summary of the Pilot Phase

Thirty-two unselected subjects participated in a pilot study designed to explore the new methodology and to determine whether blood source (own blood cells vs. another person's blood cells) was an important factor. An experimental session involved hemolysis measurements for 10 blood tubes. The subject attempted to retard the rate of hemolysis of 5 of these tubes, mentally and at a distance. The remaining 5 tubes served as control tubes that the subject did not attempt to influence. The 5 influence and 5 control tubes were scheduled according to a random sequence that was prepared by a third party and that was unknown to the experimenter who made the hemolysis measurements. Light transmission through each tube (which is proportional to hemolysis) was measured for each second of a 2-minute sampling period; the difference between the mean of the initial 5 seconds and the final 5 seconds of light measurements yielded a change score that served as the hemolysis measure.

Following the completion of the experimental session, 10 additional blood-containing tubes were measured for hemolysis rate. It was intended that these 10 measurements would provide additional "nonlocal" baseline data and would also be useful in comparing remote action (RA) versus intuitive data sorting (IDS) predictions of the experimental outcome. According to the RA hypothesis, the mean of the 5 "local" control tubes should be equivalent to the mean of the 10 nonlocal baseline tubes, and the mean of the 5 influence tubes should be lower (i.e., in the direction of less hemolysis or greater protection of the cells) than both of the former means. According to the IDS hypothesis, the mean of the local controls should be above, and that of the influence tubes should be below, that of the nonlocal baseline tubes; the *grand mean* of the *10* tubes for the experimental session should not differ from the mean of the 10 nonlocal baseline tubes.

An analysis of variance of the hemolysis scores indicated extremely great and highly significant variability among the subjects, but no other significant main effects or interactions. Therefore, significant evidence for a remote influence of the blood cells was not obtained in the pilot study. There was, however, a nonsignificant tendency for a slight "protection" effect in the "another's blood" condition, whereas the opposite effect (i.e., *less* protection during the influence trials) occurred in the "own blood" condition. The non-local baseline measurements were found to be inadequate for their intended purpose because in every case but one the mean percent light transmittance change score for the 10 nonlocal baseline tubes was lower than both experimental sessions means (i.e., consistently lower than both the control and the influence tube means). It was determined that this consistent reduction in hemolysis for the nonlocal baseline tubes (and to a lesser extent, for the tubes later in the experimental sessions as well) was due to a progressive change in the blood cells contributed by several environmental factors that increased the "noise" level of the experiments and that included higher apparatus (i.e., spectrophotometer tube holder) temperatures during later tests, and increasing exposure of the blood cells to temperature changes, air, and mechanical trauma (i.e., mechanical agitation) during the course of repeated tests of a single blood sample (i.e., multiple tests of the contents of a single Vacutainer blood collection tube). As a result of the pilot sessions themselves, as well as additional tests conducted concurrently with and subsequent to the pilot experiment, the sources of these interfering factors were identified, and steps were devised that could be taken to eliminate or greatly reduce them in the formal study.

Temperature changes in the spectrophotometer tube holder were controlled through (a) the addition of an external cooling fan to the apparatus, (b) reducing the durations of the hemolysis measurement periods, and (c) turning off the apparatus except when measurements were actually being made. The substitution of a more effective anticoagulant (acid-citrate-dextrose) for that used in the pilot study (heparin) greatly diminished the effects of progressive exposure to room temperature, air, and mechanical trauma during repeated pipette samplings so that hemolysis rate remained relatively stable over the course of 20 measurements from a given main Vacutainer source tube. When between 20 and 30 samples had been taken from a main Vacutainer tube, this stability began to deteriorate.

Summary of the Intermediate Phase of Salinity Tests

Following the completion of the pilot study, salinity tests were conducted in order to determine salinity values that might mimic anticipated psi-induced hemolysis rate changes. These tests provided the basis for Monte Carlo analyses that were designed to determine appropriate parameters for an adequate differential test of the IDS versus RA predictions of psi functioning that was to be conducted in the formal study.

A total of 332 hemolysis trials were completed using whole blood samples collected from 10 different persons. For these tests, the "noise-reducing" improvements mentioned above were incorporated. Hypotonic salinity values of 0.425%, 0.429%, 0.434%, 0.442%, and 0.450% (corresponding, respectively, to 50%, 50.5%, 51%, 52%, and 53.33% of 0.85% normal physiological saline) were tested. Sampling epochs of 1-minute duration were used, rather than the 2-minute periods of the pilot study. All other procedures were identical to those of the pilot study. These were all, of course, "control" tests in which no subjects attempted to influence the hemolysis process.

As anticipated, the "noise-reducing" improvements resulted in the virtual elimination of the extreme variability seen in the pilot study and yielded much greater stability (less degradation) of the blood samples. The optimum salinity value for mimicking an anticipated psi-induced reduction of hemolysis rate of approximately 1.0 standard deviation was found to be in the vicinity of 0.429%–0.434% saline (equivalent to 50.5%–51.0% of normal 0.85% physiological saline). Monte Carlo simulation analyses conducted on these salinity data indicated that on the basis of the magnitudes of hemolysis changes observed in these intermediate phase salinity tests, the use of 2 versus 8 samples (tubes) distributed throughout equivalent "psi effort" periods would provide adequate measurements for a differential test of the IDS versus RA interpretations of any psi effects obtained in the formal experiment.

Another purpose of these salinity tests was to assure that our spectrophotometric method was actually assessing hemolysis, rather than other possible artifactual time-changing processes that could be confused with hemolysis. A number of tests were performed using identical procedures but with normal 0.85% physiological saline rather than hypotonic saline. No hemolysis should occur with such a solution, and any changes observed could be attributed to artifacts. As expected, hemolysis did not occur in these normal saline tests; the photometric readings and curves were completely stable throughout the sampling periods.

Overview of the Formal Experiment

On the basis of the findings of the preliminary study, the pilot study, and the intermediate phase experiments, a final protocol for the formal experiment was developed that included the following features:

1. Thirty-two subjects (from the same population and selected in the same manner as in the pilot study) would each participate in one experimental session. Hemolysis measurements would be made by the experimenter, W. G. B.

2. Sixteen subjects would attempt to influence (protect) their own blood cells, and 16 would attempt to influence the cells of another person. Both subject and experimenter would be blind regarding the source of the blood until all 32 sessions had been completed. This "own" versus "other" factor was retained in the formal experiment because of the trend toward different outcomes in those two conditions observed in the pilot study.

3. Blood samples would be collected in Vacutainer tubes containing acid-citrate-dextrose (ACD) anticoagulant and would be refrigerated immediately after the blood was drawn. Blood samples would be stored at 4°C and would be removed from the refrigerator only briefly, before each hemolysis trial.

4. Hemolysis trials would be conducted between 14 and 42 hours following a blood draw. The ACD anti-coagulant permits cold storage of blood cells for as long as 3 to 4 weeks with minimal deterioration of red blood cells.

5. The temperature increase of the spectrophotometer would be minimized by means of an external cooling fan, the use of shorter sampling epochs, and allowing the apparatus to remain on only during hemolysis measurement periods.

6. A session would consist of four 15-minute periods—two control (C) periods and two protect (P) periods. For half of the subjects, these periods would be scheduled in a CPPC order; for half of the subjects, a PCCP order would be used. This block-counterbalancing design was employed in order to assure that any reasonably linear potential progressive error (such as changes in hemolysis rate due to slight progressive warming of the apparatus) would contribute equally to the two (C and P) conditions and therefore not introduce a systematic bias. Whether a given subject's sequence was CPPC or PCCP would be randomly determined by an associate (M. S.) through use of the RAND table of random numbers. The experimenter doing the hemolysis measurements would, of course, be blind regarding these sequences. A subject would learn his or her proper sequence by consulting a sealed envelope delivered to the subject after the experimenter's interactions with the subject had been completed and the experimenter had returned to the equipment room.

7. The beginning of each 15-minute period would be signaled by an appropriate number of tones delivered to the subject's headphones. The subject would have been instructed to attempt to mentally *decrease* the rate of hemolysis of the distant red blood cells during the two 15-minute protect

periods. During the two 15-minute control periods, the subject would attempt not to think about the experiment and would allow the cells to hemolyze at their normal, rapid rate. During the two protect periods, the subject would view a projected color slide of healthy, intact red blood cells as an aid to visualization and intention. During the two control periods, the subject would close his or her eyes and think about matters unconnected with the experiment.

8. During each 15-minute period, either two or eight hemolysis tubes (samples) would be measured. Monte Carlo analyses conducted at SRI International indicated that curves derived from 2 versus 8 tubes (samples) would be sufficient for an adequate test of the IDS versus RA interpretations of any obtained psi effect. The subject would be blind regarding the number of tubes being measured during any 15-minute period and would be instructed to apply mental effort as steadily and as consistently as possible throughout the entire 15-minute protect periods. The experimenter would learn whether to measure 2 or 8 hemolysis tubes during each 15-minute period by consulting a sealed envelope delivered to him just before the beginning of the measurement session. This random, balanced tube sequence would have been determined earlier by M. S., again using the RAND table of random numbers.

9. Because the subjects must remain blind regarding the number of tubes being measured during each 15-minute period, it would not be possible to provide them with real-time auditory feedback of the progress of hemolysis, as we had hoped to do. Such feedback would provide subjects with information about the number of tubes and would therefore violate the blindness requirement and add a psychological confound to the experiment. However, the subject would receive numerical feedback about hemolysis outcomes at the conclusion of the session.

10. The subject's session would be preceded by 8 minutes of tape-recorded instructions for relaxation and guided imagery, designed to help reduce distractions and focus attention upon the desired goal event—viz., decreased hemolysis during effort (protect) periods.

11. Hemolysis measurements would be accomplished with a procedure identical to that used in the pilot study, with the following exceptions: (a) the recording epochs would be 1 minute rather than 2 minutes in duration, and (b) the subject would not hear tones signaling the beginning and end of each *tube* measurement (as in the pilot study), but rather would hear tones signaling the beginning of each of the four 15-minute periods.

12. Hemolysis scores would be analyzed in a manner identical to that described in the pilot study. A similar ANOVA would be used to assess the presence of a psi effect. In addition, all hemolysis percent change scores would be normalized for purposes of additional IDS versus RA analyses.

Method

Subjects

Thirty-two subjects participated in the study. Participants were selected from a pool of normal, healthy individuals and were screened to eliminate those with known allergic or immunological disorders or other illnesses and those currently taking medication (other than oral contraceptives and/or occasional cold medicines). Twenty-one of the subjects had already participated in the pilot investigation and were asked to participate again because of their familiarity with the procedure. Eleven subjects were first-time participants who substituted for pilot study subjects who were unable to take part in the formal study. The final sample consisted of 17 females and 15 males, ranging in age from 23 to 53 years. Each subject was paid $20 as a token of appreciation for the inconvenience and slight discomfort of donating a blood sample and for participating in the subsequent 1 1/2-hour laboratory session.

Procedure

On a Monday evening, the experimenter met with a group of four participants in order to explain the experiment in detail and to have the subjects complete an Informed Consent Form, donate a 10 ml venous blood sample, and schedule an appointment for an experimental session for later that same week (i.e., on either the next day [Tuesday] or the day after [Wednesday]). An attempt was made to schedule two experimental sessions on Tuesday (at 10:00 a.m. and at 2:00 p.m.) and two sessions on Wednesday (at 10:00 a.m. and at 2:00 p.m.). On the Monday evening, the participant was given a two-page written description of the procedure and was asked to read the description at home and become familiar with it.

The four blood samples were drawn by a registered nurse.[2] The blood collection tubes (Becton Dickinson Vacutainer tubes containing acid-citrate-dextrose anticoagulant) were labeled with the names of the blood donors and were placed in a small refrigerator immediately after the blood draws. The refrigerator was maintained at 4° C throughout the experiment. When all four blood samples had been drawn, the nurse switched the name labels on two of the tubes, using a randomizing schedule that had been prepared ahead of time by an associate of the experimenter (M. S.). This schedule was always kept by the nurse (and a copy kept by M. S.) and was unknown to the experimenter until the study had been completed. The purpose of switching the labels of two tubes was to permit two subjects

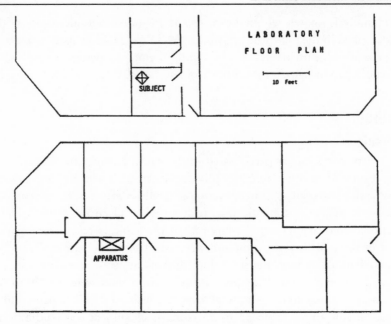

Fig. Floor plan of the laboratory rooms in which the experiments were conducted. The apparatus room and the subject room are situated in separate suites of the same building, with an exterior corridor separating the two suite areas. Interior corridors separate the rooms within each suite area. The subject room is windowless. The windows of the apparatus room were completely occluded during experimental trials.

to attempt to influence their own blood and two to attempt to influence another person's blood that week, and to keep the subjects and the experimenter blind regarding the blood source until all 32 sessions of the study had been completed.[3]

Following his or her arrival for the experimental session on Tuesday or Wednesday, the experimenter showed the subject the equipment at the target site, emphasized the spectrophotometer tube holder in which the target tubes later would be placed sequentially, and then escorted the subject to the distant subject room, located in another part of the building (see floor plan in the Figure). The subject sat in a comfortable armchair and was told that shortly after the experimenter left the subject's room, an assistant would slip an envelope under the subject's door. The subject was to retrieve that envelope and open it to find the random sequence of the four 15-minute periods of the experiment. (The 32 period-sequence envelopes had been prepared beforehand by M. S. using the RAND table of random numbers and a private algorithm. Throughout the experiment, the envelopes remained hidden from the experimenter. M. S. retained a copy of the period sequences for the 32 envelopes.) During each of the two 15-minute control periods, the subject was to attempt to keep her or his mind off the experiment and to think of other matters; if she or he could not help thinking about the experiment, the subject was asked to imagine hemolysis proceeding at its normal, rapid rate. During each of two 15-minute protect periods, the subject was to attempt to mentally retard the rate of hemolysis of the red blood cells in the tubes for that period using any of the mental strategies described on the instruction sheet. The experimenter demonstrated a slide projector that could be used by the subject during the two protect periods. The 35 mm color slide depicted healthy, intact red blood cells, and it was included as a helpful aid to the subject's visualization of the desired goal. The subject was told that the beginning of each period would be signaled by an appropriate number of tones (one for Period 1, two for Period 2, and so on) presented through headphones. The subject was also told that the first period would be preceded by an 8-minute progressive relaxation and guided imagery exercise designed to help the subject reduce distractions and focus attention upon the desired goal event, that is, decreased hemolysis during the protect periods. The exercise was accompanied by low-volume, ambient music and ocean sounds. Low-volume ambient music was also presented through the subject's headphones throughout the four periods of the experiment and was interrupted only for the four-period signaling tone presentations. The conclusion of the experiment was indicated to the subject by the cessation of the music. At that time, the subject was to sign and date his or her period sequence sheet and then be escorted back to the apparatus room by an assistant.

The experimenter returned to the apparatus room, where the equipment had already been readied for use. Just before entering this room and closing the door, he indicated to an assistant that the experiment was about to begin. The assistant gave him a sealed envelope that contained information about his 2- versus 8-tube sequence for that session, and then delivered another sealed envelope to the subject; this latter envelope contained the subject's protect versus control period sequence for that session. The experimenter started the audio tape that presented the preliminary exercises to the subject. He then conducted the 20 hemolysis measurements for the session. From his point of view, there were also four 15-minute periods of measurements; two of the periods (indicated on a sheet within his envelope) were to involve measurements of two tubes, and two of the periods were to involve measurements of eight tubes. This number of measurements factor was included to provide data for a differential test of the IDS and RA predictions of psi performance (see below). The subjects remained blind regarding the tube number schedule for the session.

Each of 20 identical 10-ml glass spectroscope tubes had been filled beforehand with 6.0 ml of 0.425% saline, and these had been kept in the refrigerator at 4° C. The saline for all tubes for all sessions came from the same stock solution of 0.85% normal physiological saline, purchased in 20-liter quantity from Fisher Scientific Supply Company and diluted with distilled water to 0.425% by the experimenter before the study began. This use of solution from the same stock eliminated variability that otherwise might have been contributed by that factor. The experimenter removed the main blood collection (Vacutainer) tube bearing that subject's name from the refrigerator, inverted the tube eight times in order to assure a homogeneous suspension of its blood cells, opened the tube, and placed it in a test tube rack on the equipment table. He then removed the first of the hypotonic saline tubes from the refrigerator and allowed it to stand at room temperature and warm slightly so that moisture from the warmer room-temperature air no longer condensed on the tube after the latter was wiped with tissue. He placed the now frost-free saline tube into the holder of the spectrophotometer and adjusted the controls of the device so that a digital reading of precisely 100.0% light transmission was obtained for this blank tube. He pressed a computer keyboard key to initiate a subroutine that signaled the subject in the distant room that a 15-minute period was about to begin. He next removed the tube from the holder and added to the tube 100 μl of whole blood from the main Vacutainer tube. He quickly stoppered the saline tube with a rubber stopper, inverted the tube twice to assure homogeneity of its contents, and quickly replaced the tube in the spectrophotometer holder. When the holder cover was closed, the chart recorder pen moved to indicate minimal light transmittance; at the point of greatest excursion of the pen, the experimenter pressed a keyboard key to initiate the 1-minute sampling epoch for that tube. The Vacutainer blood collection tube was then returned to the refrigerator, and the next hypotonic saline tube was placed in the test tube holder so that it might warm slightly for the next trial. The precise timing of all procedural events was controlled by the experimenter through the use of several procedural cues and by means of extreme stereotypy of responding. Throughout the sampling epoch, the chart recorder and the digital readout of the spectrophotometer were shielded so that they could not be observed by the experimenter. This was done in order to eliminate immediate feedback to the experimenter in the hope that this might reduce the latter's own psi contribution to the experimental outcome.

Percent light transmittance measures at a wavelength of 660 mμ (an absorbance minimum for hemoglobin) relative to the blank tube containing saline alone were taken by means of a Sequoia-Turner Model 390 spectrophotometer with digital and chart recorder readouts. The spectrophotometer provided an analog output that varied from 0 to 1.0 v DC and was linearly related to percent light transmittance (with 0 v DC = 0% T and 1.0 v DC = 100% T). This output was increased by a factor of 10 by means of a differential amplifier, and the resulting 0 to 10.0 v DC signal was fed into an analog-to-digital converter installed in an IBM PC-XT compatible computer. The A/D converter (CGRS Microtech PC DIADAC 1) uses an industry standard AD 574A 12-bit A/D chip with 0.0024 volt accuracy and 35 μsec conversion speed. A software program was written that sampled the A/D converter at the end of each second of the 1-minute trial period. Thus, the system automatically provided 60 measurements of the time course of hemolysis (i.e., percent transmittance) during each 1-minute trial. The 60 values were written to a floppy disk file and were also printed out at the end of the trial. In addition to this digital data collection, an analog chart record was obtained for each trial (using a Markson Model 1202 pen recorder).

At the end of the 1-minute sampling epoch, the experimenter removed the tube from the holder and began his preparations for the remainder of the trials. Approximately one minute elapsed between trials. If a period called for the measurement of two tubes, those two tubes were measured at the middle of the 15-minute period, that is, at times corresponding to the measurement of tubes 4 and 5 of an eight-tube period. The main Vacutainer blood collection tube and the hypotonic saline tubes remained in the refrigerator except when needed for the measurements. The completion of the hemolysis measurements for the 20 tubes of an experimental session required one hour.

When the 20th and final tube had been measured, and the results had been printed, the experimenter notified his assistant that the session was over. While the assistant went to the subject's room, the experimenter made photocopies of the data sheets and of his tube-number schedule. When the assistant returned with the subject, the assistant photocopied the subject's control/protect period sequence sheet. The assistant and the experimenter then exchanged copies of their respective sequence sheets and data printouts. These duplicate records were filed for safekeeping by the assistant and by the experimenter.

The experimenter and the subject then went to the experimenter's office, and the subject described the techniques used to attempt to influence the blood cells. After this interview, the experimenter calculated the results for the experimental session and provided the subject with information about the session outcome. This information consisted of verbal and numerical feedback about the hemolysis rates for the 20 tubes. The experimenter thanked the subject for his or her participation, and the subject left the laboratory.

Results

When all 32 experimental sessions had been completed, the blood source information was decoded so that a determination could be made of which subjects attempted to influence their own blood and which attempted to influence blood from another person. For each session, *change scores* were calculated for each of the 20 blood sample tubes (trials). For each tube (trial), the mean of the initial five A/D converter values was subtracted from the mean of the final five A/D converter values. This change score represented the change in percent light transmittance from the first 5 seconds to the last 5 seconds of the 1-minute trial and provided a quantitative measure of the rate of hemolysis for a specific blood sample tube. For each subject, change scores were available for 10 control tubes and 10 influence (protect) tubes; for each tube condition, scores were available for either 2- or 8-tube measurements during each 15-minute period. Using these change scores, a three-factor analysis of variance was used to test the major hypotheses of the study. In this ANOVA, the three factors were: blood Source (own vs. another's, between); Subjects, random and nested under Source; and Condition (protect vs. control, within). The three experimental questions explored in this formal study were the following:

1. Would the rate of hemolysis (change scores) for the protect tubes differ from that for the control tubes? Such an effect would be indicated, in the absence of a significant Condition x Subjects interaction, by a significant Condition main effect in the ANOVA. Should the Condition x Subjects interaction effect prove significant, the condition effect would be examined separately in each of the individual subjects, using appropriate within-subject error estimates.

2. Would the degree of influence of hemolysis rate differ for the two blood sources (own cells vs. another's cells)? Such an effect would be indicated by a significant *interaction* of the Source and Condition factors of the ANOVA.

3. Would results for the two-tube measuring periods differ from those for the eight-tube measuring periods, and would the function describing this two-versus eight-tube effect match more closely the RA or the IDS prediction? For the statistical tests, any probability value found to be less than or equal to .05 would be deemed significant.

The summary table for the ANOVA is given in Table 1, and means and standard deviations for the various groups and conditions are given in Table 2. The main effects for Condition and Source did not reach significance, nor did the Source x Condition interaction. However, the main effect for Subjects and the Condition x Subjects interaction were highly significant. The former effect, of course, indicates significant variability in outcome among the 32 subjects of the experiment. The significant Condition x Subjects interaction indicates that the effect of Condition (protect vs. control) differed from subject to subject; therefore, an interpretation of the Condition main effect was inappropriate, and individual subject by subject condition comparisons were called for.

These individual comparisons were made by means of matched t tests, computed for each of the 32 subjects. These were calculated by comparing the hemolysis (change) scores for a subject's 10 protect tubes with the scores for his or her 10 control tubes. Individual t tests were calculated using separate estimates of variance, instead of using the combined estimate of error variance from the ANOVA, because of the wide spread among the variances across subjects. The t scores for the individual subjects are presented in Table 3. The independently significant subjects (i.e., those with |t| [18] > 2.101, p < .05, two-tailed) are indicated by asterisks.[4] The condition effect was significant in 9 of the 32 subjects. This is to be compared with the 1.6 significant scorers expected on the basis of chance alone. The probability of observing 9 or more independently significant scorers among 32 subjects is 1.91 x 10^{-5} (exact binomial test).

Table 1. Analysis of Variance Summary Table

Source	df	SS	MS	F	p
Source (own vs another's)	1	0.3480	0.3480	0.001	.981
Condition (protect vs control)	1	2.2278	2.2278	0.461	.503
Subjects	30	18078.7369	602.6247	297.626	<10^{-16}
Source x Condition	1	2.6061	2.6061	0.539	.469
Condition x Subjects	30	145.1072	4.8369	2.389	.000063
Error	576	1166.2668	2.0248		

Table 2. Means and Standard Deviations for the Percent Change Scores for the Various Conditions

Blood Source	Control	Protect	Overall Mean
Own	$\bar{X}$ = 43.63	$\bar{X}$ = 43.36	$\bar{X}$ = 43.50
	SD = 5.07	SD = 5.10	SD = 5.08
Another's	$\bar{X}$ = 43.45	$\bar{X}$ = 43.45	$\bar{X}$ = 43.45
	SD = 5.94	SD = 5.73	SD = 5.83
Average %	$\bar{X}$ = 43.53	$\bar{X}$ = 43.41	$\bar{X}$ = 43.47
Own/Another's	SD = 5.57	SD = 5.46	SD = 5.51

In seven of those subjects, scoring was in the direction of psi hitting (i.e., slower hemolysis in the protect than in the control tubes); in two subjects, scoring was in the direction of psi missing (i.e., faster hemolysis in the protect than in the control tubes). In order to determine whether these nine independently significant subjects show a general tendency toward *hitting*, a single-mean t test may be calculated for the nine t values, comparing them with MCE = 0; such a test yields evidence for significant hitting ($\bar{X}$ = 1.63, SD = 2.23, t[8] = 2.07, p = .035, one-tailed).

Table 3. Scoring Rates (t Tests) for Individual Subjects

Subject	t	Subject	t
1	-2.17*	17	1.24
2	1.47	18	-0.98
3**	0.31	19**	2.14*
4**	-1.26	20**	-0.68
5**	-0.51	21**	-1.04
6	1.28	22	-1.14
7	0.39	23**	3.39*
8	-1.15	24	0.26
9	-0.51	25**	3.04*
10**	-0.25	26**	1.96
11	-0.84	27	-1.46
12**	2.52*	28	-0.70
13	2.96*	29	3.08*
14	0.17	30**	2.53*
15	-2.79*	31**	-1.24
16**	-1.52	32	1.10

* Independently significant ($p < .05$)
** Indicates "own blood" condition

In the overall (ANOVA) analysis, no effect was found for the Source variable: Scores for the "own" and "another's" groups were virtually identical. The Source x Condition effect, which would have indicated a dependence of the protect versus control effect upon the source of the blood, was clearly nonsignificant. Therefore, we must conclude that blood source was not an important variable in this investigation. However, an interesting trend emerges when we examine the blood source for the nine independently significant subjects. For this subgroup (for which there was evidence for a psi effect), a comparison of the t scores of the five subjects who influenced their own blood with those of the four subjects who influenced another person's blood yields a trend of greater positive scoring (i.e., psi hitting) for the "own blood" subjects (see Table 4). Because of the small number of subjects involved in the comparison, this trend is not significant ($t [7] = 1.73$, $p = .12$, two-tailed). The large magnitude of the difference, however, suggests that blood source would be an interesting variable to explore in future studies of this type.

An anonymous referee of this paper described the results of a post hoc analysis that he or she performed on the "own" versus "other" scores; that analysis pointed to a possible true scoring pattern for the two blood source conditions. The referee reported that for the 14 subjects who worked with their own blood cells, a 2 x 2 table can be constructed that divides them by positive versus negative change score x significant versus nonsignificant change. Of the 7 with positive change, 5 were significant; of the other 7, none were significant ($p < .02$). The corresponding table for the 18 subjects work-ing with someone else's blood cells yields nonsignificant differences for these same classifications.

Table 4. Comparison of Own vs. Another's t Scores
for Independently Significant Subjects

Own	Another's
2.52	–2.17
2.14	2.96
3.39	–2.79
3.04	3.08
2.53	
$\bar{X} = 2.72$	$\bar{X} = 0.27$
$SD = 0.44$	$SD = 2.76$

The raw data collected in this experiment were sent to researchers at SRI International so that they might perform certain mathematical analyses of the scores to determine which of two models of psi, the RA model or the IDS model, would provide better predictions of the obtained results for the 2-tube versus 8-tube conditions. Space limitations do not permit a detailed descrip-tion of the rationale, nature, or specific results of those analyses. However, the gist of their conclusion was that the extreme heterogeneity of the data made it impossible to make an adequate determination of which model provided a bet-ter fit of the obtained scores. They argued that psychic functioning, whatever its underlying mechanism, is highly individualized, making it difficult to test a specific theory using data combined across subjects; in the present experi-ment, insufficient data were collected for any one subject to allow tests of RA versus IDS predictions *within* individuals. Additional details regarding the IDS/RA analyses may be found in an SRI International technical report (Hubbard, Utts, and Braud, 1987).

We conducted our own analyses of the tube number factor simply by com-paring the differences between the mean change scores for the control and the protect periods for the 2-tube condition with the corresponding differences for the 8-tube condition. Matched t tests were used for these comparisons. One t test was calculated using the data from all 32 participants. A second t test was calculated using the data from only the 9 participants whose scoring rates had been independently significant. Neither analysis indicated a significant differ-ence between the two tube conditions ($\bar{X}$ difference = 0.08, t [31] = 0.25, p = .80, two-tailed, and $\bar{X}$ difference = 0.41, t [8] = 0.74, p = .51, two-tailed, respectively). According to our understanding of the IDS model, psi scoring rate should be higher for the 2-tube than for the 8-tube condition (i.e., scor-ing rate should decline with "n length"). According to a direct psychokinesis or

RA model, psi scoring rate should not depend upon the number of tubes being measured during a participant's "effort" period (i.e., scoring rate should be independent of "n length").[5] Thus, the absence of a 2-tube versus 8-tube scoring difference is more consistent with a remote action than with an IDS interpretation.

George Hansen (personal communication, May 28, 1988) raised an issue concerning the statistical analysis of the present study. The issue concerns the assumption of independence of measurements within a 15-minute period. He mentioned that substantial trial dependencies have been observed in some of the biological target systems used by other investigators and wondered whether similar dependencies might be present in this hemolysis work. If the individual "trials" (i.e., tube measurements) are not independent, then the t tests might not be appropriate. This independence issue may be addressed as follows:

1. Trial dependence would be more problematical in a situation in which *the very same target organism* participated in all trials of an extended measurement block—for example, placing a laboratory rat in a test apparatus for 15 minutes and measuring its activity 10 times (i.e., for 10 "trials") during that long period. In the present hemolysis experiment, a *different* tube of blood (with different target cells) was used for each of the "trials" (i.e., tube measurements) within a 15-minute epoch. This is analogous to making activity measurements in *10 different laboratory rats* placed sequentially in the measuring apparatus throughout a 15-minute period, all rats being selected or sampled from a common group colony cage.

2. The target tubes could not be randomized on an individual basis in the present study due to the nature of one of the experimental questions (i.e., the "few-versus-many-tubes" IDS question), which required multiple trials during a prolonged "effort" period, and therefore a blocked design. The order of the blocks (PCCP or CPPC) was randomized in order to control for any time-correlated systematic error.

3. Factors (e.g., temperature changes, handling differences, etc.) that might influence hemolysis rate were identified during the pilot and intermediate phases of the study and were eliminated or controlled so that they were no longer present or active in the formal experiment.

4. Before the formal experiment began, numerous test runs were conducted that involved multiple tube measurements under conditions identical to those of the experimental tests, but in the absence of psi influence attempts. Such tests indicated that the tubes displayed independent activities over time periods as long as those required for the completion of 20 trials.

Inspection revealed no obvious trends or dependencies in the tube data. Furthermore, autocorrelation coefficients were calculated for all tube sets of this preexperiment series for which at least 20 sequential tube measurements were available. None of these autocorrelation analyses, each carried out for lag numbers 1 through 10, provided evidence for trends or dependencies.

5. Jessica Utts (personal communication, June 1, 1988) indicated that a positive correlation of the hemolysis values across time would inflate the t tests. Although indicating that there was no easy way to check for tube independence, she did calculate correlation coefficients of hemolysis rate versus the numbers 1 through 10, for each of the 64 sets of tubes (32 control sets and 32 protect sets) for which she was supplied the raw data. She reported that the results showed an average correlation of $-.03$ for the control sets and $-.001$ for the protect sets; this indicated that there was no obvious trend or dependency.

Taken together, these various findings and considerations are more consistent with an assumption of trial (tube) independence than with one of trial (tube) dependency within time blocks.

A different statistical issue was raised by a referee who inquired as to whether the data were sufficiently close to normality to justify the use of t tests. Utts (personal communication, June 1, 1988), in checking the hemolysis data of the present experiment for normality, found that their histograms looked relatively bell-shaped and that nonparametric Mann-Whitney U tests calculated for the 32 subjects yielded results virtually identical to those of the t tests reported above (scoring rates were independently significant for 9 of the 32 individuals). She concluded that the t tests were valid (assuming the scores were independent).

Discussion

As expected, within- and between-subjects variability in this formal experiment was greatly reduced by the changes in experimental protocol that resulted from observations made in the pilot and intermediate phase experiments. This reduction in the experiment's "noise" level permitted the observation here of psi effects that could not be detected in the pilot study. Significant differences in rate of hemolysis between experimental (i.e., mentally "protected") and control blood samples were found in an extrachance number of subjects.

It may be possible to discover important differences between subjects who exhibited significant positive scoring and those who exhibited significant negative scoring or chance scoring through detailed psychological analyses that would consider both short-term ("state") and more persistent ("trait") characteristics of the subjects. State analyses could focus upon the types of *mental strategies* used by the subjects in their attempts to influence the target cells. Some subjects, for example, employed a direct strategy of visualizing the blood cells in a very realistic manner, whereas other subjects employed a more indirect strategy of visualizing objects that were similar to the blood cells and possessed characteristics similar to those that protected cells might possess. Are more direct strategies more effective than indirect strategies based upon associations and symbolic representations? Trait analyses could be accomplished by asking all participants of the formal study to return to the laboratory for various personality assessments. Such assessments would, of course, be carried out by laboratory personnel who are blind to the subjects' hemolysis results. Initial assessments might involve psychological instruments such as the Myers-Briggs Type Indicator and the Participant Information Form, which have already been shown to correlate with other types of psi performance (e.g., Berger, Schechter, and Honorton, 1985; Honorton, Barker, Varvoglis, Berger, and Schechter, 1985). Possible *interactions* between state and trait factors could also be examined.

In the present study, several factors may have interfered with the emergence of even stronger psi effects. These factors were: (a) the presence of relatively large individual differences in the characteristics of the subjects, (b) the long durations of the psi "effort" periods and of the experimental sessions as a whole, and (c) the absence of real-time feedback to the subjects concerning the state of the target system. Factor (a) could be minimized in future studies by more stringent selection of participants in terms of prior histories of successful bio-PK performance and of personality characteristics known to be correlated with psi performance. Factors (b) and (c) were necessitated by the "tube number" component of the present design (i.e., the assessment of two versus eight tubes in each 15-minute protect or control period). Lengthy "effort" periods were required to accommodate the measurement of eight tubes, and the requirement of keeping the subject "blind" to the number of tubes prevented the administration of feedback because that would have allowed the subject to keep track of the number of tubes measured. Future experiments unconcerned with a tube-number factor could include briefer psi-influence periods and could also provide feedback. It should be noted, however, that there exists a growing body of evidence that suggests that real-time sensory feedback to the subjects is not a necessary condition for the occurrence of strong psi effects (e.g., Berger, 1988; Braud, 1978b).

The major purpose of the present study was to determine whether a significant psi effect involving hemolysis could be observed using a large number of unselected subjects and an improved experimental protocol. The extra-chance number of independently significant subject performances provides an affirmative answer to this question. One of two secondary goals of the study was to explore the issue of whether an IDS interpretation or an RA interpretation provides a better explanation of the results. (The other secondary goal was to study the "own" versus "another's" blood cells factor.) According to an RA interpretation, the subjects (or the experimenter) actually retard hemolysis rate in a causal or quasi-causal manner, yielding values that would not have occurred in the absence of influence attempts. According to an IDS interpretation, the experimental personnel take advantage of already existing fluctuations of hemolysis rate among different blood samples, "sorting" those values by the scheduling and timing of their trials so as to produce an effect that simulates a causal effect. It is important to remember that an IDS effect is still a psi effect, but an informational rather than a causal one.

There were several opportunities for intuitive data sorting in the present experiment. The person (M. S.) who provided the random schedule for blood source, tube condition, and tube number used a fixed rule that involved converting published weather information into an entry point for a table of random numbers. However, arbitrary decisions were still possible in assigning odd or even digits to the various sources, conditions, and tube numbers, and such decisions by M. S. provided possible entry points for IDS that could influence both between- and within-subject effects. A second possible source of sorting involved subject scheduling for the initial blood-drawing sessions. Which subjects happened to arrive at the laboratory on a given Monday evening for blood drawing would determine their places or positions in the test schedule, and hence the particular blood samples that would be assigned to them; this could provide additional IDS entry points. Subject-scheduling IDS effects could be mediated by the experimenter (W. G. B.), by the laboratory personnel who suggested and scheduled potential subjects, and/or by the subjects themselves. A third possible source of sorting involved the experimenter's hemolysis measurements. By consciously or unconsciously altering the timing of his actions, he could determine the start point of the measured hemolysis curve, and this in turn could influence the hemolysis rate measure. The experimenter was aware of this possibility, and therefore he exercised great caution in maintaining the consistency and stereotypy of his laboratory technique in an attempt to obviate this factor. Only automated procedures could eliminate this factor, and even then, the elimination may or may not be absolute, depending upon the *completeness* of the automation. It is important to note that *deliberate* timing changes could influence only the tube number effect in the present experiment, because this was the only independent variable for which the experimenter was not blind. Because he was unaware of the blood source or of the scheduling of the protect versus control periods, any timing changes that could influence those two effects would necessarily have to be psi-mediated. It should be added that subjects may have influenced hemolysis rate indirectly by exerting a true, causal remote action influence upon the experimenter's timing behavior, so that what appears to be IDS by the experimenter could in reality be RA by the subject. As the reader will begin to appreciate, a truly definitive test of the IDS model, or even a definitive identification of IDS and RA components, is an exceedingly complex and difficult task.

The issue of whether the results of this experiment are described more closely by an IDS model or by an RA model was not resolved unequivocally by the mathematical analyses carried out at SRI International. Our own statistical analysis of the tube number factor did not indicate the presence of the significant difference between the 2-tube and 8-tube conditions that the IDS model would have predicted. The absence of a performance difference between the two tube conditions is more consistent with an RA (psychokinesis) interpretation than it is with an IDS (informational) interpretation of the obtained results.

In the blood experiments reported here, the hemolysis process occurred *in vitro* and was produced by osmotic stress. Caution should be exercised in the generalization of the results of this study to *in vivo* hemolysis. In the body, red blood cell lysis can occur through osmotic stress, but is more often contributed by other factors (Hillman and Finch, 1974; Ponder, 1971).

The rationale for selecting blood cells was that perhaps material that had once been part of the body might be more susceptible to distant mental influence than would be the case for more "alien" biological materials. At the very least, the use of such "familiar" material would be expected to increase the participants' motivational levels and hence increase the likelihood of positive results. Red blood cells were chosen as "targets" for these initial bio-PK investigations because their rate of hemolysis could be measured by means of the equipment and facilities available to us. However, this choice was not without its difficulties because the biological status of red blood cells is somewhat peculiar. On the one hand, human mature red blood cells have no nucleus, cannot reproduce, and have limited lifespans (approximately 120 days). On the other hand, as Ponder (1971) notes: "On metabolic grounds, mammalian erythrocytes are living cells; although in absolute terms the rates of respiration and of glycolysis are small, from the standpoint of cellular physiology the metabolism is far from negligible" (p. 366). Red blood cells certainly qualify as biological systems. In future cellular bio-PK investigations, however, perhaps the use of white blood cells or of artificially cultured neural cells may yield more dramatic results than those obtained in the present study.

Endnotes

1. An earlier version of this paper was presented at the 31st annual convention of the Parapsychological Association in Montreal, Canada, August 17–21, 1988. Parts of the introductory remarks are repeated or paraphrased versions of sections of an article previously written for *Parapsychology Review* (Braud, 1986) and are used with the permission of the Parapsychology Foundation. I am indebted to Rick Berger, Steve Dennis, George Hansen, Scott Hubbard, Kay Mangus, Ed May, Diane Morton, Julie Nixon, Marilyn Schlitz, Helmut Schmidt, Winona Schroeter, and Jessica Utts for their important contributions to various phases of this investigation. This project was supported, in part, by a subcontract from SRI International.

2. The order in which the subjects' blood was drawn, and hence their subject numbers for the blood source factor (i.e., whether they subsequently attempted to influence their own or another person's blood) was determined by the alphabetical sequence of the surnames of the four subjects who assembled on any given Monday evening blood-drawing session.

3. On one Monday evening, a last-minute cancellation by a subject resulted in three, rather than the usual four, blood draws that evening. The next week, the subject who had cancelled for the previous week appeared; there also appeared an additional subject who was not expected, resulting in six subjects that evening. For the two subjects who did not conform to the schedule that evening, blood tubes were inadvertently switched, when they should not have been. This resulted in a total of 18 subjects in the "another's blood" condition and 14 subjects in the "own blood" condition.

4. In order to identify significant scorers, a two-tailed test was used, allowing for the possibility of significant psi missing. This was done because the pilot study had yielded a considerable number of scores in the missing direction. If one wishes to predict only psi *hitting*, a critical $t = 1.73$ ($p < .05$, one-tailed) could be used. By such a hitting-alone criterion, eight independently significant hitters may be identified (Subject No. 26 now reaches significance). The probability of observing 8 out of 32 independently significant hitters is 1.39×10^{-4}.

5. Some models of PK or RA could predict a diminished psi effect *at the level of the individual trial* (or at the level of the *individual tube*, in the present study) for long as opposed to short "*n* lengths" due to a kind of "watering down" or "spreading thin" effect of the increased number of targets at the greater *n* length. The total psi effect for the entire ensemble of large *n* events, however, should be comparable to that for the entire ensemble of small *n* events.

References

Berger, R. E. (1988). Psi effects without real-time feedback. *Journal of Parapsychology,* 52, 1–27.

Berger, R. E., Schechter, E. I., and Honorton, C. (1985). A preliminary review of performance across three computer psi games. *Proceedings of Presented Papers: The 28th Annual Convention of the Parapsychological Association,* Vol. 1, 307–332.

Braud, W. [G.]. (1978a). Allobiofeedback: Immediate feedback for a psychokinetic influence upon another person's physiology. In W. G. Roll (Ed.), *Research in Parapsychology 1977* (pp. 123–134). Metuchen, NJ: Scarecrow Press.

Braud, W. G. (1978b). Recent investigations of microdynamic psychokinesis, with special emphasis on the roles of feedback, effort, and awareness. *European Journal of Parapsychology,* 2, 137–162.

Braud, W. [G.]. (1979). Conformance behavior involving living systems. In W. G. Roll (Ed.), *Research in Parapsychology 1978* (pp. 111–115). Metuchen, NJ: Scarecrow Press.

Braud, W. [G.]. (1986). PSI and PNI: Exploring the interface between parapsychology and psychoneuroimmunology. *Parapsychology Review,* 17(4), 1–5.

Braud, W. [G.], Davis, G., and Wood, R. (1979). Experiments with Matthew Manning. *Journal of the Society for Psychical Research,* 50, 199–223.

Braud, W. [G.], and Schlitz, M. (1983). Psychokinetic influence on electrodermal activity. *Journal of Parapsychology,* 47, 95–119.

Braud, W. [G.], and Schlitz, M. (1989). A methodology for the objective study of transpersonal imagery. *Journal of Scientific Exploration,* 3, 43–63.

Eccles, Sir J. (1977). The human person in its two-way relationship to the brain. In J. D. Morris, W. G. Roll, and R. L. Morris (Eds.), *Research in Parapsychology 1976* (pp. 251–262). Metuchen, NJ: Scarecrow Press.

Hall, H. (1984a). *Hypnosis, Imagery and the Immune System: A Progress Report Three Years Later.* Paper presented at the 36th Annual Convention of the Society for Clinical and Experimental Hypnosis, San Antonio, TX.

Hall, H. (1984b). Imagery and cancer. In A. A. Sheikh (Ed.), *Imagination and Healing* (pp. 159–169). Farmingdale, NY: Baywood.

Hillman, R. S., and Finch, C. A. (1974). *Red Cell Manual* (4th ed.). Philadelphia, PA: F. A. Davis.

Honorton, C., Barker, P., Varvoglis, M. P., Berger, R. E., and Schechter, E. I. (1985). "First-timers": An exploration of factors affecting initial psi ganzfeld performance. *Proceedings of Presented Papers: The 28th Annual Convention of the Parapsychological Association,* Vol. 1, 37–58.

Hubbard, G. S., Utts, J. M., and Braud, W. G. (1987). *Experimental Protocol for Hemolysis: Confirmation Experiment.* Menlo Park, CA: SRI International.

May, E. C., Radin, D. I., Hubbard, G. S., Humphrey, B. S., and Utts, J. M. (1985). Psi experiments with random number generators: An informational model. *Proceedings of Presented Papers: The 28th Annual Convention of the Parapsychological Association,* Vol. 1, 235–266.

Ponder, E. (1971). *Hemolysis and Related Phenomena.* New York: Grune and Stratton.

Schlitz, M., and Braud, W. (1985). Reiki-Plus natural healing: An ethnographic/experimental study. *Psi Research, 4,* 100–123.

Schmidt, H. (1976). PK effect on pre-recorded targets. *Journal of the American Society for Psychical Research, 70,* 267–291.

Schmidt, H. (1981). PK tests with pre-recorded and pre-inspected seed numbers. *Journal of Parapsychology, 45,* 87–98.

Schmidt, H., Morris, R., and Rudolph, L. (1986). Channeling evidence for a PK effect to independent observers. *Journal of Parapsychology, 50,* 1–15.

Schneider, J., Smith, C. W., and Whitcher, S. (1984). *The Relationship of Mental Imagery to White Blood Cell (Neutrophil) Function: Experimental Studies of Normal Subjects.* Paper presented at the 36th Annual Convention of the Society for Clinical and Experimental Hypnosis, San Antonio, TX.

Solfvin, J. (1984). Mental healing. In S. Krippner (Ed.), *Advances in Parapsychological Research 4* (pp. 31–63). Jefferson, NC: McFarland.

Mind Science Foundation
8301 Broadway, Suite 100
San Antonio, Texas 78209

4

Mental Interactions with Remote Biological Systems

William G. Braud, Ph.D., and Marilyn J. Schlitz, M.A.

This chapter is a review of a wide range of experiments, conducted in our laboratories, in which persons were able to mentally and at a distance influence a variety of remote living systems, including other persons' physiological activities, the activities of small animals, and the rate of hemolysis (death by osmotic stress) of human red blood cells. Possible alternative explanations are addressed and ruled out, influencing factors are discussed, and implications and applications are considered.

This research originally was published in Braud, W. G., and Schlitz, M. J. (1991). Consciousness interactions with remote biological systems: Anomalous intentionality effects. Subtle Energies and Energy Medicine: An Interdisciplinary Journal of Energetic and Informational Interactions, 2(1), *1–46. The contents of this article are Copyright © 1991 by the International Society for the Study of Subtle Energy and Energy Medicine (ISSSEEM), and the material is reprinted with permission. Contact information for ISSSEEM: 11005 Ralston Road, Suite 100D, Arvada, CO 80004; phone (303) 425-4625; fax (303) 425-4685; e-mail issseem@compuserve.com; website http://www.issseem.org*
—William Braud

Abstract: This paper describes a 13 year long, and still continuing, series of laboratory experiments that demonstrate that persons are able to exert direct mental influences upon a variety of biological systems that are situated at a distance from the influencer and shielded from all conventional informational and energetic influences. The spontaneously fluctuating activity of the target system is monitored objectively during randomly interspersed influence and noninfluence (control) periods while, in a distant room, a person attempts to influence the system's activity in a prespecified manner using mental processes of intentionality, focused attention, and imagery of desired outcomes. The experimental design rules out subtle cues, recording errors, expectancy and suggestion ("placebo") effects, artifactual reactions to external stimuli, confounding internal rhythms, and coincidental or chance correspondences. Distantly influenced systems include: another person's electrodermal activity, blood pressure, and muscular activity; the spatial orientation of fish; the locomotor activity of small mammals; and the rate of hemolysis of human red blood cells. The experiments are viewed as laboratory analogs of mental healing.

Introduction

Findings from the areas of hypnosis, autogenic training, biofeedback training psychophysiological self-regulation, placebo, meditation, and imagery research indicate that mental processes, especially intentionality, can have dramatic somatic effects. These effects, which may be observed at many levels of functioning (behavioral, autonomic, neurological, immunological, and endocrinological), are typically understood and explained in terms of a network of anatomical, hormonal, biochemical, and (perhaps) electromagnetic connections that exist between the central nervous system and the various organs, tissues, and cells of the body.

Additional evidence indicates that, under special conditions, mental influence may extend even beyond the body. Under controlled laboratory conditions, the thoughts, images, and intentions of one person may influence those of a second person even under conditions of screening or isolation that preclude conventional sensorimotor interactions between the two persons. Some of the most compelling evidence for these anomalous cognition effects may be found in several recently published meta-analyses that indicate strong and consistent interpersonal mental influences in large numbers of experiments conducted under conditions of perceptual isolation (using a ganzfeld procedure),[1] hypnotic induction,[2] forced-choice precognition,[3] and extraversion/introversion[4] testing. Accumulated evidence also points to consciousness-related anomalies in physical systems; here, small but consistent intentional influences upon remote random mechanical[5] and random electronic devices[6] have been observed.

In this paper we describe a now 13 year long, and still continuing, research program in which we have extended the remote mental influence procedure to animate rather than inanimate systems. In these experiments, the ongoing, freely varying activity of a living organism is monitored objectively while a remotely situated and suitably screened individual attempts to exert a direct mental (intentional) influence upon that activity. The monitored activities have included the autonomic and muscular activity of another person, behavioral activities of fish and small mammals, and the rate of hemolysis of human red blood cells. The goals of this remote influence have included both increments and decrements in the monitored systems' activities. Our aim is to summarize the research program and describe some of the experimental protocols in detail in order to encourage independent replications of this work by other investigators.

Living systems possess a number of advantages as target systems for direct mental influence research. They provide methodological advantages in that, unlike the subjective reports used in anomalous cognition experiments,[7] the behavioral and physiological activities of biological systems may be directly monitored by physical measuring devices. The use of living target systems also provides motivational advantages. For many experimental participants, the possibility of influencing living systems such as other people, animals, or living cells is more appealing than influencing inanimate devices. Principles discovered in the course of investigations of direct mental influence of living systems (DMILS) are relevant to our understanding of the processes underlying certain forms of unorthodox healing (i.e., mental, spiritual, or absent healing) and may even enhance our understanding of at least some of the cases in which prayer is efficacious in promoting physical healing or recovery. It may also be the case that animate target systems are inherently more susceptible to direct mental influence than are inanimate systems. Discussions of additional advantages of DMILS research and further background information may be found in other publications.[8]

General Design Considerations

The experiments to be described share certain common features. One begins by arbitrarily selecting an organism and a response system in which activity changes will be assessed. Systems characterized by a moderate amount of freely varying activity are recommended. Labile systems may be more susceptible to direct mental influence than are more inert systems.[9] It is not yet known whether physical lability itself or perceived lability is the critical factor. On the one hand, freely varying activity may reflect underlying randomness (or, perhaps, chaotic activity) which may be essential to the occurrence of direct mental effects. On the other hand, the perception of varying target activity may instill confidence that the target system can indeed change, and perhaps it is this psychological factor that is favorable to success. This important distinction (of physical versus psychological lability or "randomness") has received virtually no experimental attention. It could, however, be easily investigated using appropriate analytical designs that could separate these usually perfectly confounded factors.

Both excessively sluggish and excessively active systems may be nonoptimal. The former are inappropriate because of relative insusceptibility to change and because the system's low baseline activity level would be near a "floor" below which changes could not be observed. The latter are inappropriate because excessively "driven" systems also possess a form of "inertia" that render them difficult to change and because the system's high baseline activity would be near a "ceiling" above which changes could not be observed. The most appropriate systems may be those with moderate departures from homeostasis or balanced activity. In such cases, a directional influence consistent with a return to homeostasis could be maximally effective. Here, suitable precautions would be taken to deal with possible "regression to the mean" artifactual changes.

Once a suitable response system has been selected, experimental partici-pants are instructed to attempt to exert directional influences upon the system's activity level. Incremental aim, decremental aim, and non-influence aim (i.e., control or baseline) periods may be scheduled in various combinations, but their sequence must always be determined by an accepted random process (such as a table of random numbers or a computer's pseudorandom algo-rithm). The influence and non-influence epochs during which the system's spontaneous activity is to be continuously recorded should be numerous enough to permit efficient statistical analysis yet not so brief as to produce con-fusion or discomfort in the influencer (due to a need to switch intentions or mental states too frequently). We have typically used ten 30-second influence epochs randomly interspersed among ten 30-second non-influence (control) epochs with success. If each recording epoch is preceded by a 30-second rest/preparation period, an experimental session is usually 20 to 25 minutes in duration.

We design our experiments so that, in the absence of a direct mental influ-ence, the living target system's cumulative activity for the influence and non-influence epochs should be equivalent, much as flipping a coin should, over time, produce an equal number of heads and tails. In a design in which a num-ber of decremental aim periods (i.e., periods in which the participant intends for the system to exhibit decreased activity) are randomly interspersed among an equal number of non-influence (control) periods, 50 percent of the system's spontaneous activity would be expected (by chance alone) to occur during decremental aim periods and 50 percent during control periods. Our experi-ments typically involve a sufficient number of sessions to permit conventional statistical analyses to definitively determine whether the total decremental aim activity is indeed significantly lower than control period activity.

In analyzing our experiments, we have used the conservative strategy of reducing all of the epoch measurements to a single session score and using such session scores as the units of analysis. For a given session, we calculate a *per-cent influence score* which is the percentage of total activity that occurred in the prescribed direction during the entire set of influence (decremental or incre-mental aim) periods. A percent influence score for an experiment involving ten incremental aim epochs and ten control epochs would be calculated by summing the activity scores for the ten incremental aim epochs and dividing this sum by the total activity exhibited in the session as a whole (i.e., the sum of activity scores for all twenty epochs). In the absence of direct mental influence, the expected value of a percent influence score is 50 percent. A set of percent influence scores can be compared with mean chance expectation (MCE) of 50 percent using a single-mean t-test. This is equivalent to directly comparing the influence and non-influence (control) scores for the respective sessions by means of matched t-tests. Of course, equivalent nonparametric tests (Wilcoxon matched-pairs signed ranks tests; Mann-Whitney U tests) could be used instead.

We use percent influence scores in an effort to standardize measurements for different response systems so that results for various systems can be more readily compared. We use the more conservative session score (a kind of single, "majority vote" score) in order to bypass criticisms based on possible non-independence of multiple

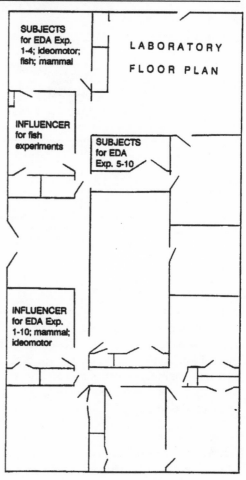

Figure 1. Floor plan of the laboratory in which the first 11 electrodermal influence experiments were conducted. This arrangement was also used for the ideomotor, fish orientation, and mammal locomotion experiments.

measures taken within a given session (and generated by the same individual organism). While it would be possible—taking a more "liberal" statistical view—to analyze individual epoch scores using, for example, a repeated measures analysis of variance (ANOVA) procedure, such an analysis assumes that the autocorrelations among the measures within each session (i.e., within each responding organism) are constant across epochs, and that the same autocorrelation applies to all sessions[10] (responding organisms), we prefer a more "conservative" approach. These assumptions cannot definitively be said to hold in all our experiments, so we prefer to use the more conservative session-based

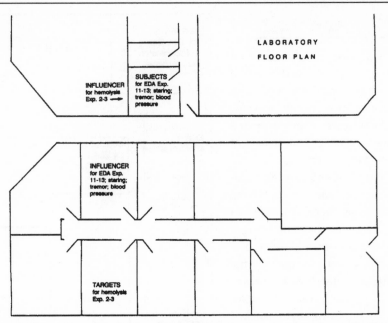

Figure 2. Floor plan of the laboratory in which the last 4 electrodermal influence experiments were conducted. This arrangement was also used for the electrodermal attention (autonomic staring detection), muscular tremor, blood pressure, and hemolysis experiments. The subject and influencer rooms are in separate suites of the same building, separated by an outside corridor and several closed doors.

(rather than epoch-based) analyses, even though the former are more wasteful of data and result in tests with reduced statistical power.

During an experimental session, it is essential that the target system be shielded from all possible conventional influences from the experimental participants. This is accomplished by placing the target system and the influencer in separate, non-adjacent rooms (with several intervening walls and closed and locked doors) and by preventing verbal or other forms of communication between the two rooms. Floor plans of laboratory spaces in which the experiments were conducted are presented in Figures 1 and 2.

If the living target system is a person, that person remains unaware of the manner in which the various epochs are scheduled (i.e., the person does not know the number, duration, or sequencing of the randomly scheduled epochs). This "blindness" eliminates the possibility of "placebo" effects or physiological self-regulation on the part of the target person based upon knowledge of the influencer's intentions. The experimental protocol specifies that no experimental personnel who interact with the target person before a session have any knowledge of the influence/non-influence epoch schedule for that session. In cases in which the target system is an animal, no one is present in the animal's test room during the actual session; scoring is accomplished automatically by equipment in another room (interconnected with the animal's test apparatus by means of shielded cables). Thus, inadvertent cueing of the animal is not possible. In cases in which the target system is a cellular preparation (i.e., in the hemolysis trials), the human measurer uses automatic equipment and is blind with respect to the influence/control epoch schedule.

Precautions against measurement errors include objective, computer-based assessments of activity scores and double- or multiple-checking of data reduction and computations. In early experiments in which measurements were made manually (i.e., in Experiments 1 through 4 of the electrodermal series described below), the measurements were blind-scored from objective chart tracings.

Deliberate cheating on the part of subjects is prevented by guaranteeing that the experimental protocol is never "broken," i.e., assuring that subjects are never allowed access to information to which they should remain blind or access to any laboratory personnel who have knowledge of such information. We also employ the strategy of working with a large sample of unselected subjects who would have minimal motivation to cheat, and the experimental protocol itself eliminates the possible effectiveness of accomplices.

In most of the experiments, the person attempting direct mental influence of the living target system (i.e., the "influencer") is provided with real-time feedback regarding the target system's activity. This feedback takes the form of a chart tracing of the target's ongoing activity. The provision of feedback allows the influencer to try different influence strategies and to have immediate knowledge of the results of those strategies. She or he may use this knowledge to continue using apparently successful methods or to shift to alternative methods. Some subjects make use of this continuously available feedback throughout a measuring epoch. Others attend to the feedback indicator only upon completing their epoch activity. Still others prefer not to use feedback at all. Our findings, and those of certain other investigators, indicate that feedback may be useful to some influencers but does not seem essential to the occurrence of DMILS and related effects.[11]

The Electrodermal Influence Series

Most of our DMILS work has involved influences by one person upon the ongoing, spontaneous electrodermal activity of another, remotely situated person. We shall review the general characteristics of these experiments; detailed information is provided elsewhere.[12]

Experimental Participants

Three categories of participants are involved in these experiments: subjects, influencers, and experimenters. The *subjects* are the persons whose ongoing physiological activities are monitored and objectively assessed. The *influencers* are individuals who use mental processes of intention, focused attention, and imagery in order to bring about prespecified changes in the physiological activities of the remotely situated subjects. The *experimenters* supervise the experiments and sometimes also function as influencers.

The subjects have been unpaid volunteers from the San Antonio community who learned about the research through announcements, newsletter and newspaper articles, lectures, media presentations, and information from previous participants. Approximately equal numbers of males and females have participated, and they have been between 16 and 65 years of age. Generally, no special inclusion or exclusion criteria were used other than interest in the studies, willingness to participate, and the ability to schedule the requisite laboratory visits. In one experiment (Experiment 5), we wished to study subjects with greater than average electrodermal activity levels (in addition to those with "average" levels), and so we chose persons who presented stress-related somatic complaints for a particular subgroup. In all experiments, all essential details of the experiments were described to the volunteers beforehand, and the volunteers gave their informed consent to participate. In all, 271 persons have served as subjects for these electrodermal influence studies.

The influencers typically have had the same characteristics as the subjects and, therefore, were also "unselected." In other experiments, the experimenters served as influencers. In still other experiments, influencers were specially selected based upon their interests and skills in unorthodox healing, mental healing, therapeutic touch, Reiki healing, meditation, and self-exploration. A total of 62 influencers participated in the entire series. An interesting finding was that results were fairly comparable for the different types of influencers.

The authors served as experimenters for this series of studies and in some experiments were assisted by experimenter J. C. (who had research experience in nursing) and experimenter H. K. (a local college student participating in a research *practicum* at the Foundation). The first author has extensive research experience in experimental psychology, physiological psychology, and parapsychology. The second author has extensive experience in anthropological and parapsychological research. Overall, four experimenters participated in the series.

Precautions Against Conventional Communication, External Stimuli, and Subtle Cues

Conventional communication between the influencer and the subject was precluded through the use of independent, nonadjacent rooms (separated by a distance of 20 meters or more and several intervening rooms, walls, and closed doors—see Figures 1 and 2) and a strict protocol that eliminated cueing possibilities. There were no active microphones through which unauthorized communication could occur and, further, the experimenter and influencer maintained silence during the experiment sessions.

Precautions Against Suggestion, Expectancy, Placebo Effects, and Confounding Internal Rhythms

The subjects remained blind regarding the nature, number, and scheduling of the influence attempts. They knew that influence attempts would be made, but they were unaware of the directions or timing of the attempts. This information was unknown even to the influencer and experimenter until after all pre-session interactions with the subject had been completed and the participants were stationed in their respective rooms. Influence periods were randomly interspersed among an equal number of non-influence (control or baseline) periods. The random scheduling of the two types of periods was accomplished through use of truly random electronic random event generators,[13] tables of random numbers, pseudorandom computer algorithms, or adequately shuffled cards. The random schedules were prepared by persons who had no further roles in the studies, and the schedules were kept in sealed envelopes in secret locations until needed. These random schedules prevented any internal rhythms or extraneous, systematic, time-varying factors from contributing in a biased fashion to either type of period.

Subjects Procedures

During a 20-minute experimental session, the subject's spontaneously fluctuating electrodermal activity was monitored while the subject (a) remained in as normal a condition as possible and made no deliberate, conscious attempts to relax or become more active, (b) observed for brief periods and then gently dismissed all thoughts, feelings, and images that spontaneously came to mind, (c) made himself or herself open to and accepting of any direct mental influences from the influencer (whom the subject had already met before the session), and (d) kept "in the back of the mind" a "gentle wish" that the experiment would have a successful outcome. Subjects followed these four instructions in every experiment. Subjects were told to avoid unnecessary movements (especially of the electroded hand and arm), but otherwise they were to maintain an everyday, ordinary state of consciousness.

Influencers Procedures

Before each of twenty 30-second electrodermal activity recording epochs, the influencer was issued instructions about what to do during the epoch. Epochs were signalled to experimenter and influencer (through headphones) by special tones audible only to them and not to the distant target person (the subject). During non-influence (control or baseline) epochs, the influencer attempted not to think about the subject or about the experiment. During decremental aim periods, the influencer created and maintained a strong intention for the remote target person to be calm and relaxed and to exhibit very little electrodermal activity. The influencer supplemented this decremental intention by calming herself or himself, visualizing the target person in calming settings, and visualizing polygraph tracings indicative of relaxation or lowered arousal (i.e., infrequent pen deflections, low amplitude pen deflections). Complementary strategies were used for incremental aim epochs, with increased activation and physiological arousal substituted for calmness and quietude.

Immediate, sensory, analog feedback regarding the subject's electrodermal activity was provided to the influencer in the form of a chart recorder tracing of the activity on a polygraph before which the influencer was seated. Influencers used this available feedback of the chart recording in various ways (see page 77). Results of some electrodermal influence sessions and, more importantly, from several entire experiments involving different target activities (see the following sections) indicate that such feedback is not essential to the occurrence of the direct mental influence effect.

In some experiments, ten epochs with calming (decremental) orientation were compared with ten non-influence, control epochs within each session. In other experiments, ten epochs with activating (incremental) orientation were compared with ten non-influence, control epochs within each session. In still other experiments, ten decremental aim epochs were compared with ten incremental aim epochs within each session.

In the various experiments of this series, periods ranging in duration from 15-seconds to 2-minutes separated the 30-second recording epochs. During these intervening periods, the influencer could rest and prepare for the next epoch.

Physiological Measurements

We chose electrodermal activity fluctuations as our physiological measure because such measurements are readily made, are sensitive indicators, and are known to be useful peripheral measures of the activity of the sympathetic branch of the autonomic nervous system. We measured the phasic, AC component of the fluctuating electrical resistance of the skin, known technically as *skin resistance reactions* (SRRs). The equipment automatically corrected for drift in baseline level (basal skin resistance) so that our measures were sensitive to changes in the subject's state and were not biased by individual differences in baseline. The occurrence of many or of high amplitude spontaneous SRRs is indicative of increased sympathetic nervous system activation or arousal, which may in turn reflect increased emotionality.[14] The occurrence of few or of low amplitude spontaneous SRRs indicates decreased sympathetic activation or arousal, which may in turn reflect decreased emotionality and, therefore, a greater degree of emotional and mental quietude and calmness. Illustrative electrodermal activity chart tracings are presented in Figure 3. The output of the skin resistance amplifier was rectified (by a diode) before it was displayed and assessed.

Figure 3. A typical chart tracing of spontaneous skin resistance reactions. The activity has been rectified, and the chart speed is 0.25 cm/sec.

Different skin resistance amplifiers, types and placements of electrodes, and chart recorders were used in the various experiments; details may be found elsewhere.[12] These changes did not appear to affect the results in a significant way.

In Experiments 1 through 4, an individual who otherwise was not involved in the research quantified the electrodermal activity by blind-scoring the pen tracings (measuring each deflection with a millimeter rule). Special precautions were taken to preclude subtle cues that might influence scoring. This was done by obscuring possible visual cues with multiple layers of opaque tape, by keeping the random influence sequence hidden, and by preventing the scorer's contact with anyone knowing the target sequence until scoring had been accomplished. In Experiments 5 through 13, scoring was completely automated by sampling the electrodermal activity at 100 msec. intervals through use of an analog-to-digital converter interfaced with a microcomputer, and averaging these values. Since electrodermal activity changes relatively slowly, this sampling rate is quite satisfactory. The computer printed a permanent paper printout of these integrated measures for each epoch at the end of each session.

Scoring of Measurements

The treatment of activity scores has already been described (see General Design Considerations section above). Percent influence scores were calculated using the electrodermal (SRR) measures for the various recording epochs of a session. The mean chance expectation for these percent influence scores was 50 percent. A significant departure from 50 percent was taken to be an indication of a direct mental influence upon ongoing, spontaneous electrodermal activity, if the direction of the departure corresponded with the aim. Therefore, one-tailed t-tests generally were used in these assessments.

Results

We have completed 13 studies of direct mental influence of electrodermal activity using the protocol just described. Some of the experiments (Experiments 1, 2, 3, 4, and 11) had only one component and were conducted simply to test the effectiveness of the method with different samples of subjects and influencers. We describe these five experiments as "nonanalytical studies" since they did not assess variables other than the influence/non-influence factor.

The remaining experiments were "analytical studies" conducted to explore the role of additional physiological and psychological variables. Since some of these analytical studies contained subcomponents, more than 13 sets of results were generated. We have used the following *a priori* rule in presenting the results: In cases in which significant differences obtained between different subconditions and/or in cases in which it had been decided in advance to evaluate certain subconditions separately, results are presented for each subcondition; otherwise, results are combined across subconditions and presented for the experiment as a whole. This rule generated 15 sets of results. For convenience, each set is called an "experiment." The number of sessions contributing to each experiment ranged from 10 to 40.

Summary statistics for the 15 experiments are presented in chronological order in Table I. For each experiment, the primary analysis was the comparison of the sessions' percent influence scores with MCE (50 percent); single-mean *t*-tests were used for these comparisons. Since percent influence scores indicate the percentage of the session's (subject's) total electrodermal activity that occurred in the expected or predicted direction, scores greater than 50 percent indicate "successful" direct mental influence outcomes while scores less than 50 percent indicate "unsuccessful" outcomes. The *t*-tests yielded independently significant ($p \leq .05$) results for 6 of the 15 experiments. This obtained 40 percent experimental success rate is to be compared with a 5 percent experimental success rate to be expected on the basis of chance alone. The overall significance of the entire series may be determined using the Stouffer (or combined *z*) method which involves converting the obtained *p*-values into *z*-scores, summing these *z*-scores, and dividing this sum by the square root of the number of studies being combined; the result is itself a *z*-score that can be evaluated by means of its associated, highly significant, *p*-value.[16] The overall *z*-score for this entire 15-part series is 4.08, which has an associated $p = .000023$. The individual *z*-scores contributing to this assessment are depicted graphically in Figure 4.

In behavioral and biomedical statistics, there is currently an increased emphasis on the effect sizes observed in experiments and the consistency of these effect sizes, rather than significance levels alone.[15] For this reason, an effect size *(r)* was calculated for each experiment according to the formula

$$r = \sqrt{\frac{t^2}{t^2 + df}}$$

These effect sizes are also given in Figure 4, next to their respective *z*-scores. The effect sizes *(r)* vary from −.24 to +.72, with a mean effect size *(r)* of +.25, which compares favorably with effect sizes typically found in behavioral and biomedical research. An appealing presentation of effect size is the *binomial effect size display* (BESD) which converts an effect size to the change in success rate (e.g., survival rate, improvement rate, etc.) that would be expected if a

Table I. Statistical Summary of Electrodermal Influence Experiments

Experiment	Influencer(s)	Number of Sessions	Mean % Influence	t	p	Type of Study
1	Experimenter	10	59%	3.07	.0065	Nonanalytical
2	Selected subject	10	59%	2.04	.035	Nonanalytical
3	10 unselected volunteers	10	58%	2.96	.0077	Nonanalytical
4	10 unselected volunteers	10	47%	-0.76	.736	Nonanalytical
5a	Experimenters	16	60%	2.40	.014	Analytical
5b	Experimenters	16	50%	-0.09	.537	Analytical
6	24 unselected volunteers	24	57%	1.77	.043	Analytical
7	Experimenters	32	53%	1.15	.13	Analytical
8	Experimenters	30	52%	0.45	.33	Analytical
9	Experimenters	30	51%	0.44	.33	Analytical
10	Experimenters	16	53%	1.31	.10	Analytical
11	3 healing practitioners	15	51%	0.62	.28	Nonanalytical
12	5 selected volunteers	40	51%	0.21	.41	Analytical
13a	8 selected volunteers	32	57%	2.41	.02	Analytical
13b	8 selected volunteers	32	48%	-0.53	.70	Analytical

Note: In this and in all subsequent tables [in this chapter], one-tailed p-values are given (for purposes of Stouffer z determinations).

treatment or procedure having that effect size were to be instituted.[16] According to a BESD, a baseline treatment which ordinarily produces, e.g., a 37.5 percent survival rate in some populations can be augmented by another treatment with an effect size of +.25 (the effect size of the mental influence in these experiments) to a 62.5 percent survival rate. This is hardly a trivial effect. It is illuminating to compare the experiment results within the series we are reporting with those of recent placebo-controlled studies of the cardiovascular effects of the drugs propranolol and aspirin conducted by the National Heart, Lung, and Blood Institute and the Physician's Health Study Research Group, respectively.[17] Both studies, each employing very large sample sizes (2,108 subjects and 22,071 subjects, respectively), were terminated prematurely because their results were deemed so favorable to the efficacy of the drugs being tested. The researchers felt that it would be unethical to continue the study and thereby deprive the placebo subjects of the benefits of the drugs. The effect size in the propranolol study was .04. A BESD analysis would indicate that a treatment with such an effect size would be expected to change success rate from 48 percent to 52 percent. The aspirin study yielded an even smaller effect size ($r = .03$).

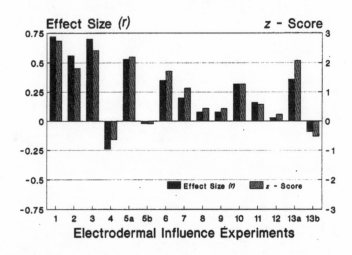

Figure 4. Effect sizes (r) and z-scores for the 15 successive electrodermal influence experiments.

Figure 5 provides still another perspective of the present results. In this figure, the mean z-score and mean effect size (r) for all 15 electrodermal influence experiments are plotted along with the corresponding scores for an electrodermal self-influence experiment and two electrodermal sham-experiments.[18] In the self-influence experiment, subjects attempted to reduce their *own* electrodermal activity, compared to baseline conditions, using psychophysiological self-regulation techniques.

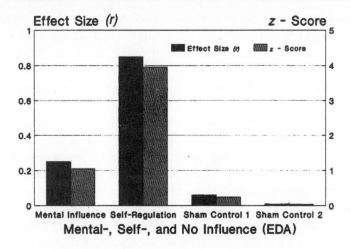

Figure 5. Mean effect sizes *(r)* and *z*-scores for electrodermal direct mental influence, self-regulation, and sham (control) experiments.

In the sham-experiments, electrodermal measurements were made under conditions comparable to those of the real direct mental influence studies, and the measurements were assigned to two conditions in a manner that mimicked that of the real experiments. The important difference was that the "influence" and "non-influence" designations were nominal only, and did not involve the participation of an actual influencer. The sham-experiments were conducted to determine whether it would be likely, by chance alone, to obtain differences as large as those found in the real experiments. For these three sets of experiments, all variables other than the source of the influence (i.e., remote mental influence, self-influence, or no influence, respectively) were virtually identical and therefore permit direct comparisons. As expected, the direct mental influence values are considerably larger than those generated under sham control (chance) conditions, but not as large as those generated through self-influence procedures.

Electrodermal Correlates of Remote Attention

In this section, and in the next six sections, we shall present brief summaries of related experimental series in which we explored DMILS effects using the same general protocol but different types of animate target systems.

Additional electrodermal studies were conducted using a protocol identical to that described in previous sections, but with two important changes. First, no polygraph feedback was supplied. Second, rather than attempt to actively influence the subject in a particular direction, the "influencer" simply devoted full attention to the distant person whose electrodermal activity was being continuously monitored. During half of the recording epochs, the "influencer" directed full attention toward ("stared at") the subject's image as it appeared on a closed-circuit television monitor. During the other (control) epochs, the influencer did not look at the monitor and did not think about the subject or the experiment. The subject, of course, was blind to the random sequence of the two types of epochs.

Four such experiments were conducted, along with a sham-experiment in which electrodermal measurements were collected during epochs that were to be analyzed as "staring" and "nonstaring" periods, but which did not involve actual staring.[19] The results are summarized in Table II. As expected, electrodermal activity scores during the "staring" and "nonstaring" periods did not differ during the *sham*-experiment. However, significant electrodermal differences between staring and nonstaring periods did emerge in each of the four "real" experiments. In three of the experiments, electrodermal activity was lower (i.e., in the direction of calming) during staring than during nonstaring periods; in one experiment, staring was associated with increased electrodermal activity (i.e., activation). These calming and activating effects are understandable in the context of the psychological conditions present in the starers and "starees" in the different experiments; however, a discussion of these patterns is beyond the scope of this paper.

Table II. Statistical Summary of Electrodermal Remote Attention Experiments

	Number of Sessions	Mean % Influence	t	p
Untrained subjects	16	59.38%	−2.66	.0089
Trained subjects	16	45.45%	2.15	.024
Replication 1	30	45.15	1.92	.032
Replication 2	16	45.66	2.08	.028
Sham Control	16	49.16	0.30	.38

Percentages > 50% indicate activation effects;
Percentages < 50% indicate calming effects
(see text for explanation of negatively-signed *t*-value)

In the meta-analysis of scientific experiments, there is a convention of assigning negative scores to results that differ in direction from the bulk of the findings. Therefore, we have assigned negative scores, *t*-tests, and effect sizes to the experiment with the "staring activation" results. The *z*-scores and effect sizes are shown in Figure 6.

Studies of Ideomotor Reactions

Ideomotor reactions are automatic reactions that are associated with thoughts; they are often subtle and unconscious. A familiar example is the

"Chevreul pendulum" in which information (typically yes/no answers to questions) is translated into subtle muscular movements of the arm, hand, and fingers, then amplified by a hand-held pendulum (a small weight suspended by a thread) to give a visible indication.[20] In three experiments, we explored the possible direct mental influence by one person of the unconscious, muscular movements of a second, remotely situated and isolated, person.[21] Circular versus linear movements of a hand-held pendulum served as the target ideomotor reactions, and circular versus linear directional aims of a distant influence were randomly scheduled.

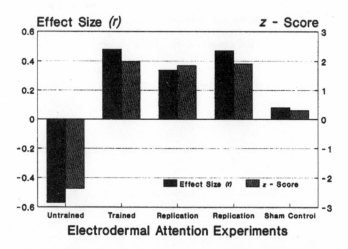

Figure 6. Effect sizes *(r)* and *z*-scores for the electrodermal attention (remote staring detection) experiments.

The pendulum movements were scored by the target persons who were blind regarding the influence aim of each recording period. The recording periods were signalled to the target persons by means of white noise (which was present during the trials, but absent during the inter-trial rests). The influencers received no immediate feedback in these experiments. Two of the three experiments yielded significant outcomes. The results are presented in Table III and in Figure 7.

Table III. Statistical Summary of Ideomotor Influence Experiments

	Number of Sessions	Mean % Influence	*t*	*p*
Experiment 1	10	55.55%	2.54	.0158
Experiment 2	15	71.65%	6.23	.000011
Experiment 3	15	47.35%	−1.29	.891

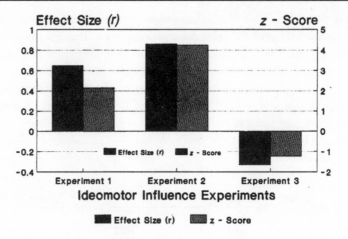

Figure 7. Effect sizes *(r)* and z-scores for the ideomotor influence experiments.

Table IV. Statistical Summary of Muscular Tremor Influence Experiments

	Number of Sessions	Mean % Influence	t	p
Experiment 1	10	47.91%	−0.92	.81
Experiment 2	9	50.09%	0.04	.48

Studies of Muscular Tremor

We conducted two experiments in which more conscious muscular responses served as the target reactions.[22] The subject (situated in a distant room and "blind" regarding the random epoch sequence) held a metal stylus within a small opening in a metal plate. The subject's aim was to be as steady as possible. Small movements of the hand (caused, for example, by nervousness) caused the stylus to contact the metal plate and were automatically registered as "errors." The subject was quite aware of these errors which could be felt and which were made even more noticeable by a small lamp that flashed each time the stylus contacted the metal. The influencer attempted to increase or decrease the number of errors (unsteadiness indications) made by the remote subject during incremental and decremental aim periods, respectively. The influencer received immediate, ongoing feedback regarding the effects of his or her intentions: The target subject's error rate was converted to gong-like tones that the influencer could hear through headphones. The pitch of the gongs was proportional to the remote subject's error rate.

These two experiments did not yield significant overall results. (Other analyses revealed significant correlations between tremor influences and nearby random event generator influences in these sessions, indicating that direct mental influences did occur in the study; a description of these analyses is beyond the scope of this paper.) Statistical summaries are given in Table IV and in Figure 8.

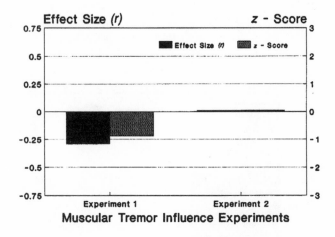

Muscular Tremor Influence Experiments

Figure 8. Effect sizes *(r)* and *z*-scores for the muscular tremor influence experiments.

Studies of Blood Pressure Influence

Two direct mental influence experiments were conducted in which blood pressure served as the targeted reaction.[23] The experimental protocol for these studies was similar to that used in the electrodermal influence series described above, with three important exceptions: (a) there were eight 2-minute epochs rather than twenty 30-second epochs, (b) blood pressure was substituted for electrodermal activity, and (c) the influencer did not receive immediate feedback regarding the remote subject's blood pressure. At the conclusion of each epoch, blood pressure was measured automatically by an IBS Model SD-700A electrosphygmomanometer (Industrial and Biochemical Sensors Corporation, Waltham, MA). The dependent variable was mean arterial pressure (MAP), calculated from the systolic (S) and diastolic (D) values according to the standard formula MAP = 1/3 (S − D) + D. Decremental aim and noninfluence (control) periods were randomly sequenced. Significant results were obtained in one of these two experiments. The results are summarized in Table V and in Figure 9.

Table V. Statistical Summary of Blood Pressure Influence Experiments

	Number of Sessions	Mean % Influence	t	p
Experiment 1	1*	51.77%**	2.45*	.025
Experiment 2	40	50.10%	0.75	.23

*The eight absolute blood pressure measurements of this pilot session of Experiment 1 were analyzed by means of a two-samples *t*-test with 6 *df*.

**The percent influence equivalent of the absolute measurements is given for Experiment 1 for comparative purposes only; they were not used in the two-samples *t* analysis.

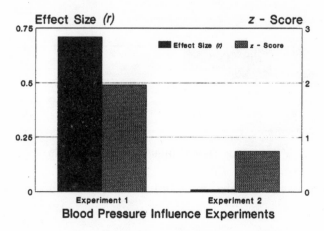

Figure 9. Effect sizes *(r)* and *z*-scores for the blood pressure influence experiments.

Spatial Orientation of Freely Swimming Fish

In addition to the studies of human physiological response systems described above, we have conducted experiments with other living organisms. In four experiments, persons attempted to influence the spontaneous swimming behavior of a small knife fish.[24,25] This fish *(Gymnotus carapo)* emits a weak electrical signal that is believed to be used for navigational purposes in its native habitat. If the fish is allowed to swim freely in a small container with metal electrodes fastened to the container's end walls, the fish's continuous AC signal arrives at those electrodes at different strengths, depending upon the fish's distance from and orientation toward the electrodes. The varying signal can be amplified, rectified, and electronically integrated. It may also be displayed on an oscilloscope screen, where it appears as a "randomly" rising and falling tracing. This oscilloscope tracing can provide immediate feedback, to an influencer, regarding the spatial orientation of a target fish that is isolated in an enclosure in another room. The integrated voltage from the fish/electrode system can be treated similarly to the changing electrical activity of the electrodermal experiments and can be compared for randomly scheduled incremental aim *versus* noninfluence (control) epochs. In this case, the incremental aim was for high amplitude oscilloscope tracings, which corresponded to a fish's perpendicular orientation toward the metal end electrodes. The results (presented in Table VI and in Figure 10) were significant for three of these four experiments.

Table VI. Statistical Summary of Fish Orientation Influence Experiments

	Number of Sessions	Mean % Influence	t	p
Experiment 1	10	52%	3.26	.00492
Experiment 2	10	53%	3.34	.00433
Experiment 3	10	54%	2.12	.0315
Experiment 4	10	51%	0.50	.314

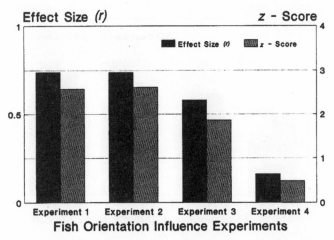

Figure 10. Effect sizes *(r)* and *z*-scores for the fish orientation influence experiments.

Locomotor Behavior of Small Mammals

In four experiments, persons attempted direct mental influence of the locomotor behavior of Mongolian gerbils *(Meriones unguiculatus)*.[24,25] The gerbil could run freely in an activity wheel. Each revolution of the wheel activated a switch that deflected an ink-writing event marker on a polygraph located in the distant room in which the influencer was stationed. Total activity (wheel revolutions) during randomly scheduled incremental aim and noninfluence (control) epochs was quantified by blind-scoring the number of pen deflections occurring during each epoch. In these experiments, the incremental aim was for frequent pen deflections (which could be viewed by the influencer and therefore provided immediate feedback), which corresponded to high activity levels (many activity wheel revolutions). The results are presented in Table VII and in Figure 11. Three of the four experiments yielded significant outcomes.

Table VII. Statistical Summary of Locomotor Behavior Influence Experiments

	Number of Sessions	Mean % Influence	t	p
Experiment 1	10	55%	1.50	.0839
Experiment 2	10	53%	2.12	.0315
Experiment 3	10	55%	2.33	.0224
Experiment 4	10	52%	2.89	.00894

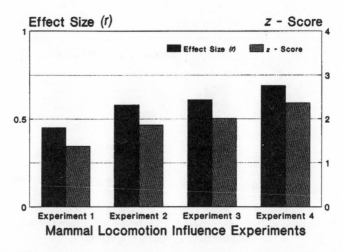

Figure 11. Effect sizes (r) and z-scores for the mammal locomotion influence experiments.

Studies of an *In Vitro* Cellular Preparation (Hemolysis)

In three experiments, we worked with an *in vitro* cellular preparation (human red blood cells) rather than an intact organism.[24,26] If red blood cells are placed in a solution having the same salinity as the blood plasma, their membranes will remain intact and the cells will "survive" for relatively long periods. If, however, the cells are osmotically stressed by a solution with too little or too much salinity, compared with blood plasma, the membranes become fragile and burst, spilling the cells' hemoglobin into the solution. The rate at which this process *(hemolysis)* occurs can be controlled by varying the salinity of the solution. Since the solution of blood cells becomes increasingly transparent to light as hemolysis proceeds, the time course and extent of hemolysis can be tracked and quantified by measuring the amount of light transmitted through the solution by means of a spectrophotometer.

We used such a measuring procedure in experiments in which persons attempted to "protect" human red blood cells by retarding their rate of hemolysis, mentally and at a distance. Rates of hemolysis of several tubes of blood were measured spectrophotometrically by a person who was unaware of which tubes were being influenced and which were noninfluence (control) tubes. The light measurements (i.e., the analog output of the spectrophotometer) were analyzed on line by means of an analog to digital converter interfaced with a microcomputer. Tubes were measured during randomly scheduled decremental aim and noninfluence (control) epochs. The influencer, who was stationed in a distant room, did not receive immediate feedback about the condition of the cells, but simply maintained a strong intention and image of the desired outcome during the decremental aim periods. The desired outcome was a reduced hemolysis rate, i.e., "healthy" cells with intact membranes that transmitted little light. Results (presented in Table VIII and Figure 12) were significant for two of the three experiments.

Overall Statistical Summary

Direct mental influences of living systems were examined in eight areas. An overall statistical summary is presented in Table IX. The data in this Table, along with the effect sizes indicated in Figures 4 through 12, provide strong evidence that persons are indeed able to exert direct mental influences upon a variety of living systems. Overall, this research program has included 37 experiments, 655 sessions, 449 different "influencees," 153 different influencers, and 13 different experimenters.

Table VIII. Statistical Summary of Hemolysis Influence Experiments

	Number of Sessions	Mean % Influence	t	p
Experiment 1	10	57.46%	8.70	.0000056
Experiment 2	32	49.55%*	−1.26*	.89
Experiment 3	32	50.01%*	4.80*	.000019

*Equivalent mean % influence scores and *t*-scores for Experiments 2 and 3 are given for comparative purposes only; actual analyses were by analysis of variance and by exact binomial tests (see original articles for details).

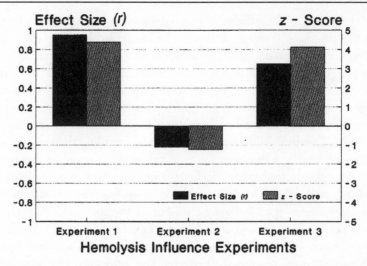

Figure 12. Effect sizes *(r)* and *z*-scores for the hemolysis influence experiments.

The reader may prefer to consider the eight subcomponents of the database separately since they involved different response systems. It is legitimate, however, to aggregate the entire body of studies if one is interested in the *general* issue of direct mental influence of living systems.[27] Indeed, the experimental protocols, analysis procedures, and mental strategies of the influencers are virtually identical across the various studies. Summary statistics for all systems combined are given in the bottom row of Table IX; overall results are depicted graphically in Figure 13.

Table IX. Overall Statistical Summary of Direct Mental Influence Experiments

Living Target System	Number of Sessions	Mean z	Stouffer z	Mean Effect Size	Percent of Experiments Significant
Electrodermal activity (influence)	323	1.05	4.08	.25	40%
Electrodermal activity (attention)	78	0.84	1.68	.18	100%
Ideomotor reactions	40	1.72	2.98	.39	67%
Muscular tremor	19	−0.42	−0.59	−.14	0%
Blood pressure	41	1.35	1.91	.36	50%
Fish orientation	40	1.88	3.78	.56	75%
Mammal locomotion	40	1.90	3.81	.58	75%
Race of hemolysis	74	2.43	4.20	.46	67%
All systems combined	655	1.34	7.72*	.33	57%

*$p = 2.58 \times 10^{-14}$ (one-tailed)

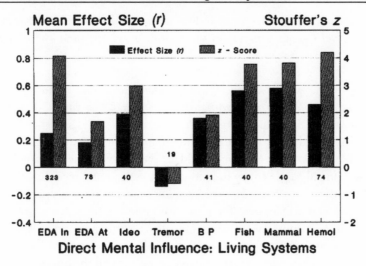

Figure 13. Mean effect sizes *(r)* and Stouffer *z*-scores for influences of all eight living target systems.

Rival Hypotheses

In order to conclude that the results described above are indeed attributable to direct mental influence it is necessary to rule out all alternative rival hypotheses, potential artifacts, and possible confounding variables. In this section, we indicate how rival hypotheses may be dismissed for the various influence series.

1. The results are attributable to coincidental or chance correspondences between the target system reactions and the influencer's intentions. The probability values of the conventional statistical tests, as well as the overall non-zero effect sizes in the intended direction, effectively rule out coincidence.

2. The results are attributable to reactions to uncontrolled external stimuli or to sensory cues. The experimental protocols eliminated these possibilities. Target systems were isolated from the influencers through the use of separate, nonadjacent rooms, distance, and intervening barriers. There were no known or obvious environmental stimuli that could have been associated differentially with the different aim conditions and that could have impinged upon the target systems.

3. The results can be attributed to common internal rhythms that could have influenced both influencer and target. Possible contributions of naturally occurring internal rhythms were ruled out by actively manipulating the influencer's aims on the basis of random schedules. The random schedules were based upon adequate randomization methods using electronic random event generators, tables of random numbers, pseudorandom computer algorithms, or well-shuffled cards.[28]

4. *The results could have been due to systematic biases contributed by recording errors or motivated misreading of records.* In 21 of the 37 experiments reviewed in this paper, target system activity was assessed by automated equipment that provided permanent numerical records that could be double-checked for accuracy. In 11 of the 37 experiments, target activity was manually assessed, by individuals who were blind regarding the nature of the trials (i.e., the scorers did not know which records were for control periods and which were for influence periods); further, the scoring was easy and straightforward (e.g., measuring permanent polygraph tracing deflections with a millimeter rule). Special precautions were taken to eliminate any subtle cues that could have compromised the blindness aspect of the protocol. In 5 of the 37 experiments, automated equipment was used but the equipment provided a numerical display that had to be read by a non-blind experimenter, rather than providing a permanent record. We believe that recording errors were effectively excluded from all of our experiments. Nonetheless, we can perform a conservative analysis to determine whether the results of only the most "rigorous" experiments (i.e., those in which permanent records were produced by completely automated equipment) remain independently significant even when contributions of less rigorous experiments are completely removed. The 21 most rigorous experiments include Experiments 5 through 15 of the electrodermal influence series, the four electrodermal attention experiments, the two muscular tremor experiments, the two blood pressure experiments, and the second and third hemolysis experiments. These combined experiments yield a mean effect size (r) = .17, a mean z-score = .84, and a Stouffer z = 3.84, which has an associated one-tailed p = 6.15 x 10^{-5}.

Because they comprise such a large component of our work, the 15 electrodermal experiments themselves were submitted to a similar rigor analysis. In order to provide a conservative analysis of that series, the first four experiments (in which polygraph records had been manually but blindly scored) were removed completely and an analysis was done to determine whether the computer-scored experiments (Experiments 5 through 15) remained independently significant when the hand-scored results were eliminated. The electrodermal direct mental influence effect remains and is significant even under these stringent conditions (Stouffer z = 2.86, p = .0021, mean effect size = .18).

5. *The "target subjects" knew, at the beginning of a session, when influence attempts were to be made and therefore could have "cooperated" by changing*

their activities to fit the influence schedule. The subjects were not told beforehand when or how many influence attempts would be made. Additionally, the experimenter who interacted with the subjects did not become aware of the influence schedule until all pre-session subject interactions had been completed and the experimenter had been isolated in a separate room. The subjects did not know of the existence of, or have access to, the envelopes containing schedule information. Thus, possible suggestion, expectancy, or "placebo" effects (of a conventional sort) were eliminated. It is not inconceivable that the "placebo effect" itself may be an everyday life manifestation of the very processes that are being studied in this research program.

6. *The "target subjects" could have learned about the non-influence/influence schedule during the session and used this information to produce artifactual activity changes.* The sensory isolation aspects of the experimental protocol prevented this.

7. *Target system activity differences for influence versus non-influence (control) periods could have reflected some systematic error, i.e., some progressive change in activity over time.* The use of randomized or counterbalanced sequences of influence and noninfluence periods assured that any time-dependent changes (e.g., due to equipment warm-up, adaptation, habituation, electrode polarization, fatigue, etc.) would not contribute differentially to the two types of periods. Additionally, the experimental protocols were deliberately designed to eliminate such time-dependent changes, and formal statistical analyses of earlier *versus* later periods indicated that the protocols were indeed effective in eliminating them.

8. *The findings are due to arbitrary selection of data.* For each experiment, the total number of trials and subjects were specified in advance, and the analyses reported in this paper include all recorded data.

9. *The findings are due to fraud on the part of the subjects.* The findings are based upon results obtained from a large number of volunteer subjects and influencers who, it would seem, would have no motive for trickery. However, even if they were motivated to cheat, no opportunities for cheating were provided. Cheating would have required knowledge of the precise sequence and timing of the various experimental epochs or the assistance of an accomplice. Both of these requirements were eliminated by the experimental protocols.

10. *The findings are due to fraud on the part of the experimenters.* No experiment, however sophisticated, can ever be absolutely safe from experimenter fraud. Even in cases in which an experiment is controlled by

outside supervisors, a hostile critic could still argue that collusion were involved. The hypothesized degree of such collusion would be limited only by the imagination and degree of paranoia of the critic. In some cases, our use of multiple-experimenter designs assured that one experimenter's portion of an experiment served as a kind of control for another experimenter's portion. Only the successful replication of this work by independent investigators in other laboratories will reduce experimenter fraud to a non-issue. One of our motives in publishing this paper is to encourage such replication attempts.

Conceptual Replications

Although, to our knowledge, there have been no exact replications of this work, several *conceptual* replications have appeared. Successful electrodermal influences were reported by researchers at Brooklyn's Maimonides Medical Center,[29] heart rate influences were reported by San Bernardino State College researchers,[30] animal behavior influences were reported by an investigator at the Institut fur Grenzgebiete der Psychologie und Psychohygiene in Freiburg, Germany,[31] and influences upon the activity of an electric fish were reported by investigators at the A. N. Severtsev Institute of Evolutionary Morphology and Animal Ecology in Moscow.[32]

In addition, in the English-language scientific literature alone, there are approximately 100 published reports of experiments in which persons have been able to influence, mentally and at a distance, a variety of biological target systems including bacteria, yeast colonies, fungus colonies, motile algae, plants, protozoa, larvae, woodlice, ants, chicks, mice, rats, gerbils, cats, and dogs, as well as cellular preparations (blood cells, neurons, cancer cells) and enzyme activity.[33] In human "target persons," eye movements, gross motor movements, electrodermal activity, plethysmographic activity, respiration, and brain rhythms have been influenced.[33]

The experiments described in this paper are, in fact, conceptual replications of human distant mental influence experiments conducted many years ago in the Soviet Union (by Bekhterev, Vasiliev, and their co-workers) and in France (by Joire, Gibert, Janet, and Richet).[34] The direct mental influence effect appears to be a widespread and robust phenomenon.

Influencing Factors

An extended discussion of the rationales and results of investigations of factors that influence the likelihood or strength of direct mental influence effects is beyond the scope of this paper. However, we felt it important to at least mention some of the physical, physiological, and psychological variables that we have studied.

Physical Factors

In our own studies of direct mental influence and in those of other investigators, physical variables have not been found to have important influences upon experimental outcome. The effects do not appear to be modulated significantly by physical distance, physical barriers or screens, or the physical constitution of the target system. There are two possible exceptions to this general conclusion.

One of these is the degree of free variability of the target system. Direct mental influence appears to occur more readily in labile systems than in inert ones. As alluded to in an earlier section of this paper, it is not yet clear whether this is truly a physical effect in which a target's susceptibility depends upon its degree of randomness, chaotic behavior, or variability, or rather, a psychological effect in which *perceived* system change encourages facilitating psychological factors of enhanced belief, confidence, and expectation of success.[35]

The other physical factor is the character of the earth's geomagnetic field (GMF) at the time of an experimental session. We have found preliminary indications that the susceptibilities of certain of our target systems (i.e., electrodermal activity, rate of hemolysis) to direct mental influence, as well as the activity levels of these systems, are higher during periods of high GMF activity than during periods of low GMF activity.[36] Complementary findings relating low GMF activity and anomalous cognition effects have already been reported.[37] This is quite interesting in view of other findings that suggest that anomalous cognition and anomalous influence processes may be affected in complementary ways by certain psychological and physiological variables and that the two processes themselves may be complementary ones.[37]

Physiological Factors

We have found indications of greater susceptibility to electrodermal direct influence in persons whose spontaneous electrodermal activity levels are relatively high, compared to those with relatively low activity levels.[18] This finding is consistent with a hypothesis that living systems characterized by a greater degree of departure from homeostasis may be more susceptible to direct mental influences aimed at restoring physiological balance. This hypothesis, which implies much more than a mere "regression to the mean" artifact, has important implications for healing practice and healing research and is one that deserves more extensive investigation. The optimal physiological conditions of *influencers* in direct mental influence experiments have not yet been adequately explored.

Psychological Factors

We are more knowledgeable about some of the psychological factors that facilitate or impede direct mental influence effects. The effect appears to be enhanced by the presence of a felt need to be influenced (in an "influencee") or by the perception (by the influencer) of such a need in an influencee.[18] Immediate, trial-by-trial, analog sensory feedback regarding the momentary state of the target system (e.g., viewing polygraph tracings of an influencee's ongoing electrodermal activity) may be facilitating for certain influencers. However, the direct mental influence effect can occur in the absence of such feedback.[19,21,26] Direct mental influence may be focused upon particular subsystems of the target system through directed intention and directed attention.[38] Persons may be able to "block" unwanted direct mental influences upon their own physiological activities.[38] The magnitude of certain direct mental influence effects may be influenced by certain "personality" factors such as the influencee's degree of extraversion/introversion, degree of social avoidance and anxiety, and the degree to which one is comfortable and nondefensive about the possibility of being strongly "interconnected" with others.[19] Direct mental influences may occur in persons who are conventionally unaware that such influences upon their physiological activities are being attempted.[12]

In our experiments, persons have rarely expressed concern about the possibility of influencing another person's physiological activity or about the possibility of being influenced in this manner. Informational congruences may occur in the influencer/influencee dyad. For example, during an experimental session, an influencee may experience certain thoughts, feelings, or images that are identical or quite similar to those experienced by the spatially remote influencer. Such anomalous cognition effects may occur whether or not direct physiological influences occur in the same sessions.[12]

It has been suggested that perhaps the ostensible "changes" in activity levels in the various target systems were not really "produced" at all, compared with what would have occurred anyway in the absence of influence attempts. Rather, it may have been the case that certain anomalous cognition processes may have resulted in optimal timing of sessions so that greater target activities happened to occur during high aim than during low aim recording epochs.[39] While this possibility cannot be ruled out unequivocally at present, explicit tests in the DMILS context have yielded outcomes that are not consistent with this "intuitive data sorting" hypothesis.[26,40]

Time-Displaced Effects?

Although we have dealt exclusively, in this paper, with direct mental influences upon living target systems, there exists a considerable literature describing experiments in which persons are able to exert direct mental influences upon inanimate random systems such as electronic random event generators and random mechanical systems.[6] In a series of remarkable experiments, physicist Helmut Schmidt has found evidence that suggests that such influences may be displaced in time. In typical random event generator (REG) experiments, a subject or operator uses mental strategies to attempt to influence the behavior of the generator in real time. In a variation on these experiments,

Schmidt's subjects used similar mental methods to successfully influence the probabilities of random events generated *in the past,* i.e., random events that had been prerecorded but had not yet been consciously observed. Apparently, present mental "efforts" were able to influence past events about which "Nature had not yet made up her mind." The subjects did not *change* the past (once events had been generated, they remained in that form), but rather, they seemed able to *influence* the initial generation of one type of random outcome over another. The rationales and details of these experiments, which were suggested by certain principles and problems of quantum mechanics *(viz.,* the time-independent nature of certain laws and the "quantum measurement problem") are too complex to describe in this paper; the interested reader can consult relevant publications.[41]

The success of these time-displaced *physical* effects suggests that similar experiments could be conducted using *living* target systems, especially target systems such as electrodermal activity (mediated by the autonomic nervous system) which was not consciously observed by the target persons themselves at the time the activities initially occurred. Ongoing fluctuating electrodermal activity (of which the target person is "unconscious") could be prerecorded on magnetic tape or on floppy disks or computer memory. The data would remain unobserved until some future session at which time the data are played back to be observed for the first time by an influencer who uses direct mental influence methods similar to those we have been discussing. Would it be possible for someone to influence his or her own prerecorded physiological activity or another individual's prerecorded physiological activity? At the time of initial recording, of course, the influence aims for different recording epochs would not yet have been determined and thus would not be known to the person initially emitting the physiological reactions. Target aims for different portions of the prerecorded record would be randomly selected only after the records had initially been made, but before the later observing/influencing session. If such experiments are successful, what would be the influence of the degree of conscious awareness of the physiological reactions at the moment of their initial generation and initial recording? Would the effect occur for initially "unobserved" autonomic reactions (such as electrodermal activity), but not for strongly observed, fully conscious actions (such as visible muscular responses)?

Schmidt has added an intriguing twist to some of his REG experiments. In some cases, the prerecorded events are observed by a third party during the interval between their initial generation/recording and the session in which the mental influence is attempted. There are indications that such intervening pre-observation, if it is sufficiently intensive, may prevent subsequent direct mental influence effects. Such effects, in which initial sensory observation appears to reduce or eliminate the susceptibility of a random system to subsequent mental influence, have occurred in cases in which humans, dogs, and goldfish were the sensory pre-observers.[42] If these effects are replicable and are indeed what they appear to be, these "conscious pre-observation" experiments have striking implications. If degree of consciousness is indeed critical in eliminating a subsequent mental influence effect, this blocking phenomenon could be used as an operational measure of consciousness, and eventually could provide a useful empirical tool in a true comparative psychology of mind. We mention these curious time-displaced effects not only for their basic research possibilities, but also because of their theoretical relevance to issues of health, particularly issues of initial symptom formation and illness prevention. Carefully designed experiments involving prerecorded physiological events could be used to explore curious practical questions such as the following: Can an individual's mental processes "reach backward in time" to actually influence the initial developmental probabilities of healthful or harmful physical changes?

Thus far, we have conducted only one time-displaced, physiological mental influence experiment of the type being proposed.[24] That study, which involved attempts by one selected influencer to influence prerecorded electrodermal activity over a small number of sessions, did not yield overall significant results. Certain interesting target system changes did, however, co-vary with changes in the influencer's psychological condition during the course of the experiment. For example, scoring improved dramatically during runs immediately following a reminder to the influencer that he had, in fact, succeeded in time-displaced experiments in another lab (which he had forgotten), and scoring dropped precipitously when the influencer (contrary to the protocol) attempted to miss (i.e., intended changes in the *opposite* direction). We hope to conduct more extensive future investigations of possible time-displaced mental influences.

Uncertainties and Reproducibility

The tables and figures presented here have already provided indications of the degree to which these direct mental influence findings have been reproducible within each type of experiment and across the various sets of experiments. Of the 37 experiments conducted, 21 yielded independently significant results. This 57 percent experimental success rate (to be compared with the 5 percent rate expected on the basis of chance alone) is particularly impressive given the generally small sample sizes and resulting low statistical power of the individual experiments.

It is not clear why some experiments succeed while others do not. If target system activity and susceptibility do indeed fluctuate with changing GMF activity, variability in this physical factor could account for some error variance. We suspect, however, that fluctuating psychological and social conditions are more directly responsible for the variable results. A "successful" session may depend crucially upon the presence of certain psychological conditions in the influencer, the influencee, and perhaps even in the experimenter, and these critical psychological ingredients may not always be reproducible. We suspect that factors such as belief, confidence, expectation of a positive outcome, and absence of psychological resistances may facilitate positive outcomes. On the other hand, boredom, absence of spontaneity, poor mood, poor interactions or rapport, psychological resistances or defensiveness, and excessive egocentric striving (excessive pressure or striving to succeed, analogous to performance anxiety) may decrease the likelihood of success.

Some variability in results may be contributed by differences in the degree of lability (free variability) of the target system, i.e., the degree to which the target system is freed from external and internal constraint or structure. A complementary process—the fullness of intention and the intensity or vividness of goal imagery—on the part of the influencer may likewise be favorable to a successful outcome. Greater certainty regarding the importance of these various factors, and the eventual specification of the requisite ingredients for more highly reproducible experiments, will depend upon the satisfactory operationalizing of these factors and a great deal of additional research.

General Discussion

The 37 experiments reviewed in this paper are variations on a single theme: Persons are able to mentally influence remote biological systems, even when those systems are isolated at distant locations and screened from all conventional informational and energetic influences. The effect appears to occur in a "goal-directed" manner; i.e., the influencer need not understand or even be aware of the specific physical or physiological processes which bring about the desired outcome. Intentionality appears to be a key factor in effecting these changes in remote biological systems. Maintaining a strong intention of a desired goal event, focusing attention upon the relevant aspect of the target system, and filling oneself with strong imagery of the desired biological activity are, under certain conditions, accompanied by a shift in the target system's activity in the intended direction.

The "mechanism" through which this shift comes about is unclear. With one possible exception, conventional physical forces would appear to be adequately ruled out, since the effect survives distance and screening effects that would block or severely attenuate such forces. The possible exception is extremely low frequency (ELF) electromagnetic radiation. ELF magnetic fields could conceivably serve as physical carriers for at least some of the observed effects.[43] ELF radiation would be expected to penetrate the shielding and barrier materials in the experimental environment and is able to propagate for great distances. Problems with ELF carriers, however, are: (a) their inability to rapidly transmit signals that are rich in informational content (i.e., their bandwidths are limited), and (b) we are unaware of plausible mechanisms through which ELF carriers could be encoded ("modulated") by influencer intentions and decoded ("demodulated") into the physical forces necessary to bring about appropriate changes at the target site. However, because ELF fields interact with GMF activity (which appears to influence the direct mental influence effect), it may be unwise to rule out ELF fields prematurely.

Several theorists have suggested that remote mental influence could occur through a reorganization of the randomness or "noise" inherent in the target system.[44] The process through which such reorganization would come about, however, remains unclear.

Regardless of how the effect is mediated, its very occurrence presupposes a profound interconnectedness between the influencers and the influencees in these experiments. The mental processes of the influencers are able to have nonlocal effects. This, in turn, suggests that these mental processes themselves may be nonlocal, rather than restricted to a particular spatiotemporal locus within the brain of the influencer. These considerations have important implications for our understanding of the nature of "mind" itself.

The results of this DMILS research program suggest certain useful methodological applications. It is possible that direct mental influence effects could be used as novel operational measures of volition or intentionality. "Unconscious" physiological responses could be used in addition to or instead of the conscious verbal reports typically used in anomalous cognition experiments; the former could provide indicators of successful information transmission that are more "primitive" and perhaps more sensitive and less prone to possible filtering and cognitive distortions.

In the experiments reported in this paper, persons influenced an arbitrarily selected response system of an arbitrarily selected organism in an arbitrarily selected direction. The successful outcomes of these experiments suggest that, in principle, judiciously selected directional mental influences could be focused upon particular organs, tissues, or cells of specific persons in ways that could be medically relevant. Thus, our findings become relevant to issues of mental healing. In fact, the experiments described in this paper may be viewed as schematized mental healing analog experiments. These healing analog studies can be modified systematically to yield information which could have useful clinical applications. Evidence continues to accumulate regarding the role of mental processes in physical health and well-being. Investigations in areas of hypnosis, biofeedback and self-regulation, and psychoneuroimmunology have documented the profound somatic impact of images, thoughts, and feelings.

Typically, these mental influences are understood and explained in terms of the biochemical and anatomical interconnections among the central nervous system, the autonomic nervous system, and the immune system. The experiments reviewed in this paper suggest that mental processes such as attention and intention can have influences that are more direct and immediate than has previously been recognized. Direct mental influence may provide an additional control system that can function in parallel with anatomical, chemical, and electrical influences within the body. It may also complement these conventional physical influences by acting, in their absence, outside the body.

We hope other investigators will join us in elucidating these subtle energetic and informational influences.

Correspondence:

William G. Braud, Ph.D.
Psychology Laboratory
Mind Science Foundation
8301 Broadway, Suite 100
San Antonio, TX 78209

Acknowledgments

We are indebted to Charles Honorton for suggesting the term "direct mental influence" as a general descriptor for these experiments and to Bruce Pomeranz, M.D., Ph.D., for his suggestion that living systems that have departed from homeostasis might provide optimal targets for direct mental influences of a balancing nature.

References and Notes

1. C. Honorton, Meta-analysis of the Psi Ganzfeld Research: A Response to Hyman, *Journal of Parapsychology* 49 (1985), pp. 51–91; C. Honorton *et al.*, Psi Communication in the Ganzfeld: Experiments with an Automated Testing System and a Comparison with a Meta-analysis of Earlier Studies, *Journal of Parapsychology* 54 (1990), pp. 99–139; R. Rosenthal, Meta-analytic Procedures and the Nature of Replication: The Ganzfeld Debate, *Journal of Parapsychology* 50 (1986), pp. 315–336.
2. E. Schechter, Hypnotic Induction vs. Control Conditions: Illustrating an Approach to the Evaluation of Replicability in Parapsychology Data, *Journal of the American Society for Psychical Research* 78 (1984), pp. 1–27.
3. C. Honorton and D. Ferrari, "Future telling": A Meta-analysis of Forced-Choice Precognition Experiments, 1935–1987, *Journal of Parapsychology* 53 (1990), pp. 281–308.
4. C. Honorton, D. Ferrari, and D. Bem, Extraversion and ESP Performance: A Meta-analysis and a New Confirmation, *Proceedings of the Annual Meeting of the Parapsychological Association* 33 (1990), pp. 113–135.
5. D. Radin and D. Ferrari, Effects of Consciousness on the Fall of Dice: A Meta-analysis, *Journal of Scientific Exploration* 5, 1 (1991), pp. 61–83.
6. D. Radin and R. Nelson, Consciousness-Related Effects in Random Physical Systems, *Foundations of Physics* 19 (1989), pp. 1499–1514.
7. S. Spottiswoode, Geomagnetic Activity and Anomalous Cognition: A Preliminary Report of New Evidence, *Subtle Energies* 1, 1 (1990), pp. 91–102.
8. W. Braud, On the Use of Living Target Systems in Distant Mental Influence Research,

in *Psi Research Methodology: A Re-Examination* (L. Coly and B. Shapin, Eds., Parapsychology Foundation, New York, NY, 1991, in press); W. Braud, Remote Mental Influence of Electrodermal Activity, *Journal of Indian Psychology* (1991), in press.

9. W. Braud, Lability and Inertia in Psychic Functioning, In *Concepts and Theories of Parapsychology* (L. Coly and B. Shapin, Eds., Parapsychology Foundation, New York, NY, 1981), pp. 1–36.

10. J. Utts, personal communication, July 13, 1991.

11. R. Berger, Psi Effects Without Real-time Feedback, *Journal of Parapsychology* 52, 1 (1988), pp. 1–27; W. Braud, Recent Investigations of Microdynamic Psychokinesis, With Special Emphasis on the Roles of Feedback, Effort, and Awareness, *European Journal of Parapsychology* 2 (1978), pp. 137–162; M. Varvoglis and D. McCarthy, Conscious-Purposive Focus and PK: RNG Activity in Relation to Awareness, Task-Orientation, and Feedback, *Journal of the American Society for Psychical Research* 80, 1 (1986), pp. 1–30.

12. W. Braud and M. Schlitz, A Methodology for the Objective Study of Transpersonal Imagery, *Journal of Scientific Exploration* 3, 1 (1989), pp. 43–63.

13. H. Schmidt, Quantum-Mechanical Random Number Generator, *Journal of Applied Physics* 41 (1970), pp. 462–468.

14. R. Edelberg, Electrical Activity of the Skin: Its Measurement and Uses in Psychophysiology, in *Handbook of Psychophysiology* (N. Greenfield and R. Sternback, Eds., Holt, Rinehart, and Winston, New York, NY, 1972), pp. 368–418; W. Prokasy and D. Raskin, *Electrodermal Activity in Psychological Research* (Academic Press, New York, NY, 1973); P. Venables and M. Christie, Electrodermal Activity, In *Techniques in Psychophysiology* (L. Martin and P. Venables, Eds., Wiley, New York, NY, 1980), pp. 3–67.

15. M. Gardner and D. Altman, Confidence Intervals Rather Than P-Values: Estimation Rather Than Hypothesis Testing, *British Medical Journal* 292 (1986), pp. 746–750; R. Rosenthal, Replication in Behavioral Research, *Journal of Social Behavior and Personality* 5, 4 (1990), pp. 1–30; J. Utts, Replication and Meta-analysis in Parapsychology, *Statistical Science* (1991), in press.

16. R. Rosenthal, Designing, Analyzing, Interpreting, and Summarizing Placebo Studies, in *Placebo: Theory, Research and Mechanisms* (L. White, B. Tursky, and G. Schwartz, Eds., The Guilford Press, New York, NY, 1985), pp. 110–136.

17. G. Kolata, Drug Found to Help Heart Attack Survivors, *Science* 214 (1981), pp. 774–775; J. Greenhouse and W. Greenhouse, An Aspirin a Day . . . ?, *Chance* 1 (1988), pp. 24–31.

18. W. Braud and M. Schlitz, Psychokinetic Influence on Electrodermal Activity, *Journal of Parapsychology* 47 (1983), pp. 95–119.

19. W. Braud, D. Shafer, and S. Andrews, Electrodermal Correlates of Remote Attention: Autonomic Reactions to an Unseen Gaze, *Proceedings of the Annual Meeting of the Parapsychological Association* 33 (1990), pp. 14–28; W. Braud, D. Shafer, and S. Andrews, Further Studies of Autonomic Detection of Remote Staring: Replications, New Control Procedures, and Personality Correlates, manuscript in preparation.

20. L. Wolberg, *Medical Hypnosis* (Grune and Stratton, New York, NY, 1948).

21. W. Braud and J. Jackson, The Use of Ideomotor Reactions as Psi Indicators, *Parapsychology Review* 13 (1982), pp. 10–11.

22. W. Braud, M. Schlitz, and H. Schmidt, Remote Mental Influence of Animate and Inanimate Target Systems: A Method of Comparison and Preliminary Findings, *Proceedings of the Annual Meeting of the Parapsychological Association* 32 (1989), pp. 12–25.

23. W. Braud, Remote Mental Influence of Blood Pressure: Biological Psychokinesis or GESP-Aided Self-Regulation, submitted for publication.

24. W. Braud, G. Davis, and R. Wood, Experiments with Matthew Manning, *Journal of the Society for Psychical Research* 50, 782 (1979), pp. 199–223.

25. W. Braud, Conformance Behavior Involving Living Systems, in *Research in Parapsychology* 1978 (W. Roll, Ed., Scarecrow Press, Metuchen, NJ, 1979), pp. 111–115.

26. W. Braud, Distant Mental Influence of Rate of Hemolysis of Human Red Blood Cells, *Journal of the American Society for Psychical Research* 84, 1 (1990), pp. 1–24.

27. G. Glass, In Defense of Generalization, *The Behavioral and Brain Sciences,* 3 (1978), pp. 394–395.

28. J. Palmer, Shuffling as a Randomization Method: How Good Is It?, *Proceedings of the Annual Meeting of the Parapsychological Association* 33, 218–226 (1990). On the basis of empirical studies. Palmer concluded that 7 shuffles were sufficient to adequately randomize a card deck; in our studies, we used 20 shuffles.

29. M. Kelly, M. Varvoglis, and P. Keane, Physiological Response During Psi and Sensory Presentation of an Arousing Stimulus, in *Research in Parapsychology* 1978 (W. Roll, Ed., Scarecrow Press, Metuchen, NJ, 1979), pp. 40–41.

30. N. Khokhlov, Remote Biofeedback in Voluntary Control of Heart Rate, *Psi Research* 2, 3 (1983), pp. 66–92.

31. E. Gruber, Conformance Behavior Involving Animal and Human Subjects, *European Journal of Parapsychology* 3 (1979), pp. 36–50; E. Gruber, PK Effects on Pre-recorded Group Behavior of Living Systems, *European Journal of Parapsychology* 3 (1980), pp. 167–175.

32. V. Protosov, V. Baron, L. Druzhkin, and O. Chistyakova, "Nile Elephant" Gnathonemus Petersii as a Detector of External Influences, *Psi Research* 2, 1 (1983), pp. 31–37.

33. D. Benor, Survey of Spiritual Healing Research, *Complementary Medical Research* 4 (1991), pp. 9–32; J. Solfvin, Mental Healing, in *Advances in Parapsychological Research,* Volume 4 (S. Krippner, Ed., McFarland and Company, Jefferson, NC, 1984), pp. 31–63.

34. L. Vasiliev, *Experiments in Distant Influence* (E. P. Dutton and Company, New York, NY, 1976).

35. Julian Isaacs has also discussed the importance of distinguishing physical lability and perceived lability; see J. Isaacs, Lability and Inertia in PK Target Systems: Is it the System Properties or the Feedback Display Properties that Matter?, Paper presented at the Annual Conference of the Society for Psychical Research, Bristol, England, 1981.

36. W. Braud and S. Dennis, Geophysical Variables and Behavior: LVIII. Autonomic Activity, Hemolysis, and Biological Psychokinesis: Possible Relationships with Geomagnetic Field Activity, *Perceptual and Motor Skills* 68 (1989), pp. 1243–1254.

37. W. Braud, ESP, PK and Sympathetic Nervous System Activity, *Parapsychology Review* 16 (1985), pp. 8–11.

38. W. Braud, M. Schlitz, J. Collins, and H. Klitch, Further Studies of the Bio-PK Effect: Feedback, Blocking, Specificity/Generality, in *Research in Parapsychology* 1984 (R. White and J. Solfvin, Eds., Scarecrow Press, Metuchen, NJ, 1985), pp. 45–48.

39. E. May, D. Radin, G. Hubbard, B. Humphrey, and J. Utts, Psi Experiments with Random Number Generators: An Informational Model, *Proceedings of the Annual Meeting of the Parapsychological Association* 28, 325–366 (1985).

40. W. Braud and M. Schlitz, Possible Role of Intuitive Data Sorting in Electrodermal

Biological Psychokinesis (Bio-PK), *Journal of the American Society for Psychical Research* 83, 4 (1989), pp. 289–302.

41. H. Schmidt, PK Effect on Pre-recorded Targets, *Journal of the American Society for Psychical Research* 70 (1976), pp. 267–291; H. Schmidt, Can an Effect Precede its Cause?, *Foundations of Physics* 8 (1981), pp. 463–480; H. Schmidt, Addition Effect for PK on Pre-recorded Targets, *Journal of Parapsychology* 49 (1985), pp. 229–244; H. Schmidt, R. Morris, and L. Rudolph, Channeling Evidence for a Psychokinetic Effect to Independent Observers, *Journal of Parapsychology* 50 (1986), pp. 1–15.

42. H. Schmidt, Superposition of PK Efforts by Man and Dog, In *Research in Parapsychology* 1983 (R. White and R. Broughton, Eds., Scarecrow Press, Metuchen, NJ, 1984), pp. 96–98; H. Schmidt, Human PK Effort on Pre-recorded Random Events Previously Observed by Goldfish, in *Research in Parapsychology* 1985 (D. Weiner and D. Radin, Eds., Scarecrow Press, Metuchen, NJ, 1986), pp. 18–21; H. Schmidt, The Strange Properties of Psychokinesis, *Journal of Scientific Exploration* 1, 2 (1987), pp. 103–118; H. Schmidt, Search for a Correlation Between PK Performance and Heart Rate, *Journal of the American Society for Psychical Research* 85, 2 (1991), pp. 101–118.

43. M. Persinger, ELF Field Mediation in Spontaneous Psi Events: Direct Information Transfer or Conditioned Elicitation?, in *Mind At Large* (C. Tart, H. Puthoff, and R. Targ, Eds., Praeger, New York, NY, 1979), pp. 191–204; M. Persinger, Psi Phenomena and Temporal Lobe Activity: The Geomagnetic Factor, in *Research in Parapsychology* 1988 (L. Henkel and R. Berger, Eds., Scarecrow Press, Metuchen, NJ, 1989), pp. 121–156.

44. R. Mattuck and E. Walker, The Action of Consciousness on Matter: A Quantum Mechanical Theory of Psychokinesis, in *The Iceland Papers,* (A. Puharich, Ed., Essentia Research Associates, Amherst, WI, 1979), pp. 111–159; D. Stokes, Theoretical Parapsychology, in *Advances in Parapsychological Research,* Volume 5 (S. Krippner, Ed., McFarland, Jefferson, NC, 1987), pp. 77–189.

5

Distant Mental Influence of Physiological Activity: New Experiments and Their Historical Antecedents

William G. Braud

This chapter reviews empirical research on a psychophysiological anomaly: the influence of physiological activities through mental sugges-tion at a distance. The review emphasizes direct (distant) mental sug-gestion research carried out by early Russian physiologists (Bekhterev, Vasiliev, Platonov, Ivanov-Smolensky), telepathy's role in Hans Berger's development of the electroencephalograph, and early French experiments in distant mental influence of hypnotized participants (Joire, Gibert, Janet, Richet). The review is brought up to date through its coverage of recent empirical studies of direct (remote, distant) mental influences of electroder-mal activity and other biological activities.

This information originally was presented as an invited contribution to a panel on "Anomalous phenomena in psychophysiology," 22nd Annual Meeting of the Association for Applied Psychophysiology and Biofeedback, Dallas, TX, March 15–20, 1991, and later published as Braud, W. G. (1992). Remote mental influence of electrodermal activity. Journal of Indian Psychology, 10(1), 1–10. The contents of this article are Copyright © 1992 by, and reprinted by permission of, the Journal of Indian Psychology.

—William Braud

The research question to be addressed here is whether it is possible for men-tal activity of one person to influence the physiological activity of another per-son at a distance and under conditions that preclude conventional sensorimotor interactions and conventional physical energies. Such questions are typically asked within the domain known as "parapsychology" or "psychical research," which deals with processes such as telepathy and clairvoyance. What is not appreciated, however, is that these very questions were actively researched by some of the founders and leading investigators of the disciplines that we now recognize as psychophysiology and conditioning and learning—disciplines that contributed importantly to the development of biofeedback and self-regulation research. I'll mention some relevant projects that were undertaken in the early 1900s in the Soviet Union by researchers who were exploring the newly discovered "conditional reflexes."

Ivan Pavlov himself addressed some of these issues. Pavlov, to whom we are all indebted for his brilliant work in classical conditioning, was intensely interested in the various phenomena of hypnosis and in the unusual physiological and psychological functions manifested in psychiatric patients. In one of his lectures in physiology, after describing the extremely fine differentiations among conditional stimuli that dogs are able to make, he continued: "In us, in human beings, our higher conscious activity runs counter to these lower abilities to differentiate and hence hinders fine differentiation. That this is so is demonstrated by the fact that, in some instances, when man's normal conscious activity is altered, his ability to differentiate is sharpened. During special states of so-called clairvoyance, the differentiating ability in man reaches infinite sharpness" (Pavlov, 1952, p. 520).

Vladimir Bekhterev, who made important contributions in what we now call "instrumental conditioning," was much more actively involved with these issues. In addition to his better known work in reflexology, Bekhterev himself conducted laboratory investigations of telepathic influence in dogs and in remote hypnotic influence of humans (see Gregory, 1976). Within his Institute for Brain Research at the University of Leningrad, he established, in 1922, a Commission for the Study of Mental Suggestion. The Commission consisted of psychologists, medical hypnotists, physiologists, physicists, and a philosopher. Its charge was to investigate spontaneous cases of psychic phenomena, psychophysiological effects of magnetic fields in hypnotized subjects, and distant mental suggestion of hypnotized subjects.

Much of the distant mental suggestion work was carried out by a young physiologist, Leonid Vasiliev. The research was conducted within a physiological framework and was guided by the electromagnetic hypothesis of telepathy developed by the German neurologist/psychiatrist Hans Berger and the Italian neurologist F. Cazzamalli (Gregory, 1976). It was, indeed, Berger's own motivation to measure this posited electromagnetic carrier of telepathy that guided the investigations that led eventually to his development of the electroencephalograph and to his recording of the first human EEG tracings in 1924 (see Brazier, 1961; Roll, 1960). In this enterprise, Vasiliev was joined by other prominent Russian psychophysiologists, notably K. I. Platonov and

Bekhterev's collaborator, A. G. Ivanov-Smolensky (who performed early, important work in developing objective methods for the study of verbal or semantic conditioning and transfer or, in Pavlovian terminology, the study of "second signalling system" conditioning).

Vasiliev's work was conducted from 1921 until 1938, discontinued from 1939 until 1960, then re-established and continued until his death in 1966. Vasiliev's major work, *Experiments in Mental Suggestion*, was first published in Russian in 1962; an English translation, authorized and revised by Vasiliev, appeared in 1963 and was re-issued in 1976 under the title *Experiments in Distant Influence*. In this highly recommended monograph, Vasiliev details the methods that he and his co-workers used to study distant influence (mental suggestion) in selected subjects. In a series of careful experiments, Vasiliev's team was able to induce motor acts, visual images and sensations, sleeping and awakening, and physiological reactions (breathing changes, changes in electrodermal activity) in persons stationed at remote locations and shielded from all conventional interactions. The methodology of these experiments included: (a) the use of selected, highly hypnotizable subjects, (b) objective recording (by means of kymographs), (c) mechanical randomizers, (d) statistical analyses of results, (e) sensory isolation, (f) electromagnetic shielding, and (g) variation of the distance between the influencer and the influencee (distances from 20 meters to 1,700 kilometers were used). The general findings were: (a) the demonstration of positive results, (b) the finding that the effects survived iron-, lead-, and Faraday-chamber screening, and (c) the identification of important psychological factors that could impede or facilitate the effects.

During this same time frame, similar investigations were being carried out in other countries. There were French experiments on inducing hypnosis at a distance (by Joire, Gibert, Janet, and Richet), Dutch experiments on the remote influence of motor acts (by Brugmans at Groningen), hypnotic experiments on "community of sensation" (in which a sensory experience of the hypnotist appeared to be experienced by the hypnotized subject), and international studies of telepathy and clairvoyance (see Vasiliev, 1976, for a discussion of some of these studies).

Ever since I read Vasiliev's (1963) monograph, I have been intrigued by his experiments and curious about whether it would be possible to replicate them. I was particularly interested in his experiments of remote mental influence of physiological activity. Through the interest and support of the Mind Science Foundation, my coworkers and I have indeed been able to replicate some of Vasiliev's work, and it is these experiments that I shall now summarize for you. Although we have studied remote influence effects upon several behavioral and physiological response systems (see Braud, Schlitz, and Schmidt, 1989), I shall restrict my comments to a series of experiments on remote mental influence of phasic electrodermal activity. We have completed fifteen experiments using the same general experimental design and methodology. Since my purpose today is to describe the method itself and the overall results, I shall not present the rationales, details or specific outcomes of the individual experiments; such detailed information may be found in our published reports (see Braud and Schlitz, 1989).

In these experiments, a subject sits in a comfortable room while his or her spontaneous skin resistance responses (SRRs) are monitored continuously by means of electronic equipment interfaced with a microcomputer. These SRRs reflect the degree of activation of the subject's sympathetic nervous system and, hence, the subject's degree of emotional, cognitive, or physical activation or arousal. Higher SRR activity is, of course, associated with physiological activation, whereas lower SRR activity reflects relaxation and calmness. In a separate, distant room (typically 20 meters away), the experimenter is stationed with another person, the "influencer." Floor plans of the research areas are given in Figures 1 and 2. The ongoing SRR activity of the distant subject is displayed to the influencer by means of a polygraph (chart recorder) and also is objectively and automatically assessed by the computer system. The influencer watches the polygraph as she or he attempts to exert a remote mental influence upon the distant subject. Influence attempts are made during ten 30-second periods; these are randomly interspersed among ten 30-second control or baseline periods during which no influence is attempted. The subject, of course, is unaware of the nature, timing, and scheduling of these periods, and is physically isolated from any conventional energetic or informational signals from the influencer. Thus, the protocol completely eliminated suggestion and expectancy effects.

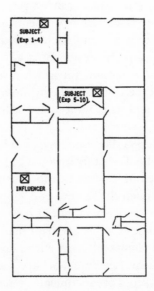

Fig. 1: Laboratory floor plan showing locations of subject and influencer for Experiments 1 through 10.

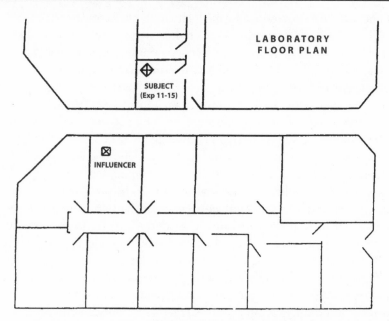

Fig. 2: Laboratory floor plan showing locations of subject and influencer for Experiments 11 through 15; subject and influencer rooms are in separate suites of the same building, separated by an outside corridor and several closed doors.

The aim of the influence is to either calm, activate, or not influence the distant subject according to a prearranged random schedule. During calming attempts, the influencer relaxes and calms himself or herself, intends and gently wishes for the subject to become calm, and visualizes or imagines the subject in a relaxing, calming setting. During activation attempts, the influencer tenses his or her own body, intends and wishes for the subject to become more active, and images the subject in activating, energizing or arousing settings and situations. During the noninfluence control periods, the influencer attempts to keep his or her mind off of the subject and to think about matters unrelated to the experiment. The influencers may use the polygraph tracings as feedback to indicate how well their influence attempts are succeeding. They may try out different mental strategies, abandon unsuccessful ones, and add variations to those that appear to be successful. Alternatively, they may proceed without such feedback and simply close their eyes and intend and visualize the desired outcomes. We have found that both feedback and nonfeedback strategies are effective.

For each experimental session, the subject's total SRR activity during each 30-second recording epoch is determined by means of an analog-to-digital converter interfaced with the microcomputer. The equipment samples the subject's SRR activity 10 times each second (which is quite adequate for a slowly changing reaction such as skin resistance) for the 30 seconds of a recording epoch and averages these measures, providing what is virtually a measure of the area under the curve described by the fluctuation of electrodermal activity over time (i.e., the mathematically integrated activity). Each session, therefore, yields ten quantities of electrodermal activity during the remote mental influence periods and ten quantities of electrodermal activity during the noninfluence, control periods. It would be possible to statistically compare the ten influence scores with the ten control scores for a given subject. However, because the scores may not be independent (i.e., may be autocorrelated), we use the more conservative strategy of reducing an entire session's activity to a single score (a type of "majority vote" score) that reflects the manner in which the subject's total electrodermal activity is distributed during the session, that is, the percentage of the subject's total activity in the predicted direction that occurs during the entire set of influence epochs; this can be contrasted with the activity occurring during the entire set of control epochs. In the absence of a remote mental influence effect, these two scores should approximate each other, that is, their expected values should be 50 percent. For a given experiment, the percent influence scores (a single score for each subject contributing to that experiment) are compared statistically with mean chance-expectation of 50 percent using single mean t tests. A schematic representation of the events of an experimental session is given in Figure 3.

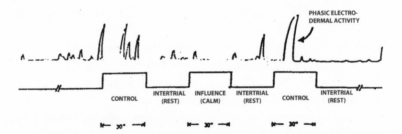

Fig. 3: Schematic representation of the events of an experimental session, along with a representative phasic electrodermal activity tracing.

Thus far, we have completed 15 electrodermal remote influence experiments, with the number of subjects in each experiment ranging from 10 to 40. In all, there have been 323 sessions conducted with 271 different subjects, 62 influencers, and 4 experimenters. The experiments have yielded evidence consistent with the hypothesis that one person may exert a remote mental influence upon another person's physiological activity. Thirteen of the 15 studies yielded overall results in the expected direction. Six of the 15 experiments (40 percent) were independently significant statistically (i.e., had p values less than .05); this is to be compared with the 5 percent experiment success rate expected on the basis of chance alone. Fifty-seven percent of the individual sessions were successful (i.e., yielded results in the expected direction); this is to be compared with the 50 percent session success rate to be expected on the basis of chance. When the series as a whole is analyzed using a recommended method for combining z scores of similar experiments (Rosenthal, 1984), an overall Stouffer $z = 4.08$, with an associated $p = .000023$, was obtained. Effect sizes were calculated using the "Cohen d" measure (in which the value of the significance test is divided by the square root of the number of scores contributing to that test)—a method recommended by those interested in the meta-analysis of scientific experiments. The mean effect size is 0.29, which compares favorably with effect sizes typically found in biomedical and behavioral research. Results are summarized and depicted graphically in Figure 4.

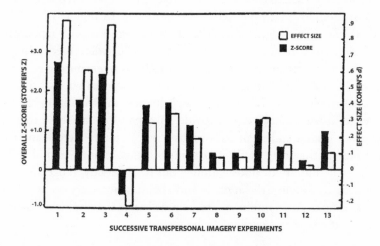

Fig. 4: Overall z scores and effect sizes (Cohen's d measures) for the successive remote mental influence experiments.

The overall results indicate that significantly more phasic electrodermal activity of the prescribed type (i.e., more activity during activation-aim periods, less activity during calm-aim periods) occurred during remote mental influence periods than during comparable control periods. The experimental design guaranteed that the obtained effects could not be attributed to conventional sensorimotor cues, common external stimuli, common internal rhythms, or chance coincidence. Neither can the results be explained in terms of various other potential artifacts or confounds; these are considered in detail and dismissed in Braud and Schlitz (1989). Thus, the results reflect an anomalous psychophysical interaction between two individuals separated from one another in space.

It is important to note that the experiments I've just reported were not carried out with special subjects. Unlike Vasiliev, we did not work only with carefully selected, highly hypnotizable subjects. Rather, we worked with anyone in the community who was interested enough to volunteer to participate in the studies. Similarly, the influencers represented a cross section of the community and were not selected on the basis of special skills or experiences. The fact that we were able to observe significant results in these unselected subjects, and often in persons attempting the task for the very first time, suggests that we are dealing with a common, widespread ability—one that is, perhaps, normally distributed within the general population. It would be of great interest, however, to conduct special experiments with influencers and subjects selected for high hypnotizability in order to approximate more closely the remarkable results reported by Russian and French investigators during the early decades of this century.

These electrodermal experiments may be viewed as successful conceptual replications of the distant mental suggestion studies reported by Vasiliev (1976). In turn, there have been several recent conceptual replications of our work (Gruber, 1979, 1980; Kelly, Varvoglis and Keane, 1979; Khokhlov, 1983). To our knowledge, however, there have been no exact replications by other laboratories of the electrodermal studies just described. One of my motives for summarizing this research here today is to encourage such replications by independent investigators. I am pleased to note that several members of this Association already have expressed interest in attempting to replicate these experiments. This would provide an expanded data base that would increase our understanding of the various physical, physiological, and psychological factors that may facilitate or impede this remote mental influence effect, provide important information about its generality, range and possible limits, help us develop the models and theories necessary to an improved understanding of these findings, and help delineate the implications and possible applications of this curious phenomenon.

Notes

1. Thirteen experiments were conducted. In those experiments, there were 15 opportunities for the remote mental influence effect to be tested. For simplicity, in this presentation, the term "experiment" is used to describe these 15 test opportunities. For Figure 4, mean z scores and mean effect sizes are shown for the 13 experiments themselves; Experiment 5 contained two test opportunities, as did Experiment 13.

2. The analog-to-digital converter was used in Experiments 5 through 15; in Experiments 1 through 4, the chart recordings were manually scored in a comparable manner by an assistant under blind conditions.

3. Using a single mean t test in this fashion is identical to using a matched (pairwise) t test to directly compare the influence versus control scores. Distribution tests indicated that the score distributions were appropriate for the use of parametric tests. When the scores were analyzed using comparable nonparametric tests, results were virtually identical to those described here.

References

Braud, W., and Schlitz, M. (1989). A methodology for the objective study of transpersonal imagery. *Journal of Scientific Exploration*, 3, 43–93.
Braud, W., Schlitz, M., and Schmidt, H. (1989). Remote mental influence of animate and

inanimate target systems : A method of comparison and preliminary findings. *Proceedings of Presented Papers, 32nd Annual Parapsychological Association Convention, San Diego, CA,* pp. 12–25.

Brazier, M. A. B. (1961). *A History of the Electrical Activity of the Brain: The First Half-Century.* London: Pitman Medical Publishing Co., pp. 110–115.

Gregory, A. (1976). Introduction. In L. L. Vasiliev, *Experiments in Distant Influence.* New York: Dutton, pp. vii–xl.

Gruber, E. R. (1979). Conformance behaviour involving animal and human subjects. *European Journal of Parapsychology,* 3, 36–50.

Gruber, E. R. (1980). PK effects on pre-recorded group behavior of living systems. *European Journal of Parapsychology,* 3, 167–175.

Kelly, M. T., Varvoglis, M., and Keane, P. (1979). Physiological response during psi and sensory presentation of an arousing stimulus. In W. G. Roll (Ed.), *Research in Parapsychology 1978.* Metuchen, NJ: Scarecrow press, pp. 40–41.

Khokhlov, N. (1983). Remote biofeedback in voluntary control of heart rate. *Psi Research,* 2, 66–92.

Pavlov, I. P., (1952). Lektsii po fiziologii (Lectures in physiology), Vol. V of his Polnoe sobrani sochinenly (Complete Collected Works), p. 520. cited in L. L. Vasiliev (1965). *Mysterious Phenomena of the Human Psyche.* New York: University Books, p. 158.

Roll, W. G. (1960). Book review of *Psyche* by Hans Berger. *Journal of Parapsychology,* 24, 142–148.

Rosenthal, R. (1984). *Meta-analytic Procedures for Social Research.* Beverly Hills, CA: Sage Publications.

Vasiliev, L. L. (1962). *Experiments in Mental Suggestion.* Leningrad: Leningrad State University. (in Russian)

Vasiliev, L. L. (1963). *Experiments in Mental Suggestion.* London: Institute for the Study of Mental Images. (English translation)

Vasiliev, L. L. (1976). *Experiments in Distant Influence.* New York: Dutton. (Revised and updated English translation)

6

On the Use of Living Target Systems in Distant Mental Influence Research

William Braud

This chapter is one of the most thorough reviews of the process of direct mental interactions with living systems. It addresses the phenomenon itself, evidence for its existence and nature, influencing factors, its historical precursors, and its implications and possible practical applications.

This information originally was published in Braud, W. G. (1993). On the use of living target systems in distant mental influence research. In L. Coly and J. D. S. McMahon (Eds.), Psi Research Methodology: A Re-examination *(pp. 149–181). New York: Parapsychology Foundation. The Parapsychology Foundation is the copyright holder of* Psi Research Methodology: A Re-examination, *Proceedings of an International Conference Held in Chapel Hill, North Carolina, October 29–30, 1988. Edited by Lisette Coly and Joanne D. S. McMahon, published in 1993. Used with permission.* —*William Braud*

The more it moves, the more it yields.

—Lao Tsu

For several years, my co-workers and I have been exploring the use of living target systems in our research on distant mental influence. We have found that living systems possess many characteristics that make them exceedingly attractive and useful to the psi researcher. I would like to share with you some of my thoughts about the positive features of such systems and their many advantages for psi research in all of its aspects—experimental, theoretical, and practical.

The Theme and Its Variations

In a typical experiment of the type to be discussed, one selects a living organism and isolates it, usually at a distance, from all conventional sensori-motor or energetic influences of the "influencer." In principle, any living organism may serve as this "target." Next, one selects some readily measured aspect of the target organism's activity, objectively monitors that activity over a period of time, and generates a permanent record of that activity. It is, per-haps, desirable to choose an activity which occurs with moderate frequency or intensity and which is relatively stable over time, although this is not an essen-tial requirement.[1] An "influencer" then attempts to influence the organism's activity, mentally and at a distance, in a prescribed fashion and according to a predetermined (and, ideally, random) schedule. Conventional statistical meth-ods are used to compare the organism's activity during periods of attempted mental influence with activity levels during comparable non-influence, control periods.

The procedure may be illustrated more concretely by an experimental technique that we have used over 300 times in our laboratory. A subject sits in a quiet, comfortable room for approximately 25 minutes while his or her sym-pathetic autonomic activity is continuously assessed by computer-monitoring of the subject's electrodermal activity (EDA). In another room, typically 20 meters away, an influencer attempts to mentally "calm" the distant subject dur-ing ten 30-second influence periods but not during ten interspersed control periods. Total EDA during the influence periods is compared statistically with total EDA during the control periods in order to determine whether the exper-iment was successful.

This is merely the latest version of an experimental procedure for which there is a long and interesting history. Some of the earliest variations of the par-adigm occurred in the contexts of "animal magnetism," "mesmerism," and hypnosis. As early as 1775, Franz Anton Mesmer described informal experi-ments in which he claimed successful action-at-a-distance effects, explaining them in terms of propagations in a "universal fluid" connecting all things (Mesmer, 1775, 1779, 1799). Similar distance and nonsensory effects were reported to occur in the *somnambules* of Amand-Marie-Jacques de Chastenet, Marquis de Puysegur, one of the chief French disciples of Mesmer. According to Puysegur, his "artificial somnambulism" effects, which he published in his 1784 memoirs, depended importantly upon the mesmerist's firm belief, faith, and confidence in his own powers, and in his strong wanting or willing of the desired effects. Distant "mesmeric" effects were reported by James Esdaile (1852) in India, and by John Elliotson (1843)[2] and Chauncey Hare Townshend (1844) in England.[3]

Perhaps the best known, and best controlled, of these early distant influ-ence attempts were the remarkable *le sommeil a distance* experiments carried out in Le Havre in 1885 and 1886 by Joseph Gibert and Pierre Janet with the special subject, Leonie B. (Janet, 1886a, 1876b), some of which were witnessed by psychical researchers Frederic W. H. Myers, Charles Richet, and Julian Ochorowicz. Richet (1888) later reported his own similar successful experiments with Leonie and with other subjects.[4] P. Joire (1897), in Lille, conducted experiments on mental suggestions of specific motor acts.

Between 1920 and 1922, the once heralded but now, unfortunately, neglected experiments of Brugmans, Heymans, and Weinberg were conducted at the University of Groningen in the Netherlands. These experiments, which involved distant mental influence of motor actions in a special subject, A. S. van Dam, have been re-assessed by Schouten and Kelly (1978).

Controlled experiments were conducted at the Institute for Brain Research of the University of Leningrad in an attempt to determine whether complex motor acts in dogs (specially trained for "will-less" obedience) could be influenced by mental suggestion in the absence of sensory cues. The subjects for these experiments, which yielded suggestive but not entirely satisfactory results, were the trained dogs of Vladimir Durov, a celebrated circus clown and dog trainer; the experimenters were the respected reflexologists, Vladimir M. Bechterev (1920) and A. G. Ivanov-Smolensky (1920).[5]

Between 1921 and 1938, the mental influence research at the Institute for Brain Research shifted its emphasis from dogs to humans, and experiments were carried out, primarily under the direction of Leonid L. Vasiliev, which constitute what is perhaps the most impressive, systematic research on distant mental influence ever to be conducted. Vasiliev and his colleagues found positive evidence that "sleeping" and "waking," swaying, and a variety of motor reactions could be mentally influenced from a distance. A complete account of these experiments was not published until 1962 (Vasiliev, 1962).

This book was translated into English, under the aegis of C. C. L. Gregory and Anita Kohsen (Gregory), and published in 1963 under the title, *Experiments in Mental Suggestion*. A revised edition appeared in 1976 as *Experiments in Distant Influence*. It is noteworthy that in each and every instance of the research reviewed thus far, distant mental influence was attempted *only with special subjects*. The "hypnosis at a distance" trials were carried out with subjects who were already known to be especially susceptible to hypnotic influence, and the dogs studied in Leningrad had been specially trained, beforehand, for "will-less" obedience.

Following a hiatus of about a decade, additional reports of distant mental influence began to appear. These experiments almost inevitably involved the distant mental influence of more "primitive" biological systems and were typically carried out in two new contexts: psychokinesis and healing. These investigations now become too numerous to be mentioned individually. Fortunately, they have been well reviewed by Solfvin (1984) and by Benor (1984, 1985, 1988). Both selected and unselected participants attempted to mentally influence the growth or viability of bacteria, fungus colonies, yeast, and plants, or to influence the movements of protozoa, larvae, woodlice, ants, chicks, mice, rats, gerbils, and cats. Some experiments involved attempts to influence cellular preparations (blood cells, neurons, cancer cells) or enzyme activity.

A very small number of researchers continued to conduct human influence experiments in the French and Russian traditions. Douglas Dean (1964) reported successful attempts to influence the direction of eye movements (recorded electrophysiologically) in dreaming "target persons." Hiroshi Motoyama (1977) reported successful attempts by one person to influence physiological activities (EDA, plethysmographic activity, respiration) of another person while the two persons were isolated in separate, lead-shielded rooms. In our own laboratory, we have observed successful distant mental influences by one person of the EDA, pulse rate, muscular tremor, ideomotor activity, and mental imagery of another person when deliberate attempts were made to influence these specific reactions (Braud, Davis, and Wood, 1979; Braud and Jackson, 1982, 1983; Braud and Schlitz, 1988). Elmar Gruber (1979, 1980) reported successful attempts to influence the locomotor behaviors of "target persons." Jule Eisenbud (1983) reported the tantalizing results of his attempts to issue mental commands for distant individuals to telephone him.

Of relevance to the mental influence theme are those investigations in which physiological reactions were used as psi indicators—e.g., Dean's (1962) plethysmography studies, Tart's (1963) EEG and GSR studies, the EEG influence studies of Lloyd (1973) and of Targ and Puthoff (1974), etc. Reviews of these studies may be found in Beloff (1974), Millar (1979), Morris (1977), and Tart (1963).

This brief survey was presented in order to indicate the range of phenomena to be addressed in this paper. Common to all of the observations is an influence by one person upon another person (or upon another living system). A number of terms have been offered as descriptions (or, unfortunately, even as explanations) of these interactions. When the experimental outcome involved symptoms of hypnosis, the terms "le sommeil a distance" and "telepathic hypnotization" naturally suggested themselves. When hypnosis was no longer the desired outcome, those terms gave way to "telepathy at a distance" or "active agent telepathy." "Telergy" and "living target psychokinesis" followed in the wake of PK experimentation. Rex Stanford (1974) suggested "mental or behavioral influence of an agent (MOBIA)." In our own work, we first suggested the term "allobiofeedback" (see Braud, 1978), since, in our initial experiments, one person received feedback for an attempted biological influence of another person. We subsequently observed that the provision of feedback was not necessary to the occurrence of the effect and coined the term "bio-psychokinesis (bio-PK)" to describe what we were interpreting as psychokinetic influences upon living systems (see Braud and Schlitz, 1983). In presenting this work at an international conference devoted to imagery (Braud and Schlitz, 1987), we used the term "transpersonal imagery effect," following the lead of psychologist Jeanne Achterberg (1985) who suggested "transpersonal imagery" for a possible effect upon one person of the imagery of another. However, imagery does not seem to be required for the effect; non-imagistic *intention* may suffice. Hence, I considered "transpersonal imagery or intentionality effect" (Braud, 1987). This term is much too cumbersome, and I am abandoning it in favor of the much more straightforward "distant mental influence," which seems to convey the essence of the interaction with minimal surplus meaning. In a later section of this paper, I shall consider in more detail the various processes or mechanisms that might underlie these phenomena.

Advantages

I shall discuss four important advantages of using living target systems in distant mental influence research.

• The findings of research with living target systems potentially have great *relevance* to important and meaningful human processes such as healing and social influence.

• *Motivation* is heightened in participants in living target experiments.

• Distant mental influence of living systems has a certain *plausibility* that experiments on influence of inanimate systems do not possess.

• Living systems may be particularly *appropriate* as detectors of psi influence.

Relevance

The findings of distant mental influence research with living target systems may have important implications for healing and for social influence. If arbitrarily selected physiological processes of living organisms can be influenced mentally in an arbitrary manner in experimental contexts, does this not suggest that, in principle, similar mental influences could be directed to bodily organs, tissues, or cells in a manner favorable to health and well-being? If behavioral tendencies can be influenced in the laboratory, does this not suggest that decisions, behaviors, and social actions could be influenced psychically in everyday life? There are a number of "pathways" through which distant mental influence might bring about healing or other practical effects.

1. Physiological or biochemical processes in one person might be directly influenced by another person. Such an influence could correct an imbalanced or diseased process or could forestall possible medical problems.

2. The mental influence of one person might trigger the self-healing capabilities of another person; the influence might instill in the latter an awareness of a problem, an increased wish to initiate or increase self-healing, or simply provide an opportunity to engage in more efficient self-healing.

3. The mental influence might provide increased motivation for self-healing through instilling greater feelings of self-worth, increased knowledge of reasons for self-healing, or simply an increased awareness that significant others truly desire and expect one's improvement.

4. Conceivably, distant mental influence could remove physical or psychological impediments to the physical healing process.

5. Especially in situations in which two or more decisions have approximately the same likelihood, a distant mental influence could bias the decision-making process toward or away from one of the choices. Such processes as walking or driving patterns, purchasing behavior, voting tendencies, and so on might be influenced in this way. The consequences of such influenced decisions could be trivial or profound.

6. Paradoxically, distant mental influence could be used to instill or strengthen attitudes of self-responsibility or internal locus of control which could minimize the subsequent suggestive influences of others. This would be analogous to hypnotizing someone, then suggesting that he or she could no longer be hypnotized.

7. Distant mental influence conceivably could influence the accessibility of memories, alter perceptions, feelings, or the timing and sequencing of actions; such alterations could, in turn, have trivial or profound individual or social effects.

Motivation

The motivation to succeed in living target distant influence experiments is likely to be greater than would be the case in inanimate target experiments. To many people, living targets themselves would appear more interesting or appealing than inanimate targets. The implications of success might be clearly or dimly perceived, and, provided such implications are not construed as threatening, this knowledge could heighten motivation in subjects and experimenters alike. Motivation would be especially enhanced if the experiment were conducted in a psychic healing context, since the benefits or implications of success would be readily apparent. The perceived relevance of a living target experiment immediately would endow it with meaningfulness and importance and lift it far above its possible consideration as a mere laboratory game or curiosity.

Plausibility

In discussing the plausibility of living target studies, we may distinguish two types of plausibility: (a) plausibility to the subject, and (b) theoretical plausibility.

Subject plausibility. All of us have had considerable experience in influencing living systems through ordinary means. We continually influence other people and animals through our words and actions. We are aware that we can influence our muscular movements, breathing, feelings, and perhaps even our autonomic reactions through volition, imagery, and the generation of specific kinds of thoughts. We have heard that bodily control may be enhanced through hypnosis, biofeedback, and meditation. We are familiar with the notion that people can become aware of being stared at by an unseen person and may even have personally experienced this phenomenon. Our families, our friends, and the media have exposed us to the notion of telepathic influence. Given these sorts of experiences, the prospect of mentally influencing another person or an animal in a psi experiment does not seem excessively alien or implausible.

We do not have such a network of supporting experiences in the case of awareness or influence of inanimate objects, and it is not surprising that we would be filled with feelings of confusion, uncertainty, doubt, and pessimism when confronted with the task demands of clairvoyance or inanimate-target PK experiments. To the extent that attitudes influence psi performance, living target experiments would be expected to have advantages over inanimate target experiments.

Theoretical plausibility. To paraphrase George Orwell, all psi tasks are impossible, but some are more impossible than others. The simple exercises of wiggling our fingers or reviving specific memories will convince us that our minds influence our own brains. We have no genuine understanding of how these mind-body interactions come about. They are impossible. Nonetheless, they happen all the time, and their familiarity has pushed far from our awareness any disturbing thoughts or concerns about their impossibility. If my mind can influence my own brain, perhaps it can influence other, similar brains as well, even if those similar brains are outside my physical body. Stated somewhat differently, if my mind can influence my brain while the latter is inside my skull, perhaps my mind can continue to influence this neural tissue *or similar biological material* even if the latter is removed and maintained outside my skull or body. Further, the degree to which my mind can continue to exert its influence on a distant target system may depend upon the similarity of that system to my familiar brain. On the basis of theoretical speculations such as these, distant mental influence of living targets becomes more plausible than the influence of inanimate systems. Could the same process underlie both familiar volitional actions (such as muscular movements and memory constructions) and the less familiar volitional actions that we know as "psychokinesis"?

This idea is not a novel one; it has been advanced on several occasions in the history of psychical research. Louisa Rhine (1970) and John Beloff (1979) presented the idea as it was formulated in the 1940s and 1950s by J. B. Rhine (1943, 1947), Thouless and Wiesner (1946, 1947), and J. C. Eccles (1953). Rhine and Thouless were seeking to understand the place in nature of the PK "force" or process which had just been demonstrated by the freshly reported dice-influence experiments; Eccles was attempting to develop a neurophysiological explanation of the action of the will. D. Scott Rogo (1980) reminded us that the idea of PK as a "force" that normally regulates events within the body had been proposed as early as 1909 by Hereward Carrington. Carlos Alvarado (1981) traced the idea back to 1874 and to Serjeant-at-Law E. W. Cox, the pre-S.P.R. psychical researcher who assisted William Crookes in the latter's investigations of the "psychic force" (a term coined by Cox) exhibited in physical mediumship. The following sampling conveys the flavor of these thoughts on the possible identity of PK and "ordinary" volition.

• John Beloff (1979): ". . . can PK be regarded as the extrasomatic (and hence paranormal) extension of what, in ordinary volitional activity, is endosomatic (and hence normal)?" (p. 99)

• Evan Harris Walker (1975): ". . . the action of the consciousness to secure the collapse of the state vector has the physical consequence of determining the subsequent states of that system in a manner that corresponds to the concept of the 'will.' . . . Since the brain is responding to sensory input from events external to the body, physically the brain is tied to and is thus a part of a larger physical system incorporating the external world. Whichever state the brain goes into, it must be one consistent with the state the external world . . . enters. As a result, specification of the w_i [will] variables can effect a change in the state of both the brain and events external to the body." (p.8, 9)

• John Eccles (1953): ". . . a special property . . . is exhibited by the dynamic patterns of neuronal activity that occur in the cerebral cortex during conscious states, and the hypothesis is developed that the brain by means of this special property enters into liaison with mind, having the function of a 'detector' that has a sensitivity of a different kind and order from that of any physical instrument . . . at any instant the 'critically poised neurones' would be the effective detectors and amplifiers of the postulated action of the 'will' . . . 'will' modifies the spatio-temporal 'fields of influence' that become effective through this unique detector function of the active cerebral cortex. . . . It will be agreed with Rhine (1948) that, if the so-called psi capacities (psychokinesis and extrasensory perception) exist, they provide evidence of slight and irregular effects which may be similar to the effects which have here been postulated for brain-mind liaison, where they would occur in highly developed form." (p. 267, 275, 277, 284)

- Thouless and Wiesner (1947): "We wish to suggest . . . that these [paranormal cognition and psychokinesis] are merely unusual forms of processes which are themselves usual and commonplace, and that in their usual and commonplace form, they are to be found as elements in the normal processes of perception and motor activity. . . . I control the activity of my nervous system (and so indirectly control such activities as the movements of my body and the course of my thinking) by the same means as that by which the successful psychokinetic subject controls the fall of the dice or other object." (p. 195, 197)

- J. B. Rhine (1943): "The mind or subjective self in its domination of the body exercises a causal influence which cannot be otherwise than kinetic. Thus psychokinetic action . . . is the basis on which every man interprets his routine experience of daily life." (p. 70)

- Hereward Carrington (1909): "Now, if mind exists apart from the brain and merely utilizes it to manifest through, it is acting upon it by a species of telekinesis all the time! Every mental state and change—accompanied, as it doubtless is, by molecular action, chemical changes, etc.—is the result of a telekinetic action! There should be no very great difficulty in imagining consciousness capable of affecting the outside material world, therefore." (p. 295)

- Sydney Alrutz (1909): "What is happening in the brain . . . when we move an arm by means of an act of will? . . . Are these entirely electrical and chemical forces? . . . Might there not be . . . some form of energy more closely allied to the psychic acts, constituting a sort of bridge or transition between psychic phenomena . . . and electrical and chemical phenomena?" (cited in Carrington, 1921, p. 114–115)

- F. W. H. Myers (1886, 1903): ". . . perhaps when I attend to a thing, or will a thing, I am directing upon my own nervous system actually that same force which, when I direct it on another man's nervous system, is the 'vital influence' of mesmerists, or the 'telepathic impact.' . . ." (1886, p. 127–128); ". . . the telekinetic force . . . is generally . . . a mere extension to a short distance from the sensitive's organism of a small part of his ordinary muscular power." (1903, vol. 2, p. 208)

- E. W. Cox (1874): "The theory of *Psychic Force* is in itself merely the recognition of the now almost undisputed fact that under certain conditions . . . a Force operates by which . . . action at a distance is caused. . . . As the organism is itself moved and directed within its structure by a Force which either is, or is controlled by, the Soul, Spirit, or Mind . . . , it is an equally reasonable conclusion that the Force which causes the motions beyond the limits of the body is the same force that produces motion within the limits of the body." (p. 101)

Honorton and Tremmel (1979) and Varvoglis and McCarthy (1986) have recently begun to develop potentially useful empirical methods for exploring the volition/PK theory.

Appropriateness

Living target systems possess a number of characteristics that make them especially susceptible to distant mental influence and hence quite useful as "psi detectors."

Lability. Early empirical PK work (see Rush, 1976), the various quantum mechanical and noise-reorganization models of PK (see Oteri, 1975; Puharich, 1979), Stanford's (1978) conformance behavior model, and my own lability/inertia model (Braud, 1981) all predict greater psi influences upon random or labile systems than upon non-random or inert systems. Since living systems possess a great deal of lability or free variability, they would seem to be excellent candidates for sensitive and effective detectors of distant mental influence. I was delighted to find the following unexpected passage in one of Carrington's (1921) volumes. The passage is from a presentation of Sydney Alrutz to the Sixth Psychological Congress which met in Geneva in August, 1909.

> When we wish to study the electrical charge contained in any body, we obtain exactitude only when we succeed in transferring this charge to another body; we may then study the nature of the charge under varying circumstances, and establish the influence of the two charges upon one another. It is only in this way that experimentation becomes truly fertile. Should we not apply the same laws to the phenomena of the nervous system, and institute a similar mode of experiment for the nervous energies? Under what conditions can we conceive this transference?
>
> The most natural supposition seems to be that it would occur, if at all, in labile organizations; in those subjects which, according to Janet . . . possess an excessively unstable personality; and whose psychic life is characterized by great suggestibility, by instability, and a certain peculiar mobility. Such individuals are also characterized by the great facility with which the functions vary and react upon one another. Binswanger has said that the nervous system of these individuals is characterized by the variability of the dynamic cortical functions; that is to say, by the fact that the nervous segments of their cerebral cortex present a *melange* of greater or lesser irritability. (p. 115)

It was pleasing to find such an early, yet relatively accurate, statement of the lability idea. An important warning should be inserted at this point. The term "lability" is sometimes used in psychology and psychiatry to indicate extreme reactivity or extreme variability of mood or behavior. My own use of "lability" should not imply such an extreme case. I use the term to indicate flexibility, ease of expression, freedom to change, free variability. A system that is excessively active is too "driven" to be modulated in an efficient manner; it is constrained by its own overactivity. Such excessively active systems might be as insusceptible to psi influence as would be excessively sluggish or inert systems.

Perhaps the usage of lability that comes closest to my own is that of I. P. Pavlov, who used the expression in his classification scheme of the "types of nervous systems" of his experimental animals. Pavlov and his co-workers classified their dogs on the basis of three major dimensions: *strength, equilibrium,* and *mobility.* "Strength" referred to the "working capacity" of the cerebral cells, their resistance to powerful external disruptors and stress, and their resistance to the development of a kind of brain-protecting "transmarginal inhibition" due to excessive stimulation or excessive environmental demands. "Equilibrium" referred to the balance of excitatory and inhibitory tendencies. "Mobility" or "lability" referred to the ease and speed with which behavior and brain processes could shift from one state to another to keep pace with changing environmental demands. A labile nervous system changed rapidly in response to a stimulus and rapidly returned to its prior state upon the removal of the stimulus. A good synonym for lability might be "speedy appropriateness" of responding (this synonym was not used by Pavlov). The various combinations of these three dimensions yielded several "types" of nervous systems, and these types were observed to respond quite differently to environmental stimuli, conditioning demands, pharmacological agents, spontaneous stressors, etc. Although developed initially to help understand the varied reactions of dogs in the experimental study of the physiology of "higher nervous activity," the typology concept was subsequently extended to human behavior and to the area of psychopathology. Relevant information may be found in Pavlov (1927, 1957), Gray (1964), Kaplan (1966), Lynn (1966), and Sargant (1957). The Pavlovian typology issue is an exceedingly complex one, and, perhaps for that reason, it has suffered unfortunate neglect. Many of the Pavlovian concepts have important similarities to the introversion/extraversion and neuroticism constructs of Hans Eysenck (e.g., Eysenck, 1967a, 1967b). Eysenck's personality theory is, in a way, a blending of the typologies of Ivan Pavlov and of Carl Jung. In view of recent parapsychological interest in introversion/extraversion (e.g., Palmer, 1977) and in the Myers-Briggs Type Indicator, which is, of course, based upon Jung's typology (see Berger, Schechter and Honorton, 1986; Honorton, Barker, Varvoglis, Berger and Schecter, 1986; Schmidt and Schlitz, 1988), a careful and systematic exploration of Pavlov's typology in relation to psi influence might prove quite productive.

We have been considering the possible psi-conduciveness of the *physical* lability of target systems. Physical lability is almost inevitably accompanied by *perceived* lability, and this latter factor may have an important *psychological* influence upon an investigation's outcome. In any experiment which provides trial-by-trial feedback to the subject, the subject is necessarily aware of the ongoing changes in the state of the target system. The knowledge that develops during the course of the experiment that the target system *can* change, and indeed is changing, may increase the subject's belief or confidence that a distant influence upon the target system is indeed possible; these attitudinal shifts may, in turn, affect psi scoring. Even in distant influence experiments in which immediate feedback is not provided, the simple *knowledge* that a changing living system is involved may have psi-favorable psychological effects. It would be possible to disentangle the usually confounded effects of physical lability and perceived or known lability through the use of special experimental designs in which these factors are manipulated independently and blindly; however, such investigations have not yet been carried out.

Living systems as detectors/amplifiers. There is a tendency to think of *detectors* as inanimate, physical devices that respond to the presence of particular materials or energies. However, living organisms themselves can function as exquisitely sensitive detectors of subtle energies and of extraordinarily low concentrations of materials. In biology and in medicine, biological preparations are sometimes used to detect the presence or amount of a substance for which physical detectors have not yet been developed. Such "bioassays" may continue to be used even after the development of appropriate physical detectors. Bioassays have been, and continue to be, useful in the discovery of various hormones, vitamins, and neurotransmitters. Otto Loewi's (1921) use of heart muscle preparations in the detection of "vagus substance" (later identified as acetylcholine) is one of the best known applications of the bioassay.[6] The technique could be extended to yield a "behavioral bioassay" in which observations of changes in the behavior of intact organisms indicate the presence of some agent or energy. Examples of the behavioral bioassay include observations of changes in the aggressive behavior of Siamese fighting fish in the evaluation of tranquilizing drugs (e.g., Walaszek and Abood, 1956) and observations of changes in web-spinning behaviors in spiders in response to minute quantities of LSD, psilocybin and mescaline (Weckowicz, 1967). There have been claims that behavioral bioassays may be used to detect specific memories; however, those claims remain controversial (Braud, 1970; Braud and Braud, 1972; Smith, 1974; Stewart, 1972: Ungar, Desiderio and Parr, 1972).

This material is offered as background to the conjecture that *the human brain may function as a bioassay for mind, and living systems may function as bioassays for psi.* This conjecture is not essentially different from the proposals of Eccles (1953), Dobbs (1967), and Walker (1975) that the brain may be an especially sensitive detector of small influences due to its vast number of interconnecting neurons, some of the synapses of which may be poised at critical levels of excitability. A slight influence at the synapse of a single "critically poised" neuron could lead to a cascade of subsequent neuronal firings that could in turn lead to gross behavioral, physiological, or subjective reactions. Thus, the system could function not only as a detector, but also as an amplifier of subtle mental influence.

It may be possible to construct complex, interactive inanimate systems whose similarity to the brain might allow them to function as mental influence detectors. As early as 1947, Thouless and Wiesner described the requirements of such a physical system.

> . . . the ideal mechanism for studying [psychokinesis] would be one in which very minute forces could start processes in systems of small size, which processes could act as triggers for subsequent processes involving sufficiently large forces to be easily observable. If indeed a physicist could construct for us a mechanism in which there were delicately balanced systems of very small size, which balance could be upset by small forces yet was protected from being upset by small forces accidentally impinging from outside, and if, moreover, any change in these small systems could be automatically magnified to a large energy change in some larger system, then we might hope to have the ideal mechanism for the experimental demonstration of psychokinesis. We have not succeeded in devising such a mechanism in our laboratories. (p. 198)

Today, there exist several physical systems that may indeed satisfy the Thouless and Wiesner requirements. One such system consists of computer-based *neural networks* (see Kelly, 1979; Radin, 1988). Other possibilities would include physical systems of the sort being explored in the new discipline of "chaos" theory and which exhibit unusually sensitive dependence upon initial conditions (see Gleick, 1987; Jantsch, 1980).

Multiple psi channels. It is likely that distant mental influence experiments with living target systems will yield better psi results than similar experiments with inanimate targets because of the greater number of "psi channels" available in the former. Living target systems may "cooperate" in bringing about the desired outcome through the aid of their own telepathic or clairvoyant abilities, combined with intentional or unintentional self-regulation. This would seem especially likely when other persons serve as target systems. In a "bio-PK" experiment, for example, in which my aim is to reduce the EDA of Person A during certain periods, I may achieve that goal by means of a direct PK influence upon A's EDA. However, it is also possible that A will "scan" the experimental environment and discern the pattern of activity that I expect of him or her. Person A could become aware of that pattern through telepathic access to my influence attempt or through clairvoyant access to a physical record of the influence/non-influence schedule. Person A could then produce the desired EDA patterns in herself or himself through autonomic self-regulation. Of course, both PK and telepathy/clairvoyance may be occurring at once, or the two forms of psi may alternate throughout the trials of an experiment. For one of these "channels," the locus of the psi is the influencer; for the other channel, the psi locus is the ostensible "target person."

We are planning a blood pressure bio-PK study that may cast light on this issue of the true psi locus in such experiments. We plan to assess subjects' abilities to self-regulate blood pressure in an initial screening phase of the study. In the next phase, all of the screened subjects will serve as target persons in bio-PK sessions in which an influencer will attempt to influence their blood pressures. We are quite interested in learning whether and how the self-regulation and hetero-regulation scores of these two phases are correlated. If successful bio-PK is due primarily to receptive psi plus self-regulation, one would expect that the most successful sessions would involve subjects who are very good at self-regulation; a strong positive correlation between the Phase 1 and Phase 2 scores would be consistent with this hypothesis. On the other hand, if no correlation or a strong negative correlation between Phase 1 and Phase 2 scores obtains, such an outcome would be more difficult to explain on the basis of this hypothesis. A strong negative correlation or no correlation would be of greater interest than a strong positive correlation, since the latter could also be interpreted as merely an indication that the blood pressure activity of certain subjects is more labile and *generally* influenceable than that of other subjects. It is recognized that no experimental outcome will point conclusively to one or the other of these two interpretations of "true psi locus," and my suspicion is that successful bio-PK experiments include elements of both processes, and that the real locus of the effect is in neither the influencer nor the subject, exclusively, but in an interactive field in which both participate.

Another approach to the self-regulation issue would involve experiments with response levels that vary in the ease with which they can be self-regulated. Motor activities and breathing are relatively easy to consciously self-regulate, while certain autonomically mediated functions (such as foot temperature or blood pressure) are more difficult to voluntarily control. Would bio-PK studies involving the former be more successful than those involving the latter? Again, outcomes will not be conclusive because of the problem of possible skeletal or cognitive "artifacts" (see Katkin and Murray, 1968), but would be of interest nonetheless.

A third approach to this issue would involve varying the phylogenetic or ontogenetic status of the target organism, or testing organisms under conditions that would be expected to influence self-regulation ability. What would be the outcomes of bio-PK experiments in which one attempts the distant mental influence of skeletal responses in: (a) infants who have not yet manifested a great deal of motoric self-control, or (b) persons in REM sleep in which most motoric activity is inhibited, or (c) organisms that are only distantly related to human beings?

We have been assuming that in distant influence experiments, there are indeed changes in the activities of the living target systems that are produced "causally" or "psychokinetically" by the influencer and that would not occur otherwise. In situations in which the *a priori* probability of a particular target reaction is quite low or in cases in which the reaction is relatively complex, this seems to be a quite reasonable assumption. However, in *statistical* experiments in which the target activity has a relatively high probability of occurring naturally, there arises the additional possibility that the influencer *psychically perceives* the present or future activities of the system and schedules his or her "influence" attempts so that they happen to coincide with the system's activities, thus producing an illusion of a causal effect. The precognizing of "favorable" segments of ongoing random events was suggested many years ago by W. E. Cox (Cox, personal communications; Hansen, 1987) as a possible alternative explanation of most radioactivity-based REG "PK" effects. This interpretation has recently been revived, with the new name "intuitive data sorting" or "intuitive data selection" (see Weiner and Nelson, 1987, pp. 136–144).

It certainly is possible that IDS may play some role in some distant mental influence experiments; however, the extent of such influences remains to be determined. Two recent explicit tests of the IDS hypothesis in our own laboratory (Braud and Schlitz, 1987; Braud, 1988) did not yield results consistent with the IDS explanation.

Levels of influence. Another advantage of living target systems in distant mental influence research is that such systems possess multiple "levels" of activity which may be targeted for possible influence. The relative susceptibility of those levels could then be compared. For example, experiments could be designed in which one attempts distance mental influence of the thoughts, images, feelings, behavior, gross physiological activity, molecular physiological activity, biochemical activity, or immunological activity of another person. Would influence attempts be equally successful at those various levels of response? One could even attempt to exert a distant mental influence upon the *psychic* activity of another person.

Experiments could be designed to study the degree of co-variation or dissociation among various levels. What happens at Level X when distant mental influence is directed toward Level Y? Are the outcomes symmetrical or asymmetrical? What happens at Level Y when attempts are made to influence Level X?

Other influences upon the target. Living target systems permit the study of several nonpsi factors that might influence the psi susceptibility of the system. I was tempted to say that living systems are susceptible to more influences than are inanimate systems, but this is not necessarily true (see below). It would be more accurate to say that living and nonliving systems are susceptible to *different* nonpsi influences. It would be possible to study the influence of a certain nonpsi factor upon both influencer and target system. Would the factor have the same influence in these two cases? According to certain models of psi functioning (e.g., my lability/inertia model [Braud, 1981]; Roll's "systems theoretical" model [Roll, 1985]), certain variables (e.g., level of arousal, degree of cognitive constraint, etc.) are expected to have opposite effects upon different psi processes or upon systems with different roles in psi interactions. Living target system research can provide a testing ground for some of these ideas.

Disadvantages

The use of living systems in distant mental influence research is not without its disadvantages.

Logistical Difficulties

Experiments with living target systems may be more complicated than inanimate target system experiments in that the target systems themselves require additional scheduling and maintenance. In planning an REG-PK session, one has only to schedule an influencer. In a bio-PK session involving a human target person, one has to schedule two people, and if one of these persons fails to appear at the laboratory, the session must be cancelled or postponed. If the experiment involves animals, plants, or cellular systems, one must have additional facilities for their housing, maintenance, and preparation. The experimenter also will have to become familiar with the living system and learn its requirements, sensitivities, habits, preferences, etc. All of this makes life more complicated, but also more interesting, for the experimentalist. And what does one do with one's experimental organisms when the experiments have been completed? The most ideal and most humane solution is to borrow one's target creatures from nature for a while, then return them unharmed when the study has been completed.

Manipulation of Life Forms

One must deal with the ethical issue of whether it is proper to influence the actions of other people or of other life forms. We have worked with over four hundred people in our various bio-PK studies, and less than a half-dozen of these expressed any concern about the possibility of influencing or being influenced by another person. If other persons serve as target systems, there would seem to be no ethical problems as long as (a) subjects give informed consent, (b) the planned influences are not deleterious to the target person, and (c) there is proper debriefing in which the likelihood and extent of distant influence are placed in a proper context for the subjects. Some formal or informal screening might be useful in order to eliminate subjects who might deal with the issue of distant influence in an imbalanced manner. As in any other experiment, clinical interaction, or everyday life situation, problems may be avoided if one uses good judgment and common sense. In our own work, we have dealt with possible ethical issues by choosing target reactions that are generally beneficial to our subjects. In the use of animals or plants in distant influence experiments, one could choose target reactions or activities which are not harmful to those organisms. In healing analog studies, participants are sometimes asked to destroy "harmful" organisms such as bacteria or cancer cells. The same arguments and considerations used by those who use pharmacological or other treatments to destroy these organisms in other, usually medical, contexts would be relevant here. Cost/benefit analyses, "greater good" judgments, and personal attitudes will govern final decisions.

Those who are troubled by the "manipulative" aspects of distant mental influence studies might consider whether a psychic "command" to make a particular movement really differs from a similar "command" of an agent to a subject to *think about* a particular target in a card guessing, ganzfeld, or remote viewing experiment. The major difference seems to lie in which influence might be considered to have a stronger possible impact upon the external, physical world. To assert that the psychic production of a muscle twitch is more powerful or more coersive or manipulative than is the production of a "mere" mental image is to ascribe a greater reality status to the former than to the latter. This is certainly questionable. Images are no less real than are muscular movements and may have even more profound environmental and social consequences under certain conditions.

Psi-Missing in Healing Studies

It is my impression that psi-missing occurs less frequently in living target studies than in inanimate target studies. Still, psi-missing has been reported in healing analog studies (e.g., Grad, 1967; Wells and Klein, 1972). This possibility should be kept in mind by anyone considering practical applications of distant mental influence; the procedure could backfire. This possibility is not really surprising, since any treatment (e.g., drugs) can have reversed or "paradoxical" effects under certain conditions, and no treatment is entirely without possible negative side effects. It would seem especially important to explore the conditions which tend to produce psi-missing in living target influence experiments so that those conditions could be avoided in practical application attempts.

Experimental Control

It might seem that controlling extraneous variables would be more difficult for living than for inanimate target systems and more difficult for complex organisms than for primitive ones. This is not necessarily true. The extraneous variables are simply different in these cases and not necessarily more or less numerous. One has but to read Hubbard, Bentley, Pasturel, and Isaacs' (1987) account of the development of their monitoring, isolation, and artifact detection systems for piezoelectric strain gauge PK targets to realize how difficult controlling extraneous variables can be in the case of inanimate target systems.

In the case of living systems, "higher" organisms can filter or screen themselves or compensate for environmental variables which would exert strong influences upon more "primitive" organisms. For example, a subtle temperature change could have a marked influence upon *in vitro* cellular preparations, while even large temperature fluctuations might go completely unnoticed by human laboratory participants. What is a signal at one level of biological development becomes noise at another level, and adaptive pressures have produced and perfected quite efficient noise-cancelling mechanisms.

Statistical Issues

Possible statistical problems unique to living target distant influence research have been discussed by Solfvin (1984) and by Rush (1986). An issue that has not received adequate treatment is the possible *lack of independence* of repeated measurements of the activity of a biological target organism. In electrodermal bio-PK experiments, for example, external conditions or endogenous rhythms could result in relatively long "bursts" of nonindependent activity or inactivity. If successive "samples" of this activity are treated as independent when they are in fact positively correlated over time, statistical tests (such as *t* tests) that assume independent units would be artificially inflated. We have dealt with this possibility in our EDA bio-PK experiments by not treating the many trials of our sessions as units for statistical analysis, but, rather, have collapsed all of the trial activities into a single score for the subject; i.e., the entire session becomes either a hit or a miss (depending upon whether there was more or less overall activity, respectively, in the prescribed direction in the influence trials, compared with the noninfluence, control trials). This amounts to a "majority vote" procedure which eliminates possible statistical dependency problems but is extremely wasteful of data. Alternative solutions would be: (a) to attempt to determine empirically the nature of the correlation among the data points and include the value of such a correlation as a correction in computing *t* scores, or (b) to attempt to show that successive target activity measures are in fact independent. If the data can be reduced to binary form, a number of statistical tests for intertrial independence or randomness are available (e.g., Davis and Akers, 1974; Dudewicz and Ralley, 1981). For analog data, correlations of activity with trial number (Utts, personal communication, 1988) or the use of autocorrelation techniques (see Braud, 1988) would be helpful.

Trial dependence is problematical in situations in which the very same target organism participates in all trials of an extended measurement block—e.g., placing a laboratory rat in a test apparatus for 15 minutes and measuring its activity 10 times (i.e., for 10 "trials") during that long period. The problem can be reduced or obviated by using different organisms or different biological samples for the different trials. This is analogous to making activity measurements in 10 different laboratory rats placed sequentially in the measuring apparatus throughout a 15-minute period, all rats being selected or sampled

from a common group colony cage. One would still have to be careful to eliminate external or internal factors that could bias subsets of organisms or trials
in different directions for experimental and control treatments, respectively.

Resistance

Researchers who explore the distant mental influence of living systems will
encounter *resistance* in all of its manifestations. The living target system itself, at a
physiological level, may resist a distant mental influence, especially if that influence
opposes a strong homeostatic tendency. Psi influence attempts may be most successful when they are directed in a manner that would assist the organism's return
to a balanced condition. Assisting homeostasis should not be confused with the
statistical artifact of regression to the mean, about which Child (1977, 1978) has
warned us, and against which experimental precautions should be taken.

Of equal interest are the various forms of *psychological* resistance that may
be encountered in the subjects and experimenters of distance influence experiments, as well as in the reactions of one's colleagues and critics. Success at distant mental influence may trigger conscious or unconscious thoughts or feelings
about the possible abuse of such "powers" which in turn may activate certain
psychological mechanisms of defense against the resultant threatening impulses
or fears. These issues have been discussed by Eisenbud (1963, 1972, 1977,
1983), Tart (1984), Braude (1986), Inglis (1981), and Braud (1984). Fears of
the possibility of "evil" mental influence may indeed be responsible for the
dearth of studies of distant mental influence of human subjects, even in the
context of healing or healing analog investigations. It has also struck me as curious that the most explicit treatments of the possibility of harmful psi influences
in everyday life, i.e., the theories of the Greek psychical researcher Angelos
Tanagras (1949, 1967), have been almost totally ignored by parapsychologists.[7]

The issue of psychological resistance is a complicated one and one for
which there would seem to be no easy solution. One method of countering
defenses would be the provision of a nonthreatening context for one's distant
influence experiments, such as healing or another positive application (see
Braud, 1984; Benor, 1985). Another method of dealing with defenses would
be to attempt to assess the presence and degree of these defenses in various subjects and experimenters (through use of a specially constructed version of the
"Defense Mechanism Test," for example) and to study the manner in which
this assessed factor interacts with psi performance or ways in which such
defenses might be reduced.

A Converging Strategies Approach

In a classic 1968 *Psychological Review* paper, Stoyva and Kamiya proposed
a "converging operations" approach to the study of consciousness and illustrated that strategy in the contexts of the experimental study of dreaming and

the waking mental activity associated with EEG alpha control. The strategy utilizes the convergence of different types of indicators (i.e., psychophysiological, behavioral, and verbal) in the definition of a hypothetical construct such as a particular state of consciousness. Recently, Rex Stanford has been using a similar approach in his studies of the psi-conduciveness of ganzfeld stimulation (e.g., Stanford, Kass, and Cutler, 1988). I would like to suggest a multi-component strategy in which converging operations of three kinds may be used in elucidating the problems of "consciousness" and "life." The strategy may be illustrated in the context of psychokinesis. Research would be conducted in three areas in order to determine: (a) whether animate and inanimate systems differ in their *susceptibility* to a PK influence, (b) whether animate and inanimate systems differ in their ability to *produce* psychokinetic effects in other systems (or, better, to produce "conformance behavior" in other systems; see Stanford, 1977, 1978; Edge, 1978; Braud, 1980; Varvoglis, 1986), and (c) whether "pre-observations" of random events by animate versus inanimate systems differentially influence the susceptibility of such events to later psychokinetic influence (see Schmidt, 1984, 1985, 1986, 1987). Ideally, many studies would be carried out in parallel in these three different areas by many different investigators (preferably by investigators with different belief systems regarding the studied phenomena). The studies could be done using a variety of life forms (of different phyletic and ontogenetic status) and a variety of inanimate systems (differing, perhaps, in their degree of complexity and the degree of interconnectedness of their component elements). Similar parallel experiments could be conducted with human influencers, influencees, and pre-observers who are in various states of consciousness during their experimental sessions. Outcomes of these studies that would be of great interest and theoretical importance would be: (a) the discovery of specific graded or discontinuous curves relating outcome likelihood to the life- or consciousness-status of the experimental participants in each of the three research areas, and (b) a *similarity* of the three obtained functions. Throughout this endeavor, great care would have to be exercised to assure that comparison tests were carried out under identical psychological conditions, using the proper multiple blinds and design considerations. While findings in any one area would be far from definitive, the *convergence* upon the same conclusion of evidence from three different research domains would be more compelling and could lead ultimately to a true comparative psychology of consciousness or mind.

Let me illustrate the use of this strategy more concretely. Let us suppose that we carry out REG-PK experiments with alert humans, drowsy humans, dolphins, chickens, earthworms, protozoa, plants, complex machines, and simple machines as the ostensible subjects or influencers. In some cases, the REG "hit" feedback would have to be transformed into environmental events that satisfy the organisms' needs or allow the execution of some strong predisposition. It is also important that the various experiments be given sufficiently fair tests; i.e., experiments should not be tried only once or a few times and abandoned prematurely because they "didn't work." Next, we conduct distant influence experiments in which these same respective organisms and devices serve as targets. Finally, we have the respective organisms or devices "pre-observe" REG events before the latter are subsequently displayed as PK targets for a human influencer. Let us suppose that the strength of the PK effect (as assessed by some appropriate standardized measure such as effect size) differs for alert *versus* drowsy human influencers, and that a similar functional relationship is found in the case of alert *versus* drowsy human pre-observers. Or, suppose all three effects tend to occur for dolphins, humans, and chickens, but not for earthworms, protozoa, plants, or machines. Such convergent outcomes might point to interesting gradients or discontinuities among the systems which then could be explored more incisively.

Other Contexts

We have been considering distant mental influence as it occurs in the context of quasi-experimental or experimental psi studies. In these studies, attempts are made to study the process *in isolation,* without the possibility of conventional sensory or motor accompaniments. A sufficient number of such experiments have been successful, and have yielded sufficiently impressive results, to lead us to conclude that direct distant mental influences upon living systems are possible. There is, therefore, an even greater likelihood that direct distant mental influences upon living systems may occur frequently and strongly in everyday life situations and may be intertwined with more "conventional" control modalities. We influence others by means of our words, expressions, and actions. We influence our own bodies through various neural and hormonal processes. It is not unreasonable to assume that we also influence other persons or our own physiological functioning through direct psychic means, acting in parallel with more conventional means. Perhaps psychic influences modulate or orchestrate the more familiar physical and chemical processes that support and govern our everyday actions.

When we succeed in influencing ("self-regulating") our somatic functioning in contexts of auto-hypnosis, autogenic training, biofeedback, visualization effects, or rehabilitation training to recover or compensate for lost muscular functioning, perhaps we are exerting direct mental ("psychokinetic") influences upon our somatic systems. When we attempt to help restore the mental and physical health and well-being of other persons in contexts of medical treatment, nursing care, therapy, counseling, and teaching, perhaps we are exerting direct mental influence upon our patients, clients, and students. Similar suggestions may be found in a previous paper of mine (Braud, 1986) and in the writings of R. A. McConnell (see McConnell, 1983, 1987). Tanagras (1949, 1967) discusses the possibility of direct psychic influences in mundane contexts, as well as in more exotic contexts involving "the evil eye" and the possible effects upon others of negative thoughts and feelings. A more contemporary treatment of possible interactions of sorcery, psi phenomena, and stress among certain Amerindian groups may be found in Lake (1987). On the more positive side, direct mental influence may be implicated in extraordinary athletic or martial arts accomplishments, such as those described by Murphy and White (1978).

Harmful or Unwanted Influences

Is it possible to prevent harmful or unwanted distant mental influences, and if so, how can this be done? This is an important issue, and one that is difficult to address adequately because of the absence of necessary research findings. In the various experiments on the distant mental influences of human subjects, which have been discussed in this paper, it might be argued that influence is possible provided the subjects give explicit or tacit consent to be influenced. In the various "bio-PK" experiments on physiological influence that we have conducted in our laboratory, the subjects knew the nature of the influences which were to be attempted and agreed to let such influences occur. The experiments could be viewed as social agreements or "contracts" in which the experimenter, the influencer, and the subject all agree to play certain roles having psychic components; each participant plays his or her proper role in order for the experiment to succeed. Allowing one's body to be influenced is part of the task demand with which the subject willingly complies. It is in everyone's best interest for the experiment to succeed. It could be argued that even in experiments in which the subject is conventionally "unaware" that influence attempts will be made, the subject may be *psychically* aware of the possibility, and that there are tacit understandings of appropriate roles and useful outcomes. Perhaps such tacit consent to be influenced and resultant compliance would not occur in situations in which the attempted influence is a deleterious one.

There is actually no compelling evidence that bears directly on the issue of whether external psi influences can produce undesirable effects in a person. If the influences under consideration are direct, causal, "psychokinetic" ones, unwanted influence may have a greater likelihood than if the influences are really unconsciously self-produced and merely aided or triggered by telepathic or clairvoyant knowledge of what actions are expected. To the extent that psi manifestations in a "target person" make use of the images, thoughts, and feelings of that person as "vehicles" for their expression, psi influences could be allowed or prevented through the use of the same self-control techniques by which the target person customarily modulates his or her own thoughts, images, feelings, and behaviors in more conventional contexts. Processes of intention and acts of will should be just as effective in the psi realm as they are in more ordinary domains.

Psychological and, possibly, psychic techniques could be used to prevent unwanted influences. These techniques involve reminders of self-control, self-responsibility, internal locus of control, and ultimate "veto power" over what one does upon the suggestion of others. Confidence enhancing images of barriers, screens, shields, or other symbols of protection might be used to effectively block unwanted psi influences. We have used such techniques, with initial indications of success, in some of our EDA bio-PK experiments. More extensive studies of "psi blocking" techniques in other contexts are still in progress, and we hope to report their results soon.

The demand characteristics of laboratory experiments make difficult or impossible any final resolution of the issue of whether unwanted or harmful psi influences can ever really occur. The situation is similar to the one that obtains in hypnosis or compliance research: Subjects may, at some level, recognize that experimenters would not allow anything that is truly harmful to occur; i.e., subjects may discern that some experiments may be dramatic instances of play-acting designed to prove a particular point of view of the investigators. Ethical constraints would not allow more realistic experiments or tests in everyday life that are not "play-acting."

Perhaps the most valid evidence bearing on this issue will come from careful anthropological observations in natural settings. Even here, however, certain complexities and alternative interpretations will remain. For example, consider a well-authenticated case in which Person X becomes seriously ill or even dies shortly after being "hexed" or "cursed," *without his or her knowledge,* by Person Y. How could we exclude the possibilities that (a) Person Y precognized Person X's illness or death and then engineered the ostensible "magical" influence in order to convince others of unusual causal powers that Person Y really does not possess, or (b) that Person X actually injured himself or herself as self-punishment for some real or imagined crime, sin, or taboo-violation—i.e., that Person X committed a sort of socially approved suicide, using psi-provided knowledge of an actually ineffective "curse" as an opportunity for this action? Are these reasonable alternative explanations, or are they continuing manifestations of psychological defense against the possibility of truly causal external psi influences of a harmful nature? Could our hope to assign a definite form and definite locus to these effects be a misguided one? Perhaps the most satisfactory interpretation will be one in which psi influence effects are understood as field-like effects contributed by *all* participants and involving several "forms" of psi.

The complexity of designing research or of interpreting findings relevant to this issue soon becomes apparent. It becomes more understandable why so few researchers have grappled with the issue of possible harmful or unwanted psi influence effects.

Findings

I have presented various considerations which favor the use of living target systems in distant mental influence research. I will conclude with a brief summary of some of the findings and conclusions that are emerging from our work with living target systems in our Mind Science Foundation laboratories.

1. Based upon overall statistical results, the distant mental influence effects are relatively reliable and robust.

2. The magnitudes of the effects are not trivial and, under certain conditions, may compare favorably with the magnitudes of self-regulation effects.

3. The ability to manifest the effect is apparently widely distributed in the population. Sensitivity to the effect appears to be normally distributed in the volunteer subjects who have participated in our various experiments. Many persons are able to produce the effect, with varying degrees of success, including unselected volunteers attempting it for the first time. More practiced individuals seem able to produce the effect more consistently. There are indications of improvement with practice in some influencers.

4. The effect can occur at a distance, typically 20 meters; greater distances have not yet been explored.

5. Subjects with a greater need to be influenced (i.e., those for whom the influence is more beneficial) seem more susceptible to the effect.

6. Immediate, trial-by-trial analog sensory feedback is not essential to the occurrence of the effect; intention and visualization of the desired outcome is effective.

7. The effect can occur without the subject's knowledge that such an influence is being attempted.

8. It may be possible for the subject to block or prevent an unwanted influence upon his or her own physiological activity; psychological shielding strategies in which one visualizes protective surrounding shields, screens, or barriers may be effective.

9. Generally, our volunteer participants have not evidenced concern over the idea of influencing or being influenced by another person.

10. The effect can be intentionally focused or restricted to one of a number of physiological measures; it may also take the form of a generalized influence of several measures, if that is the intent of the influencer.

11. A number of target systems have been found to be susceptible to the effect, including the spatial orientation of fish, the locomotor activity of small mammals, the autonomic nervous system activity of another person, the muscular tremor and ideomotor reactions of another person, the mental

imagery of another person, and the rate of hemolysis of human red blood cells *in vitro*.

12. The living target systems can be influenced bi-directionally; i.e., their activity levels can be either increased or decreased.

13. The activity levels of at least some of the target systems (i.e., electrodermal activity, rate of hemolysis) and their susceptibility to distant mental influence appear to be influenced by geomagnetic field (GMF) activity; i.e., the systems are more active and more susceptible to influence when the earth's geomagnetic field activity is more "stormy" than during more "quiet" GMF periods.

14. Distant mental influence, in the expected direction, seems more successful when the intentions and images of the influencer are focused *specifically on the desired target activity*, rather than directed toward the target in a more general or global manner.

15. The effect does not always occur. The reasons for the absence of a significant effect in some experiments of a series which is otherwise successful are not clear. We suspect that the likelihood of a successful distant mental influence effect may depend upon the presence of certain psychological conditions, in both influencer and subject (and perhaps even in the experimenter), which are not always present. Possible success-enhancing factors may include belief, confidence, positive expectation, and appropriate motivation. Possible success-hindering factors may include boredom, absence of spontaneity, poor mood of influencer or subject, poor interactions or poor rapport between influencer and subject, and excessive egocentric effort (excessive pressure or striving to succeed) on the part of participants. We suspect that the effect occurs most readily in subjects whose nervous systems are relatively labile (i.e., characterized by free variability) and are momentarily free from external and internal constraints. Perhaps fullness of intention and intensity or vividness of visualization in the influencer facilitate the effect. Additional research, of course, is needed to determine the validity of these conclusions and to explore more thoroughly the various physiological and psychological factors that are favorable or antagonistic to the occurrence of the effect.

We are continuing our laboratory studies of distant mental influence of living systems, being especially interested in exploring the possible limits of such effects and whether the effects can be *strong* and *consistent* enough to yield possible practical applications (e.g., in the area of healing). We hope that this presentation will encourage others to carry out similar investigations.

Endnotes

1. Julian Isaacs (1983) has argued that physical systems with very low spontaneous activity levels would be ideally suited to the direct detection of subtle PK effects.

2. It was Elliotson who introduced the stethoscope into hospital practice.

3. Inspired by these sorts of reports (especially those of Townshend), Edgar Allan Poe featured distant mesmeric influences in two of his short stories, "A Tale of the Ragged Mountains" (1844) and "The Facts in the Case of M. Valdemar" (1845). Distant mesmeric influence also was featured in Robert Browning's poem, "Mesmerism," written during this same time period (see Schneck, 1956).

4. It is not generally known that, in 1889, Charles Richet published a sensational novel, *Sister Marthe*, which featured hypnosis and dual personality; he published the novel under the pseudonym, Charles Epheyre.

5. It was Bechterev, of course, who pioneered the learning paradigm which later came to be known as "instrumental" or "operant" conditioning, and which was later so well explored and exploited by B. F. Skinner and his co-workers; Ivanov-Smolensky specialized in "semantic" conditioning and his investigations are important in the understanding of the Pavlovian "second signalling system" (i.e., the experimental study of language and thinking).

6. Loewi's findings were not immediately accepted because of the difficulties encountered by other investigators in replicating his work. The vagus nerve of the frog also contains a sympathetic accelerating component. The nerve's action is therefore mixed, sometimes accelerating the heart and sometimes decelerating it. Which particular action predominates depends upon the frog and varies with the season of the year. Opposite seasonal variations occur in the toad. Eventually, when these initially occult interactions were realized, Loewi's discoveries were confirmed and are now universally accepted (see Goodman and Gilman, 1956). Perhaps these events will provide encouragement to psi researchers who continue to experience difficulties in their replication attempts.

7. I have been able to find reference to Tanagras' work in the writings of only one psychical researcher, Jule Eisenbud. Tanagras' name does not appear in the indices of the major parapsychological reference works. The one exception is Wolman's (1977) *Handbook*, which gives a single page reference to Tanagras and this is merely to the definition of his term "psychoboly" in the glossary at the end of the volume; interestingly, the page reference is incorrect and Tanagras is not to be found even on the single page for which he is referenced. Could this be still another indication of psychological resistance to the possibility of powerful negative psi influences?

References

Achterberg, J. (1985). *Imagery in healing.* Boston: Shambala.

Alrutz, S. (1921). Cited in H. Carrington, *The problems of psychical research.* New York: Dodd, Mead.

Alvarado, C. (1981). PK and body movements: A brief historical—and semantic—note. *Journal of the Society for Psychical Research,* 51, 116–118.

Bechterev, V. M. (1920). Experiments on the effects of "mental" influence on the behavior of dogs. In *Problems in the study and training of personality* (pp. 230–265). Petrograd.

Beloff, J. (1974). ESP: The search for a physiological index. *Journal of the Society for Psychical Research,* 47, 403–420.

Beloff, J. (1979). Voluntary movement, biofeedback control and PK. In B. Shapin and L.

Coly (Eds.), *Brain/mind and parapsychology* (pp. 99–109). New York: Parapsychology Foundation.

Benor, D. J. (1984). Fields and energies related to healing: A review of Soviet and Western studies. *Psi Research,* 3(1), 21–35.

Benor, D. J. (1985). Research in psychic healing. In B. Shapin and L. Coly (Eds.), *Current trends in psi research* (pp. 96–112). New York: Parapsychology Foundation.

Benor, D. J. (in press). *The psi of relief: A review of research in psychic healing and related topics.*

Berger, R. E., Schechter, E. I., and Honorton, C. (1986). A preliminary review of performance across three computer psi games. In D. H. Weiner and D. I. Radin (Eds.), *Research in parapsychology 1985* (pp. 1–3). Metuchen, NJ: Scarecrow Press.

Braud, L. W., and Braud, W. G. (1972). Biochemical transfer of relational responding (transposition). *Science,* 176, 942–944.

Braud, W. G. (1970). Extinction in goldfish: Facilitation by intracranial injection of "RNA" from brains of extinguished donors. *Science,* 168, 1234–1236.

Braud, W. G. (1978). Allobiofeedback: Immediate feedback for a psychokinetic influence upon another person's physiology. In W. G. Roll (Ed.), *Research in parapsychology 1977* (pp. 123–134). Metuchen, NJ: Scarecrow Press.

Braud, W. G. (1980). Lability and inertia in conformance behavior. *Journal of the American Society for Psychical Research,* 74, 297–318.

Braud, W. G. (1981). Lability and inertia in psychic functioning. In B. Shapin and L. Coly (Eds.), *Concepts and theories of parapsychology* (pp. 1–28). New York: Parapsychology Foundation.

Braud, W. G. (1984). The two faces of psi: Psi revealed and psi obscured. In B. Shapin and L. Coly (Eds.), *The repeatability problem in parapsychology* (pp. 150–175). New York: Parapsychology Foundation.

Braud, W. G. (1986). PSI and PNI: Exploring the interface between parapsychology and psychoneuroimmunology. *Parapsychology Review,* 17(4), 1–5.

Braud, W. G. (1987). Studies of transpersonal imagery and intentionality effects. Paper presented at Esalen Institute's Invitational Conference on Healing, Big Sur, CA.

Braud, W. G. (1988). Distant mental influence of rate of hemolysis of human red blood cells. *Proceedings of the 31st Annual Parapsychological Association Conference,* Montreal, Canada, 1–17.

Braud, W. G., Davis, G., and Wood, R. (1979). Experiments with Matthew Manning. *Journal of the Society for Psychical Research,* 50, 199–223.

Braud, W. G., and Jackson, J. (1982). Ideomotor reactions as psi indicators. *Parapsychology Review,* 13(2), 10–11.

Braud, W. G., and Jackson, J. (1983). Psi influence upon mental imagery. *Parapsychology Review,* 14(6), 13–15.

Braud, W. G., and Schlitz, M. (1983). Psychokinetic influence on electrodermal activity. *Journal of Parapsychology,* 47, 95–119.

Braud, W. G., and Schlitz, M. (1987a). Possible role of intuitive data sorting in electrodermal biological psychokinesis (bio-PK). *Proceedings of the 30th Annual Parapsychological Association Convention,* Edinburgh, Scotland, 18–30.

Braud, W. G., and Schlitz, M. (1987b). A methodology for the objective study of transpersonal imagery. Paper presented at the Second World Conference on Imagery, Toronto, Canada.

Braud, W. G., and Schlitz, M. (in press). *Distant mental influence: A systematic research program.*

Braude, S. (1986). *The limits of influence.* New York: Routledge and Kegan Paul.

Carrington, H. (1909). *Eusapia Palladino and her phenomena.* New York: B. W. Dodge.

Carrington, H. (1921). *The problems of psychical research.* New York: Dodd, Mead.

Child, I. L. (1977). Statistical regression artifact in parapsychology. *Journal of Parapsychology,* 41, 10–22.

Child, I. L. (1978). Statistical regression artifact: Can it be made clear? *Journal of Parapsychology,* 42, 179–193.

Cox, E. W. (1874, January). Quoted in W. Crookes. Notes of an enquiry into the phenomena called spiritual. *Quarterly Journal of Science.*

Davis, J. W., and Akers, C. (1974). Randomization and tests for randomness. *Journal of Parapsychology,* 38, 393–408.

Dean, E. D. (1962). The plethysmograph as an indicator of ESP. *Journal of the Society for Psychical Research,* 41, 351–353.

Dean, E. D. (1964). A statistical test of dreams influenced by telepathy. *Journal of Parapsychology,* 28, 275–276.

Dobbs, A. (1976). The feasibility of a physical theory of ESP. In J. R. Smythies (Ed.), *Science and ESP* (pp. 225–254). New York: Humanities Press.

Dudewicz, E. J., and Ralley, T. (1981). *Handbook of random number generation and testing with TESTRAND computer code.* Syracuse, NY: American Sciences Press.

Eccles, J. C. (1953). *The neurophysiological basis of mind.* Oxford: Clarendon Press.

Edge, H. (1978). A philosophical justification for the conformance behavior model. *Journal of the American Society for Psychical Research,* 72, 215–232.

Eisenbud, J. (1963). Psi and the nature of things. *International Journal of Parapsychology,* 5, 245–273.

Eisenbud, J. (1972). The psychology of the paranormal. *Journal of the American Society for Psychical Research,* 66, 27–41.

Eisenbud, J. (1977). Perspectives on anthropology and parapsychology. In J. Long (Ed.), *Extrasensory ecology: Parapsychology and anthropology* (pp. 28–44). Metuchen, NJ: Scarecrow Press.

Eisenbud, J. (1983). *Parapsychology and the unconscious.* Berkeley, CA: North Atlantic Books.

Elliotson, J. (1843). *Numerous cases of surgical operations without pain in the mesmeric state.* London: H. Bailliere.

Epheyre, C. [Charles Richet]. (1889). Soeur Marthe. *Revue des Deux Mondes,* 93, 384–431.

Esdaile, J. (1852). *Natural and mesmeric clairvoyance, with the practical application of mesmerism in surgery and medicine.* London: H. Bailliere.

Eysenck, H. J. (1967a). *The biological basis of personality.* Boston: Thomas.

Eysenck, H. J. (1967b). Personality and extra-sensory perception. *Journal of the Society for Psychical Research,* 44, 55–70.

Gleick, J. (1987). *Chaos.* New York: Viking.

Goodman, L. S., and Gilman, A. (1956). *The pharmacological basis of therapeutics* (2nd ed.). New York: Macmillan.

Grad, B. (1967). The "laying on of hands": Implications for psychotherapy, gentling and the placebo effect. *Journal of the American Society for Psychical Research,* 61, 286–305.

Gray, J. A. (1964). *Pavlov's typology.* New York: Macmillan.

Gruber, E. R. (1979). Conformance behavior involving animal and human subjects. *European Journal of Parapsychology,* 3, 36–50.

Gruber, E. R. (1980). PK effects on pre-recorded group behavior of living systems. *European Journal of Parapsychology,* 3, 167–175.

Hansen, G. P. (1987). Striving toward a model. In D. H. Weiner and R. D. Nelson (Eds.), *Research in parapsychology 1986* (pp. 140–141). Metuchen, NJ: Scarecrow Press.

Honorton, C., Barker, P., Varvoglis, M., Berger, R., and Schechter, E. (1986). First-timers: An exploration of factors affecting initial psi ganzfeld performance. In D. H. Weiner and D. I. Radin (Eds.), *Research in parapsychology 1985* (pp. 28–32). Metuchen, NJ: Scarecrow Press.

Honorton, C., and Tremmel, L. (1979). Psi correlates of volition: A preliminary test of Eccles' "neurophysiological hypothesis" of mind-brain interaction. In W. G. Roll (Ed.), *Research in parapsychology 1978* (pp. 36–38). Metuchen, NJ: Scarecrow Press.

Hubbard, G. S., Bentley, P. B., Pasturel, P. K., and Isaacs, J. D. (1987) Instrumentation and protocol for a remote action experiment. *Proceedings of the 30th Annual Parapsychological Association Convention,* Edinburgh, Scotland, 451–474.

Inglis, B. (1981). Power corrupts: Skepticism corrodes. In W. G. Roll and J. Beloff (Eds.), *Research in parapsychology 1980* (pp. 143–151). Metuchen, NJ: Scarecrow Press.

Isaacs, J. (1983). A twelve-session study of micro-PKMB training. In W. G. Roll, J. Beloff, and R. A. White (Eds.), *Research in parapsychology 1982* (pp. 31–35). Metuchen, NJ: Scarecrow Press.

Ivanov-Smolensky, A. G. (1920). Experiments in mental suggestion on animals. In *Problems in the study and training of personality.* Petrograd.

Janet, P. (1968a). Report on some phenomena of somnambulism. *Journal of the History of the Behavioral Sciences,* 4, 124–131. (Reprinted from *Revue Philosophique de la France et de l'Etrangere,* 22 [1886], 190–198.)

Janet, P. (1968b). Second observation of sleep provoked from a distance and the mental suggestion during the somnambulistic state. *Journal of the History of the Behavioral Sciences,* 4, 258–267. (Reprinted from *Revue Philosophique de la France et de l'Etrangere,* 22 [1886], 212–223.)

Jantsch, E. (1980). *The self-organizing universe.* Oxford: Pergamon Press.

Joire, P. (1897). De la suggestion mentale. *Annales des Sciences Psychique,* 4, 193.

Kaplan, M. (Ed.). (1966). *Essential works of Pavlov.* New York: Bantam.

Katkin, E. S., and Murray, E. N. (1968). Instrumental conditioning of autonomically mediated behavior: Theoretical and methodological issues. *Psychological Bulletin,* 70, 56–68.

Kelly, E. F. (1979). Discussion. In B. Shapin and L. Coly (Eds.), *Brain/mind and parapsychology* (pp. 50–51). New York: Parapsychology Foundation.

Lake, R. G (1987/88). Sorcery, psychic phenomena, and stress: Shamanic healing among the Yurok, Wintu, and Karok. *Shaman's Drum,* 11, 38–46.

Lloyd, D. H. (1973). Objective events in the brain correlating with psychic phenomena. *New Horizons,* 1, 69–75.

Loewi, O. (1921) Uber humorale Übertragbarkeit der Herznervenwirkung. *Arch. f. d. ges. Physiol.,* 189, 239–242.

Lynn, R. (1966). *Attention, arousal, and the orientation reaction.* Oxford: Pergamon.

McConnell, R. A. (1983). *An introduction to parapsychology in the context of science.* Pittsburgh: Author.

McConnell, R. A. (1987). *Parapsychology in retrospect.* Pittsburgh: Author.

Mesmer, F. A. (1980a). Letter from M. Mesmer, Doctor of Medicine at Vienna, to A. M. Unzer, Doctor of Medicine, on the medicinal uses of the magnet, 1775. In G. J. Bloch (Ed.), *Mesmerism* (p. 28). Los Altos, CA: William Kaufmann.

Mesmer, F. A. (1980b). Dissertation on the discovery of animal magnetism, 1775. In G. J. Bloch (Ed.), *Mesmerism* (pp. 52–53). Los Altos, CA: William Kaufmann.

Mesmer, F. A. (1980c). Dissertation by F. A. Mesmer, Doctor of Medicine, on his discoveries, 1779. In G. J. Bloch (Ed.), *Mesmerism* (pp. 115–130). Los Altos, CA: William Kaufmann.

Millar, B. (1979). Physiological detectors of psi. *European Journal of Parapsychology,* 2, 456–478.

Morris, R. L. (1977). Parapsychology, biology, and anpsi. In B. B. Wolman (Ed.), *Handbook of parapsychology* (pp. 687–715). New York: Van Nostrand Reinhold.

Motoyama, H. (1977). Physiological measurements and new instrumentation. In G. W. Meek (Ed.), *Healers and the healing process* (pp. 147–155). Wheaton, IL: Theosophical Publishing.

Muftic, M. A. (1959). A contribution to the psychokinetic theory of hypnotism. *British Journal of Medical Hypnotism,* 10, 21–26.

Murphy, M., and White, R. A. (1978). *The psychic side of sports.* Reading, MA: Addison-Wesley.

Myers, F. W. H. (1903). *Human personality and its survival of bodily death.* Volume 2. New York: Longmans, Green.

Oteri, L. (Ed.). (1975). *Quantum physics and parapsychology.* New York: Parapsychology Foundation.

Palmer, J. (1977). Attitudes and personality traits in experimental ESP research. In B. B. Wolman (Ed.), *Handbook of parapsychology* (pp. 175–201). New York: Van Nostrand Reinhold.

Pavlov, I. P. (1927). *Conditioned reflexes.* London: Oxford University Press.

Pavlov, I. P. (1957). *Experimental psychology and other essays.* New York: Philosophical Library.

Poe, E. A. (1844, April). A tale of the ragged mountains. *Godey's Magazine and Lady's Book.*

Poe, E. A. (1845, December). The facts in the case of M. Valdemar. *The American Review.*

Puharich, A. (Ed.). (1979). *The Iceland papers.* Amherst, WI: Essentia Research Associates.

Puységur, A. M. (Marquis de). (1784). *Mémoires pour à l'histoire et à l'établissement du magnétisme animal.* Paris.

Radin, D. I. (1988). Searching for "signatures" in human interaction data: A neural network approach. *Proceedings of the 31st Annual Convention of the Parapsychological Association,* Montreal, Canada, 61–73.

Rhine, J. B. (1943). The mind has real force. *Journal of Parapsychology,* 7, 69–75.

Rhine, J. B. (1947). *The reach of the mind.* New York: William Sloan.

Rhine, L. E. (1970). *Mind over matter.* New York: Macmillan.

Richet, C. (1888). Relation de diverses expériences sur la transmission mentale, la lucidité, et autres phénomènes non explicables par les données scientifiques actuelles. *Proceedings of the Society for Psychical Research,* 5, 18–168.

Rogo, D. S. (1980). Theories about PK: A critical evaluation. *Journal of the Society for Psychical Research,* 50, 359–378.

Roll, W. G. (1985). A systems theoretical approach to psi. In B. Shapin and L. Coly (Eds.), *Current trends in psi research* (pp. 47–86). New York: Parapsychology Foundation.

Rush, J. (1976). Physical aspects of psi phenomena. In G. Schmeidler (Ed.), *Parapsychology: Its relation to physics, biology, psychology and psychiatry* (pp. 6–39). Metuchen, NJ: Scarecrow Press.

Rush, J. H. (1986). Findings from experimental PK research. In H. Edge, R. Morris, J. Palmer, and J. Rush, *Foundations of parapsychology* (pp. 237–275). Boston: Routledge and Kegan Paul.

Sargant, W. (1957). *Battle for the mind.* New York: Doubleday.

Schmidt, H. (1984). Superposition of PK efforts by man and dog. In R. A. White and R. Broughton (Eds.), *Research in parapsychology 1983* (pp. 96–98). Metuchen, NJ: Scarecrow Press.

Schmidt, H. (1985). Addition effect for PK on prerecorded targets. *Journal of Parapsychology,* 49, 229–244.

Schmidt, H. (1986). Human PK effort on pre-recorded random events previously observed by goldfish. In D. H. Weiner and D. I. Radin (Eds.), *Research in parapsychology 1985* (pp. 18–21). Metuchen, NJ: Scarecrow Press.

Schmidt, H. (1987). The strange properties of psychokinesis. *Journal of Scientific Exploration,* 1, 103–118.

Schmidt, H., and Schlitz, M. (1988). A large scale pilot PK experiment with prerecorded random events. *Proceedings of the 31st Annual Parapsychological Association,* Montreal, Canada, 19–35.

Schneck, J. M. (1956). Robert Browning and mesmerism. *Bulletin of the Medical Library Association,* 44, 443–451.

Schouten, S. A., and Kelly, E. F. (1978). The experiment of Brugmans, Heymans, and Weinberg. *European Journal of Parapsychology,* 2, 247–290.

Smith, L. T. (1974). The interanimal transfer phenomenon: A review. *Psychological Bulletin,* 81, 1078–1095.

Solfvin, G. F. (1982). Expectancy and placebo effects in experimental studies of mental healing. Unpublished doctoral dissertation, University of Utrecht, The Netherlands.

Solfvin, J. (1984). Mental healing. In S. Krippner (Ed.), *Advances in parapsychological research* (Vol. 4, pp. 31–63). Jefferson, NC: McFarland.

Stanford, R. G. (1974). An experimentally testable model for spontaneous psi events: II. Psychokinetic events. *Journal of the American Society for Psychical Research,* 70, 321–356.

Stanford, R. G. (1978). Toward reinterpreting psi events. *Journal of the American Society for Psychical Research,* 72, 197–214.

Stanford, R. G., Kass, G., and Cutler, S. (1988). Session-based verbal predictors of free-response ESP-task performance in ganzfeld. *Proceedings of the 3lst Annual Convention of the Parapsychological Association,* Montreal, Canada, 395–411.

Stewart, W. W. (1972). Comments on the chemistry of scotophobin. *Nature,* 238, 202–210.

Stoyva, J., and Kamiya, J. (1968). Electrophysiological studies of dreaming as the prototype of a new strategy in the study of consciousness. *Psychological Review,* 75, 192–205.

Tanagras, A. (1949). The theory of psychobolie. *Journal of the American Society for Psychical Research,* 43, 151–154.

Tanagras, A. (1967). *Psychophysical elements in parapsychological traditions.* New York: Parapsychology Foundation.

Targ, R., and Puthoff, H. (1974). Information transmission under conditions of sensory shielding. *Nature,* 252, 602–607.

Tart, C. T. (1963). Possible physiological correlates of psi cognition. *International Journal of Parapsychology,* 5, 375–386.

Tart, C. T. (1984). Acknowledging and dealing with the fear of psi. *Journal of the American Society for Psychical Research,* 78, 133–144.

Thouless, R. H., and Wiesner, B. P. (1946). On the nature of psi phenomena. *Journal of Parapsychology,* 10, 107–119.

Thouless, R. H., and Wiesner, B. P. (1947). The psi process in normal and "paranormal" psychology. *Proceedings of the Society for Psychical Research,* 48, 177–197.

Townshend, C. H. (1844). *Facts in mesmerism.* London.

Ungar, G., Desiderio, D. M., and Parr, W. (1972). Isolation, identification and synthesis of a specific-behaviour-inducing brain peptide. *Nature,* 238, 198–202.

Varvoglis, M. P. (1986). Goal-directed and observer-dependent PK: An evaluation of the conformance-behavior model and the observational theories. *Journal of the American Society for Psychical Research,* 80, 137–162.

Varvoglis, M. P., and McCarthy, D. (1986). Conscious-purposive focus and PK: RNG activity in relation to awareness, task-orientation, and feedback. *Journal of the American Society for Psychical Research,* 80, 1–30.

Vasiliev, L. L. (1976). *Experiments in distant influence.* New York: Dutton. (Reprinted from *Exsperimentalnie issledovaniya mislennogo vnusheniya.* Leningrad: Zhdanov Leningrad State University, 1962.)

Walaszek, E. J., and Abood, L. G. (1956). Effect of tranquillizing drugs on fighting response of Siamese fighting fish. *Science,* 124, 440–441.

Walker, E. H. (1975). Foundations of paraphysical and parapsychological phenomena. In L. Oteri (Ed.), *Quantum physics and parapsychology* (pp. 1–44). New York: Parapsychology Foundation.

Weckowicz, T. (1967). Animal studies of hallucinogenic drugs. In A. Hoffer and H. Osmond, *The hallucinogens.* New York: Academic Press.

Weiner, D. H., and Nelson, R. D. (1987). *Research in parapsychology 1986* (pp. 136–144). Metuchen, NJ: Scarecrow Press.

Wells, R., and Klein, J. (1972). A replication of a "psychic healing" paradigm. *Journal of Parapsychology,* 36, 144–149.

7

Reactions to an Unseen Gaze (Remote Attention): Autonomic Staring Detection

William Braud, Donna Shafer, and Sperry Andrews

This chapter describes results of experiments in which physiological measurements revealed that persons can be influenced by the unseen gaze of another, distant person. In these experiments, staring (remote attention) occurred through the use of a closed-circuit television system that eliminated the possibility of sensory and other conventional cues.

This work originally was published in Braud, W., Shafer, D., and Andrews, S. (1993). Reactions to an unseen gaze (remote attention): A review, with new data on autonomic staring detection. Journal of Parapsychology, 57(4), 373–390. The contents of this article are Copyright © 1993 by the Parapsychology Press, and the material is used with permission. —William Braud

Abstract: Have you ever had the feeling that someone was staring at you from behind and, upon turning around, found you were correct? Some of these experiences may be merely coincidental or attributable to subtle cues. However, laboratory studies have demonstrated that some instances of accurate detection of remote staring cannot be attributed to such conventional factors. Rather, they suggest that at least some experiences of remote staring detection may contain valid psychic or parapsychological components. We review earlier studies in which persons could guess successfully when they

We thank George Hansen, Charles Honorton, Robert Rosenthal, and Jessica Utts for helpful comments and suggestions.

This article is partly based on a paper presented at the Thirty-Third Annual Convention of the Parapsychological Association, August 1990, at Chevy Chase, MD.

were being stared at by persons beyond the range of possible sensory cues (that is, the trials were carried out via one-way mirrors or closed-circuit television and according to an unknown, random schedule). We then report original, well-controlled research in which unconscious, autonomic reactions (electrodermal activities) were used to provide physiological, rather than conscious and verbal, indications of accurate remote staring detection. A closed-circuit video system was used in a randomized, blinded experimental design in order to eliminate the possibility of sensory cuing. Accurate and significant effects were obtained, with moderately large effect sizes [t (15) = 2.66; p = .018, two-tailed; effect size = 0.59]. The unconscious, physiological (autonomic nervous system) measure used in the present work appears to yield stronger effects than did previous, more conscious, cognitive guessing measures. Additionally, qualitatively different reaction patterns occurred for untrained starees versus starees who had experienced extensive training in becoming more sensitive to others and dealing with their own psychological resistance to being "connected" with other people [t (15) = 2.15, p = .048, two-tailed; effect size = 0.50].

Have you ever had the feeling that someone was staring at you from behind and, upon turning around, found you were correct? From time to time, most of us have had such a feeling, which appears to be a common part of the human experience. In surveys conducted in California (Coover, 1913) as early as 1913, 68% to 86% of respondents reported having had the feeling of being watched or stared at on at least one occasion, and a more recent Australian survey (Williams, 1983) placed the figure at 74%. In a survey of San Antonio respondents recently completed as part of the present project, the figure was found to be approximately 94%.

Despite its widespread occurrence and familiarity, the staring experience has been subjected to surprisingly little scientific scrutiny. Is the presumed ability to detect an unseen gaze merely a superstition, a cultural myth without real substance, an overinflation of coincidental occurrences, or a response to subtle sensory cues? Or, alternatively, could the experience be a valid indicator of an exceptional yet poorly understood human capability?

In 1898, Titchener published a short article in which he addressed the "feeling of being stared at." Titchener mentioned that he had conducted a series of laboratory experiments at Cornell University on this topic and that the experiments had yielded negative results; unfortunately, he reported no details regarding those experiments. Titchener indicated that such experiments "have their justification in the breaking down of a superstition which has deep and widespread roots in the popular consciousness" (p. 897). He attempted to provide a psychological interpretation of the prevalence of the "staring" belief based on nervousness in social situations, attracting the attention of the starer, turning, and noticing the starer's gaze.

In 1912–1913, experimental research on staring detection was carried out by Coover at Stanford University. Coover (1913) reported the results of a study in which each of 10 subjects made 100 guesses of whether he or she was being stared at by an experimenter seated behind the subject in the same room; the subject kept his or her eyes closed and "shaded with one hand." The staring versus nonstaring schedule was determined by tossing a die. The duration of a staring or nonstaring trial was 15–20 sec; the 100 trials were distributed over three to four hourly sessions that were spaced one week apart. Overall, the subjects' accuracy of guessing did not depart significantly from chance. Coover discussed qualitative differences in the subjects' imagery and subjective impressions that he thought were correlated with the degree of confidence or certainty of their guesses, but did not substantiate his conclusions with quantitative data. He interpreted his findings as support for Titchener's claim that the belief in staring detection was empirically groundless.

In 1959, Poortman of Leyden University (Netherlands) reported a preliminary staring detection study in which he himself served as a subject for 89 trials (distributed over a 13-month period) and attempted to guess whether or not he was being stared at by another experimenter. The same person served as experimenter throughout the tests. Poortman was seated in a separate room adjoining that of the starer, with his back to the starer. Staring and nonstaring trials were of 2 to 5 min duration and were randomly scheduled by means of card shuffles. Poortman achieved a 59.55% accuracy rate which he called "suggestive and highly promising." A reanalysis of Poortman's data by the present authors yields a one-tailed $p = .04$. Poortman also provided several interesting observations of psychological conditions that appeared, in his own experience, to facilitate or to impede accurate staring detection.

In the Coover and Poortman experiments, test conditions were poorly controlled. The subject and the starer were in the same room or in open adjoining rooms, and the subject could have discriminated staring from nonstaring periods by means of subtle, unintentional auditory cues. This cueing possibility was eliminated in two recent studies by Peterson (1978) and Williams (1983).

Peterson (1978) reported two preliminary pilot studies and a formal experiment conducted at the University of Edinburgh. The pilot studies were relatively informal and were conducted in order to ascertain effective procedures that would later be used in the formal experiment. The formal experiment involved nine starer-staree pairs of participants. The starer and staree occupied separate, adjacent, closed cubicles. Special lighting and the use of one-way mirrors permitted visual access in one direction only—that is, the starer could see but could not be seen by the staree. Isolation was increased further by requiring the staree to listen to sound-masking white noise through headphones. The staree pressed a pushbutton whenever he or she felt "stared at"; the button presses marked a chart recorder and provided "time on target" measures. The staring and nonstaring periods were scheduled randomly by means of special equipment. The actual test trials were preceded by brief training periods in which the staree received feedback in an attempt to develop an appreciation of internal cues that might be associated with staring detection. The members of each dyad reversed roles during the experiment so that each person had an opportunity to serve as both starer and staree. There were 36 experimental sessions overall, each of 6-min duration; each session contained three 30-sec staring periods. Analysis of results indicated significantly accurate detection of staring ($p = .012$, two-tailed).

The experimental design was improved even further by Williams (1983) at the University of Adelaide (South Australia). Williams provided excellent sensory isolation of her starers and starees by stationing them in separate, closed rooms 60 ft apart. Instead of using a one-way mirror, the starer watched the subject by means of a closed-circuit video camera/monitor arrangement. Twenty-eight starees participated in the study and indicated their staring detection guesses by means of button presses. Each staree experienced 52 twelve-second staring trials and 52 twelve-second nonstaring trials; the two trial types were scheduled randomly by means of signalling tapes created on the basis of random numbers. Conventional measures of accuracy, as well as sensitivity measures (d') derived from signal detection theory, yielded significant results ($p = .04$, one-tailed).

Three of the four empirical studies reviewed above yielded suggestive evidence that persons are able to consciously discriminate periods of staring from those of nonstaring, even in cases in which the possibility of subtle sensory cues has been eliminated. In fact, an examination of the tabulated results (see Table 1) reveals that scoring actually improved as test conditions were made more stringent, especially if success is measured in terms of effect size (defined as z score or z score equivalent divided by the square root of the number of contributing score units; see Rosenthal, 1984).

Table 1. Summary of Previous Staring Detection Experiments

Investigator	Affiliation	Design features	Scoring rate (%)	Effect size
Titchener (1898)	Cornell U.	No data reported	—	—
Coover (1913)	Stanford U.	Same room	50.20	.004
Poortman (1959)	Leyden U.	Adjoining rooms	59.55	.18
Peterson (1978)	U. of Edinburgh	One-way mirrors	54.86	.42
Williams (1983)	Adelaide U.	Closed-circuit TV	51.31	.32

However, the effect, although consistent, was not particularly striking. A plausible reason for this is that the testing method used in these studies was not the most appropriate one. The laboratory experiments were designed to encourage deliberate conscious guessing in order to identify staring periods.

Such a procedure would be expected to maximize possible cognitive interferences and distortions of subtle internal staring-related cues; it would be difficult or impossible for the staree to avoid the use of guessing strategies, response biases, intellectual analysis and interpretation, and so forth. In the everyday life context, on the other hand, staring detection frequently takes the form of spontaneous unconscious behavioral and bodily changes. Often, such changes are reported to be rich in physiological content (for example, tingling of the skin, prickling of neck hairs) and automatic movements (for example, spontaneous head-turning, unplanned glances). Higher cognitive functions seem to play minor roles in these staring detection contexts.

On the basis of these considerations, it was hypothesized that an experimental design based on more unconscious autonomic nervous system reactions might be more sensitive to staring detection than would one based on conscious motor guessing. Therefore, we designed an experiment in which we would be able to monitor sympathetic autonomic nervous system activity (using electrodermal recording techniques) in the staree during staring and nonstaring periods to determine whether those periods could be unconsciously differentiated. We used spontaneous phasic skin resistance response (SRR) activity as an indicator of the subject's degree of sympathetic autonomic arousal or activation. Our equipment automatically corrected for drift in baseline level (basal skin resistance) so that our measures would be sensitive to changes in the subject's state and would not be biased by individual differences in baseline. As in the Williams (1983) study, separable closed rooms and a closed circuit television system were used to eliminate conventional communication channels between starer and staree. We also sought to compare the autonomic staring detection ability of two groups of subjects. One group (tested in Phase 1) consisted of untrained subjects. Another group (tested in Phase 2) consisted of subjects who had undergone special training designed to help them increase their sensitivity to internal physiological reactions, to increase their understanding of what it might feel like to be "interconnected" with other persons, and to help them deal with their possible psychological resistances to such interconnectedness.

Method

Subjects

Thirty-two subjects participated as starees in this experiment. The subjects were unselected (perhaps a better description would be "self-selected") persons from the local community who were interested in our experiments and who had become aware of them through various media (local radio, newspaper, and newsletter) descriptions and through information provided by previous subjects. The participants ranged in age from 22 to 71 years; there were 24 females and 8 males. The 32 subjects were tested in two phases. Phase 1 involved 16 subjects who were "untrained." Phase 2 involved 16 subjects who had been "self-" selected from the same general subject pool; these subjects, however, were tested following their participation in a "connectedness training" program (see section on Procedure) conducted by the third author (S. A.). The second author (D. S.) played the role of both experimenter and starer throughout the experiment; she, too, participated in the connectedness training following Phase 1 and preceding Phase 2.

Apparatus

The experimental apparatus consisted of silver/silver chloride palmar electrodes (7.0 mm in diameter), a skin-resistance amplifier (Lafayette Model 76405), and an analog-to-digital converter interfaced with a microcomputer. A color video camera (Hitachi Camcorder VM-2250) in the staree's room permitted the staree to be viewed by the starer in a distant room without the possibility of sensory cueing. The camera's radio frequency (RF) output was boosted by a 10 dB amplifier, then conveyed via heavy duty 300-ohm impedance twin-lead cable to a 19-inch color TV monitor (Sony Trinitron KV-1114) situated in the starer's room. Additional details concerning equipment, room layout, and physiological monitoring are given in Braud and Schlitz (1989).

Procedure

The experimenter met with the subject in the Mind Science Foundation's library, where the subject completed the Myers-Briggs Type Indicator (MBTI, Form F: see Briggs and Myers, 1957), a 55-item personal history survey (Participant Information Form [PIF]) developed at the Psychophysical Research Laboratories (Psychophysical Research Laboratories, 1983), and a brief staring questionnaire. The staring questionnaire asked whether the subject had ever felt an unseen person staring at him or her and whether such an experience took the form of a physical sensation or a conscious thought; the questionnaire also asked that the subject describe the experience. After completing these assessments, the subject was taken to the starer's room, shown the television monitor, and informed of the details of the procedure.

Next, the experimenter led the subject to the staree's room, which was located in an entirely different suite area across an outside corridor. The two rooms were separated from each other by two inner hallways, an outer corridor, and four closed doors. Neither room had any windows. The physical separation and geography of the rooms, along with a firm experimental protocol that precluded auditory cueing, assured that conventional sensorimotor communication between these two rooms, under the conditions of the experiment, was not possible. The staree's room was brightly illuminated by means of overhead fluorescent lights. The camera, which was continuously active throughout the entire session, was mounted on a tripod 6 ft away from the subject's chair, at eye level, and at an angle of approximately 45 degrees left of center (from the subject's point of view). The camera's zoom lens was set so that the subject's shoulders, neck, and head would be visible on the monitor in the starer's room. The camera's autofocus function was disabled in order to eliminate distracting camera lens movement noises that otherwise might have resulted from automatic tracking of subject movements; this also eliminated possible distracting changes in the subject's image, from the starer's point of view.

The subject was seated in a comfortable recliner chair (which remained in an upright position throughout the experiment), and the experimenter attached two silver/silver chloride electrodes filled with partially conductive electrode gel to the subject's left palm by means of adhesive electrode collars. The subject was asked to sit quietly for the next 20 min, to refrain from unnecessary movements (especially of the left hand and arm), and to think about whatever he or she wished during the experiment. The subject was told that the camera would be on throughout the 20-min session, but that the experimenter would watch the monitor only at certain randomly determined times during the experiment; at those times, the experimenter would stare intently at the subject's image on the monitor and would attempt to gain the subject's attention. The subject was asked not to try to guess consciously when those periods (of which the subject was, of course, kept blind) might be occurring, and was told that we were exploring whether any unconscious physiological reactions might be associated with remote staring. The experimenter then left the subject alone in the staree room and went to the distant starer's room, closing all doors behind her.

In the starer's room, the experimenter recorded the subject's basal skin resistance and then, prior to starting the microcomputer that controlled the session events, retrieved from a hidden location a sealed opaque envelope that contained the random sequence of staring and nonstaring periods that would be used for that session. Thirty-two such envelopes had been prepared previously by W. B., who had used a computer's random algorithm to generate the random sequence of the 10 staring and 10 nonstaring periods for each session. In a hidden location known only to him, W. B. kept his own copies of the 32 random sequences. The microcomputer program controlled the timing of the various events of the experiment and recorded the subject's electrodermal activity during each of 20 thirty-second recording periods. Each of the 20 recording periods was signalled by a low-pitched tone (audible only to the experimenter, through headphones); a 30-second rest period followed each recording period. The experimenter consulted the contents of the session envelope to learn which of the 20 recording periods were to be devoted to staring and which were to serve as the nonstaring, control periods. In consulting her sheet of random epoch sequences, the experimenter used a method of occluding her view of all epoch instructions, other than the present one, so that she could devote full attention to her assignment for that epoch without being distracted by instructions for subsequent or earlier epochs. If the random sequence indicated a staring period, the experimenter silently swiveled her chair around so that it faced the TV monitor, she stared intently at the subject's monitor image throughout the 30-sec recording period. During nonstaring periods, she kept her chair turned away from the monitor so that she could not see the monitor's screen; she busied her mind with matters unrelated to the experiment. All reflective surfaces had been carefully covered so that inadvertent glimpses of the monitor screen were not possible.

Throughout the session, the experimenter received no information about the subject's ongoing electrodermal activity; the latter was continuously and automatically assessed by the computer system. The equipment sampled the subject's spontaneous phasic skin resistance responses (SRR) 10 times each second for the 30 seconds of a recording epoch and averaged these measures, providing what is virtually a measure of the area under the curve described by the fluctuation of electrodermal activity over time (that is, the mathematically integrated activity). At the end of the experimental session, the computer printed the electrodermal results for each of the 20 recording periods. The experimenter filed away the printout, without looking at the electrodermal measures, then went to the staree's room and discussed the experiment with the subject in general terms. Neither experimenter nor subject had any knowledge of the numerical results for the session. Only after all 32 sessions had been completed did W. B. analyze the results and give the experimenter feedback. The experimenter later provided feedback to those subjects who requested it.

The first 16 subjects who participated in Phase 1 of the study had no special preparation. The 16 starees of Phase 2, however, had participated in approximately 20 hours of "connectedness training" provided by the third author (S. A.), in which the participants engaged in intellectual and experiential exercises involving feelings of interconnectedness with other people. The training began with a group viewing of a videotape based on Peter Russell's book, *The Global Brain* (1983). This was followed by discussions of the videotape, lectures by S. A., and experiential exercises in which participants became increasingly comfortable and adept at "connecting" with each other. The latter took the form of staring into another person's eyes for long periods of time, becoming comfortable with this, observing how one's physiological reactions came to more closely resemble those of the other, and conversing and retrieving information while maintaining eye contact (rather than averting the gaze upward or sideways, as would usually occur during memory retrieval and cognitive processing; see Bakan, 1980). Individual and group discussions were devoted to learning about and dealing with psychological resistances that interfered with the process of connectedness or with feelings of "merging" with another person. The experimenter/starer for the present study (D. S.) actively participated in all training sessions. The participants were aware that the training would be followed by an experiment involving physiological detection of staring, but were not aware of any more details of the study than were the 16 untrained subjects of Phase 1.

In the present study, we simply explored the possible effects of the connectedness training. We did not make any predictions about scoring direction in the two phases and therefore planned to use two-tailed tests in their evaluations. On the one hand, it could be argued that the training could increase the sensitivity of Phase 2 subjects to any effects that might occur in Phase 1. Alternatively, the training might result in a qualitatively different reaction pattern in Phase 2.

Results

For each subject, electrodermal activity was measured during 10 staring and during 10 nonstaring periods. Rather than compare these multiple scores within a given subject, we reduced the activities for an entire session to a single score for each subject and performed statistical tests using subjects, instead of multiple period scores, as the units of analysis. We used the more conservative session score (a kind of single, majority-vote score) in order to bypass criticisms based on possible nonindependence of multiple electrodermal measures taken within a given session. Although it would be possible to analyze individual epoch scores using, for example, a repeated measures analysis of variance procedure, such an analysis assumes that the autocorrelations among the measures within each session (i.e., within each subject) are constant across epochs, and that the same autocorrelation applies to all sessions (subjects). Because these assumptions may not be met in these experiments, we preferred to use the more conservative session-based (rather than epoch-based) analyses, even though the former are more wasteful of data and result in tests with reduced statistical power.

For each of the 32 sessions, a total score was calculated for all 20 recording periods (10 staring and 10 nonstaring). This total score was divided into the sum of the electrodermal activity scores for the 10 staring (S) periods; the process was repeated for the 10 nonstaring (N) periods. In the absence of a remote staring effect, these two ratios $[S/(S + N), N/(S + N)]$ should approximate 50%. A remote staring effect would be indicated by a significant departure of the scores from the 50% mean chance expectation (MCE). Single mean t tests were used to assess the departure of the ratios from MCE (50%). This is approximately equivalent to calculating dependent (matched) t tests to compare the raw scores for each subject for staring versus nonstaring periods. We have consistently used ratio scores in our various projects as a method of "standardizing" scoring so that scoring magnitude could be more meaningfully compared for the different dependent measures (response systems) with which we work.

First, an analysis was performed on the staring/total-activity ratios of the 16 untrained subjects of Phase 1. A single mean t test indicated that the 16 untrained starees exhibited significantly greater spontaneous electrodermal activity during staring periods than during nonstaring, control periods. The mean percent electrodermal activity for staring periods was 59.38%, rather than the 50.00% expected by chance. The single mean t test comparing the 16 percentages with 50% MCE was 2.66 which, with 15 degrees of freedom, has an associated two-tailed $p = .018$, and a calculated effect size = .59; the 95% confidence interval is bounded by the values 51.86% and 66.90%. Thus, these subjects were significantly more activated (in terms of sympathetic autonomic activity) by remote staring than by the nonstaring, control periods.

Next, a parallel analysis was performed on the scores for the 16 Phase 2 subjects who had experienced connectedness training prior to their experimental sessions. A single mean t test indicated that the 16 trained starees exhibited significantly less spontaneous electrodermal activity during staring periods than during nonstaring, control periods. The mean percent electrodermal activity for staring periods was 45.45%, rather than the 50.00% expected by chance. The single mean t test comparing these 16 percentages with 50% MCE was 2.15 which, with 15 degrees of freedom, has an associated two-tailed $p = .048$, and a calculated effect size of .50; the 95% confidence interval is bounded by the values 40.94% and 49.95%. Thus, these trained subjects were significantly more calmed (in terms of sympathetic autonomic activity) by remote staring than they were by the nonstaring, control periods.

If the scores for the subjects of the two phases are directly compared by means of an independent-groups t test, a significant difference is found between the untrained and the trained subjects ($t = 3.39$, $df = 30$, $p = .002$, two-tailed). It should be noted that the latter test was post hoc and was carried out simply to quantify the obvious difference in scoring patterns observed for the two phases.

Secondary analyses were performed to test the equivalence of the Phase 1 and Phase 2 subjects in terms of their personality (MBTI) and physiological (electrodermal activity) characteristics; a summary of these analyses is presented in Table 2. For the MBTI scores, group means are presented for the continuous scores of the extraversion/introversion (E/I), sensing/intuition (S/N), thinking/feeling (T/F), and judging/perceiving (J/P) dimensions. A score of 100 represents the midpoint of each continuum. Scores less than 100 indicate tendencies toward extraversion, sensing, thinking, and judging; scores greater than 100 indicate tendencies toward introversion, intuition, feeling, and perceiving. For electrodermal activity scores, group means are given for the sum of spontaneous skin resistance responses integrated over all 20 recording epochs (total SRR) and for the subjects' initial basal skin resistance (BSR) in ohms. High total SRR scores and low BSR scores are associated with increased sympathetic autonomic arousal, whereas low total SRR scores and high BSR scores are associated with decreased sympathetic arousal. Analyses indicated that the Phase 1 and Phase 2 groups did not differ significantly on any of these six measures.

We are now able to supplement the findings previously summarized in Table 1 with the results of the present investigation at the Mind Science Foundation, using closed-circuit TV and autonomic measures (see Table 3).

A more detailed statistical summary of all relevant staring detection research is presented in Table 4. If effect size is taken to be the most appropriate measure of the strength of an obtained outcome (Rosenthal, 1984, 1990), it appears that the autonomic recording method of the present study does indeed yield stronger results than do the conscious-guessing measures of staring detection used in previous studies.

Table 2. Group Means and Statistical Comparisons of Personality and Physiological Characteristics of Phase 1 and 2 Subjects

Phase	MBTI Continuous Scores[a]				Electrodermal activity	
	E/I	S/N	T/F	J/P	Total SRR	BSR *ohms*
1 (untrained)	87.12	123.00	106.75	117.00	605.19	343,506
2 (trained)	98.75	133.69	98.06	103.50	656.06	289,047
t	1.49	1.44	1.23	1.52	0.54	0.82
p	.15	.16	.28	.14	.60	.42

Note. All *p*s are two-tailed.
[a]E/I denotes extraversion/introversion; S/N, sensing/intuition; T/F, thinking/feeling; J/P, judging/perceiving.

Table 3. Summary of Present Autonomic Staring Detection Experiments

Subjects	Scoring rate	Effect size
Untrained	59.38%	.59
Trained	45.45%	−.50

The statistical values of Table 4 may be used in a preliminary meta-analysis of all staring-detection studies reported to date. The table lists the statistical test presented in the original report, the one-tailed *p* value associated with that test, the *z*-score equivalent of the one-tailed *p* value, the number of units contributing to the analysis, and the effect size (calculated by dividing the equivalent *z* score by the square root of *n*). Combining all six tabulated entries yields a mean $z = 1.07$, a Stouffer $z = 2.62$ (with associated one-tailed $p = .0044$), and a mean effect size $= .17$. The Stouffer z procedure, an accepted method for combining probabilities of several studies testing essentially the same hypothesis, is described by Rosenthal (1984); this source also provides an excellent discussion of various effect size measures.

Table 4. Statistical Summary of All Staring Detection Experiments

Study	Test	p^a	z	n	Effect size
Coover (1913)	$z = 0.126$	.4499	0.126	1000	.004
Poortman (1959)	$z = 1.70$	.044	1.70	89	.18
Peterson (1978)	$t = 2.648$	.006	2.51	36	.42
Williams (1983)	$t = 1.77$	.044	1.70	28	.32
Braud, et al. (1993)					
Untrained subjects	$t = 2.66$	.009	2.37	16	.59
Trained subjects	$t = -2.15$	.976	-1.98	16	-.50

^{a}one-tailed

A comment is necessary regarding the effect size for trained subjects given in Tables 3 and 4. In the present research, we sought to determine whether the subjects would autonomically discriminate the staring from the nonstaring (control) periods; indeed, they were able to do this in both phases. No prediction was made regarding the direction of their differential autonomic response, that is, whether their electrodermal activity would be greater or less during staring periods (compared with nonstaring periods). For this reason, two-tailed tests were used for each phase, and results for both phases were "successful" (i.e., both scoring rates departed significantly from chance expectation). However, for purposes of meta-analysis, it is customary to use only one-tailed tests and p values in the tabulations. It is also customary to use a negative sign for a result that is inconsistent with the bulk of the results (see Rosenthal, 1984, p. 95). We have followed this convention when entering the results for the trained subjects (Phase 2) of this study. This provides a conservative estimate of overall results, because the autonomic discrimination of the trained Phase 2 subjects was just as effective as that of the untrained Phase 1 subjects, but happened to be in the calm as opposed to the active direction. This reversal of direction becomes understandable when considered in relation to the nature of the training experienced by these Phase 2 subjects (see the Discussion section). If the results of the Phase 2 subjects are also considered "positive," then the alternative values for summarizing Table 4 become: mean $z = 1.73$, Stouffer $z = 4.24$, $p = .000011$, mean effect size = .34.

It should also be pointed out that the units of analysis for the effect sizes reported in Tables 1 and 4 are not comparable for all studies. Effect sizes for the first two studies (those of Coover and of Poortman) are based on trial units, whereas those of the remaining studies are based on subject units. These differences should be kept in mind when evaluating these effect sizes.

Discussion

Prior research yielded suggestive evidence that persons were able to discriminate staring and nonstaring periods by means of deliberate, conscious guesses. The aim of the present project was to determine whether staring and nonstaring periods could be differentiated by means of more "unconscious" physiological reactions. The electrodermal activity differences between staring and nonstaring periods indicated that such differentiation could indeed occur. We chose to measure spontaneous electrodermal fluctuations (that is, changes in skin resistance reactions) because such measurements are easy to make, are sensitive indicators, and are known to be useful peripheral measures of the activity of the sympathetic branch of the autonomic nervous system. The occurrence of many or of high amplitude skin resistance reactions (SRRs) is symptomatic of increased sympathetic activation or arousal, which may in turn reflect increased emotionality (see Edelberg, 1972; Prokasy and Raskin, 1973; Venables and Christie, 1980). On the other hand, the occurrence of few or of low amplitude SRRs indicates decreased sympathetic activation or arousal, which may in turn reflect decreased emotionality and, therefore, a greater degree of emotional and mental quietude or calmness.

The results of both phases of the present study indicate reliable autonomic discrimination of staring and nonstaring periods, and the relatively large effect sizes suggest that autonomic detection may be a more powerful method than conscious guessing for the detection of staring effects. Phase 1 findings suggest that the starees were more activated during the staring than during the nonstaring epochs. Phase 2 findings suggest that those starees were more calm during staring than during nonstaring periods. The latter finding does in fact make sense in view of the training experienced by the Phase 2 subjects. That training was designed to allow persons to become more comfortable with staring and with connecting with other people; and it permitted the trainees to reduce at least some of their defenses or resistances to staring, being stared at, and sharing a mutual gaze. In the course of their training, many trainees reported that their staring encounters became quite positive and pleasant interactions, and they expressed disappointment when the encounters ended (see Kellerman, Lewis, and Laird, 1989, for the sometimes powerful effects of sharing a mutual gaze). We speculate that similar processes may have occurred during remote staring: The trained subjects of Phase 2 may have missed the contact with, and the attention of, the starer (with whom they had become increasingly familiar during the course of the training), and may have become more relaxed and calm when that attention was provided, albeit in remote form, during the staring epochs of the experiment. A useful analogy for the reader (from the domain of animal behavior studies) might be the alarm and distress that occur upon the removal of an imprinted object from the environment of an imprinted precocial fowl or other organism, and the distress-reduction that occurs when the imprinted object is reintroduced (see Bateson, 1966; Ratner and Denny, 1964). For the Phase 1 starees, who did not have the benefit of the connectedness training, being stared at (even in its remote form) may have been experienced in a more typical way, that is, as threatening (see Argyle, 1975, pp. 229–250; Argyle and Dean, 1965) and sympathetically activating (rather than calming).

These comments apply, not only to the starees, but also to the starer (D. S.). Although she attempted to behave identically and maintain identical attitudes in the two phases, when she began Phase 2, she had of course participated actively in the connectedness training and may well have been more comfortable and relaxed about attending to and connecting with her subjects in Phase 2 than she had been in Phase 1. This increased comfort and relaxation could have been reflected in the calming direction of the Phase 2 results. We deliberately included the starer in the connectedness training because the training was directed at changing dyadic relationships and the starer was a critical part of the dyadic effects we wished to explore in Phase 2.

It could be hypothesized that the different patterns of findings for Phase 1 and Phase 2 may have been contributed by a nonequivalence of initial characteristics of the subjects in the two phases. This hypothesis is not a convincing one inasmuch as the subjects for the two phases came from the same general participant pool and did not appear to differ importantly in terms of PIF characteristics, MBTI profiles, or overall electrodermal reactions (either basal skin resistance or total electrodermal activity for the session, both of which reflect general arousal level, nervousness, etc.). It would appear that the participants in the two phases were sufficiently similar in their initial characteristics to rule out differences attributable to those factors alone. The participants' training (that of the starees and of the starer in Phase 2) appears to have been more critical than possible initial differences in determining the qualitatively different outcomes of Phase 2. Further research will help clarify these issues.

The results of this physiological investigation, along with those of previous behavioral studies, provide evidence that persons are indeed able to respond to instances of remote attention (such as that provided by an unseen gaze) even under conditions that preclude conventional sensorimotor communication. The specific mechanism underlying these anomalous manifestations is unknown, although several physical and other models that do accommodate nonlocal interconnectedness of consciousness have been proposed (see, for example, Stokes, 1987). The present findings are consistent with a vast body of similar evidence collected within the field of parapsychology or psi research. Some of the most compelling evidence is provided in several recent meta-analyses (Harris and Rosenthal, 1988; Honorton et al., 1990; Radin and Nelson, 1989; Rosenthal, 1986; Utts, 1991), reviews (e.g., Child, 1985), and handbooks (Krippner, 1977–1982 and 1984–1990; Rao, 1984; Schmeidler, 1988; Wolman, 1977). Critical discussions of this evidence may be found in

Kurtz (1985). The studies reviewed in the present paper indicate that these anomalous phenomena are not merely laboratory curiosities, but may be reliable and robust enough to have important influence in everyday life and may have important impacts upon, and implications for, a wide variety of psychophysiological and social psychological investigations.

References

Argyle, M. (1975). *Bodily communication*. New York: International Universities Press.

Argyle, M., and Dean, J. (1965). Eye-contact, distance, and affiliation. *Sociometry, 28*, 289–304.

Bakan, P. (1980). Imagery, raw and cooked: A hemispheric recipe. In J. Shorr, G. Sobel, P. Robin, and J. Connella (Eds.). *Imagery. Its many dimensions and applications* (pp. 35–53). New York: Plenum.

Bateson, P. P. G. (1966). The characteristics and context of imprinting. *Biological Review, 41*, 177–220.

Braud, W. G., and Schlitz, M. J. (1989). A methodology for the objective study of transpersonal imagery. *Journal of Scientific Exploration, 3*, 43–63.

Briggs, K. C., and Myers, I. B. (1957). *Myers-Briggs Type Indicator Form F.* Palo Alto, CA: Consulting Psychologists Press.

Child, I. L. (1985). Psychology and anomalous observations. *American Psychologist, 40*, 1219–1230.

Coover, J. E. (1913). The feeling of being stared at. *American Journal of Psychology, 24*, 571–575.

Edelberg, R. (1972). Electrical activity of the skin: Its measurement and uses in psychophysiology. In N. Greenfield and R. Sternbach (Eds.), *Handbook of psychophysiology* (pp. 368–418). New York: Holt, Rhinehart, and Winston.

Harris, M. J., and Rosenthal, R. (1988). *Interpersonal expectancy effects and human performance research*. Washington, DC: National Academy Press.

Honorton, C., Berger, R., Varvoglis, M., Quant, M., Derr, P., Schechter, E., and Ferrari, D. (1990). Psi communication in the ganzfeld. *Journal of Parapsychology, 54*, 99–139.

Kellerman, J., Lewis, J., and Laird, J. D. (1989). Looking and loving: The effects of mutual gaze on feelings of romantic love. *Journal of Research in Personality, 23,*145–161.

Krippner, S. (1977–1982). *Advances in parapsychological research: Volumes 1–3*. New York: Plenum.

Krippner, S. (1984–1990). *Advances in parapsychological research: Volumes 4–6*. Jefferson, NC, and London: McFarland.

Kurtz, P. (1985). *A skeptic's handbook of parapsychology*. Buffalo, NY: Prometheus Books.

Peterson, D. M. (1978). *Through the looking glass: An investigation of the faculty of extra-sensory detection of being stared at*. Unpublished thesis. University of Edinburgh, Scotland.

Poortman, J. J. (1959). The feeling of being stared at. *Journal of the Society for Psychical Research, 40*, 4–12.

Prokasy, W. F., and Raskin, D. C. (1973). *Electrodermal activity in psychological research*. New York: Academic Press.

Psychophysical Research Laboratories (1983). *1983 PRL annual report*. Princeton, NJ: Psychophysiological Research Laboratories.

Radin, D. I., and Nelson, R. D. (1989). Evidence for consciousness-related anomalies in random physical systems. *Foundations of Physics, 20* (1), 1499–1514.

Rao, K. R. (1984). *The basic experiments in parapsychology.* Jefferson, NC: McFarland.

Ratner, S. C., and Denny, M. R. (1964). *Comparative psychology: Research in animal behavior* (pp. 351–368). Homewood, IL: Dorsey Press.

Rosenthal, R. (1984). *Meta-analytic procedures for social research.* Beverly Hills, CA: Sage Publications.

Rosenthal, R. (1986). Meta-analytic procedures and the nature of replications: The ganzfeld debate. *Journal of Parapsychology,* 50, 315–336.

Rosenthal, R. (1990). Replication in behavioral research. *Journal of Social Behavior and Personality,* 5 (4), 1–30.

Russell, P. (1983). *The global brain.* Los Angeles, CA: J. P. Tarcher.

Schmeidler, G. R. (1988). *Parapsychology and psychology: Matches and mismatches.* Jefferson, NC: McFarland.

Stokes, D. M. (1987). Theoretical parapsychology. In S. Krippner (Ed.), *Advances in parapsychological research, Volume 5* (pp. 77–189). Jefferson, NC, and London: McFarland.

Titchener, E. B. (1898). The feeling of being stared at. *Science,* 8, 895–897.

Utts, J. (1991). Replication and meta-analysis in parapsychology. *Statistical Science,* 6, 363–403.

Venables, P. J. and Christie, M. H. (1980). Electrodermal activity. In I. Martin and P. Venables (Eds.), *Techniques in psychophysiology* (pp. 3–67). New York: Wiley.

Williams, L. (1983). Minimal cue perception of the regard of others: The feeling of being stared at. Paper presented at the 10th Annual Conference of the Southeastern Regional Parapsychological Association. West Georgia College, Carrollton, GA, February 11–12 [See also Williams, L. *The feeling of being stared at: A parapsychological investigation.* Unpublished manuscript, n.d.]

Wolman, B. B. (1977). *Handbook of parapsychology.* New York: Van Nostrand Reinhold.

Institute of Transpersonal Psychology
744 San Antonio Road
Palo Alto, CA 94303

8

Additional Studies of Bodily Detection of Remote Staring

William Braud, Donna Shafer, and Sperry Andrews

This chapter describes extensions of the work presented in chapter 7.
The experiments on physiological detection of remote staring are repli-
cated, new controls are added, and the phenomenon is related to certain
personality characteristics of the research participants.

This information originally was published in Braud, W., Shafer, D., and
Andrews, S. (1993). Further studies of autonomic detection of remote star-
ing: Replication, new control procedures, and personality correlates.
Journal of Parapsychology, 57(4), 391–409. The contents of this article are
Copyright © 1993 by the Parapsychology Press, and the material is used
with permission.   —William Braud

Abstract: In a previous paper, we reviewed early experimental attempts to assess subjects' accuracy in consciously detecting when they are being watched or stared at by
someone situated beyond the range of their conventional senses. We also reported new
results of our own experiments in which a more "unconscious" autonomic nervous system reaction (spontaneous electrodermal activity) was used to assess accuracy of detection of staring (remote attention). In our experiments, one subject (the starer) directed
full attention to another distant subject's (staree's) image on the monitor of a closed-
circuit television system used to eliminate the possibility of subtle sensory cues. The
staree's spontaneous electrodermal activity, meanwhile, was monitored objectively by a
computer system during randomly interspersed staring and nonstaring periods; the
staree was blind regarding the number, timing, and sequencing of the two types of

This article is based on a paper presented at the Thirty-Fifth Annual Convention of the
Parapsychological Association in Las Vegas, NV, in August, 1992.

period. We found evidence for significant blind autonomic discrimination between the staring and nonstaring episodes. In the present paper, we report evidence for autonomic discrimination of staring versus nonstaring periods in two replications—one involving the same starer who had participated in the earlier studies ($t[15] = 2.08$; $p = .05$, two-tailed; effect size $r = .47$), and the second involving three new starers ($t[29] = 1.92$; $p = .06$, two-tailed; effect size $r = .34$). Chance results were found, as expected, in a new, improved control condition (a "sham control") in which the data were treated as they were in a true staring study, but staring did not, in fact, occur. We also found that the magnitude of the remote autonomic staring detection effect was significantly related to the starees' degree of introversion (Myers-Briggs Type Indicator) and to their degree of social avoidance and distress (social anxiety).

In a previous paper (Braud, Shafer, and Andrews, 1993), we reviewed the scientific literature dealing with the purported ability to detect when one is being watched or stared at by someone situated beyond the range of the conventional senses. Surveys indicated that between 68% and 94% of various samples reported having had staring detection experiences in their everyday lives. Previous investigations provided suggestive evidence that persons were indeed able to detect, consciously, when they were being stared at under conditions in which precautions were taken to eliminate possible subtle sensory cues. In particular, positive conscious-guessing results were obtained in two studies in which sensory cueing was eliminated through use of one-way mirrors (Peterson, 1978) and use of a closed-circuit television system (Williams, 1983).

We hypothesized that stronger effects might be obtained if relatively "unconscious" autonomic nervous system activity were used as the indicator of staring detection, rather than conscious guessing. Our reasoning was that autonomic reactions might be less distorted by higher cognitive processes and therefore might provide a purer and more sensitive indicator. We presented the results of two original experiments in which sympathetic nervous system activation was assessed by means of electrodermal monitoring during randomly interspersed remote-staring and nonstaring (control) periods. The monitored participant was unaware of the number, timing, or scheduling pattern of these two types of periods. The possibility of sensory cueing was eliminated through the use of a closed-circuit television system for staring: the starer devoted full attention to the staree's image on the television monitor. In the first experiment, 16 untrained participants evidenced significant autonomic discrimination, becoming more activated during staring than during nonstaring periods. In the second experiment, 16 subjects who had been extensively trained to become more aware of their interconnections with other people and less defensive about their connectedness also evidenced significant autonomic discrimination, but became more calm during staring than during nonstaring periods; the starer had been similarly trained. As judged by effect sizes, unconscious autonomic detection did indeed appear to yield stronger effects than did previous conscious verbal or motoric detection assessments.

In the present paper, we present our attempts to replicate and extend our previous findings. Identical equipment, basic procedures, and analysis methods were used. The first replication involved 3 new starers and 30 new starees. The second replication involved the same starer who had participated in the earlier experiments reported in 1993, but employed 16 new starees. We made two additions in these studies. One of these was the introduction (into Replication 2) of an additional empirical control condition. This was a "sham control" in which we treated sessions and data as we did for real staring sessions, but staring did not, in fact, occur; this provided an empirical assessment of the likelihood of obtaining chance discriminations of otherwise equivalent session segments. The second improvement was the introduction of a new personality assessment for the starees in both replications. In addition to the Myers-Briggs Type Inventory (MBTI) we had been using in the original studies, we included an assessment of social anxiety or discomfort in a social situation (a Social Avoidance and Distress scale), in order to explore the possible interrelationships of these personality characteristics with the autonomic staring detection effect.

Method

Subjects

Thirty volunteer participants (22 females and 8 males) served as "starees" for Replication 1, and 16 volunteers (5 males and 11 females) participated as starees for Replication 2. In Replication 1, half of the starees were persons already known by the starers (relatives, friends, or familiar undergraduate classmates), whereas half were unknown at the time of the laboratory session (i.e., they were unfamiliar undergraduates); only one of the starees had participated previously in laboratory psi experiments. (Later results did not differ for the known versus unknown starees.) It had been decided in advance that each starer was to work with 10 starees and that results for all 30 starees were to be pooled for purposes of analysis. In Replication 2, 13 of the starees were previously unknown undergraduate students from a local college, and 3 were friends or relatives of the starer; only 2 of the starees had participated previously in laboratory psi studies. Participants were selected on the basis of availability during planned laboratory session times and on the basis of interest in participating in a study exploring the "feeling of being stared at." Across both replications, staree age ranged from 17 years to 40 years.

The starers of Replication 1 were three undergraduate psychology students (two females and one male) from a local college who were participating in independent studies internships at the Mind Science Foundation. None of these starers had prior laboratory psi research experience. The starers were trained for the experiment by the second author (D. S.), who had served as starer in our original (1993) staring detection experiments. D. S. served as starer for Replication 2. She herself had participated previously in extensive "connectedness" training that had been provided by the third author, S. A. This training (which is described in Braud, Shafer, and Andrews, 1993) took the form of approximately 20 hours of intellectual and experiential exercises designed to help individuals become more adept at and comfortable with experiencing interconnections with others, and to become more aware of, and to deal more effectively with, psychological resistances to such connectedness. It is important to note that D. S. was very comfortable with "connecting with" (i.e., having feelings of "merging with") others when these replication experiments began, and that it is likely that she communicated this ease and comfort to the three Replication 1 starers during the course of their training by her. (This training involved discussions of the rationale for the studies, previous results, and procedural information about the experiments; the three Replication 1 starers experienced no formal "connectedness" training, although that training was discussed in general terms by D. S.)

Apparatus

The experimental apparatus was identical to that described in Braud, Shafer, and Andrews (1993) and consisted of silver/silver chloride palmar electrodes (7.0 mm in diameter) attached with semi-conductive electrode gel, a skin-resistance amplifier (Lafayette Model 76405), and an analog-to-digital converter interfaced with a microcomputer. A color video camera (Hitachi Camcorder VM-2250) in the staree's room permitted the staree to be viewed by the starer in a distant room without the possibility of sensory cueing. The camera's radio frequency output was boosted by a 10-dB amplifier, then conveyed via heavy duty 300-ohm impedance twin-lead cable to a 19-in. color TV monitor (Sony Trinitron KV-1914) situated in the starer's room. Additional details concerning equipment, room layout, and physiological monitoring are given in Braud and Schlitz (1989).

Procedure

With the exceptions to be noted later, procedural details for the two replications were identical to those of the original experiments (see Braud, Shafer, and Andrews, 1993). For both replications, the starer (who was also the experimenter) greeted the staree in the starer's room, explained the experiment, and showed the staree the television monitor on which the latter's image would appear during the session.

Next, the starer led the staree to the staree's room, which was located in an entirely different suite area across an outside corridor. The two rooms were separated from each other by two inner hallways, an outer corridor, and four closed doors. Neither room contained any windows. Conventional sensorimotor communication between these two rooms, under the conditions of the experiment, was not possible. The staree's room was brightly illuminated by means of overhead fluorescent lights. The camera, which was active continuously throughout the entire session, was mounted on a tripod 6 ft away from the staree's chair, at eye level, and at an angle of approximately 45 degrees left of center (from the staree's point of view). The camera's zoom lens was set so that the staree's shoulders, neck, and head would be visible on the monitor in the starer's room. The camera's autofocus function was disabled in order to eliminate distracting camera lens movement noises that otherwise might have resulted from automatic tracking of staree movements; this also eliminated possible distracting changes in the staree's image, from the starer's point of view.

The staree was seated in a comfortable recliner chair (which remained in an upright position throughout the experiment), and the experimenter attached two silver/silver chloride electrodes filled with partially conductive gel to the staree's left palm by means of adhesive electrode collars. The staree was asked to sit quietly for the next 20 min and to refrain from unnecessary movements (especially of the left hand and arm). In order to more closely simulate a naturalistic staring-detection situation, the staree occupied his or her mind during the session by studying, reading a magazine, or thinking about and planning the day's activities (for Replication 1), or by completing the personality assessments (in Replication 2). The staree was told that the camera would be on throughout the 20-min session, but that the experimenter would watch the monitor only at certain randomly determined times. At those times, the starer would stare intently at the staree's image on the monitor and would attempt to gain the staree's attention. The staree was asked not to try to guess consciously when those periods (of which the staree was, of course, kept blind) might be occurring and was told that we were exploring whether any unconscious physiological reactions might be associated with remote staring. The experimenter then left the subject alone in the staree room and went to the distant starer's room, closing all intervening doors.

In the starer's room, the experimenter/starer recorded the staree's basal skin resistance and then, prior to starting the microcomputer that controlled the session events, retrieved from a hidden location a sealed opaque envelope that contained the random sequence of staring and nonstaring periods that would be used for that session. Forty-six such envelopes had been prepared previously by W. B., who had used a computer's random algorithm to generate the random sequence of the 10 staring and 10 nonstaring periods for each session. In a hidden location known only to him, W. B. kept his own copies of the 46 random sequences. The microcomputer program controlled the timing of the various events of the experiment and recorded the staree's electrodermal activity during each of the twenty 30-sec recording periods. Each of the 20 recording periods was signaled by a low-pitched tone (audible only to the experimenter, through headphones); a 30-sec rest period followed each recording period. The experimenter/starer consulted the contents of the session envelope to learn which of the 20 recording epochs were to be devoted to staring and which were to serve as the nonstaring control periods. If the random sequence indicated a staring period the experimenter/starer silently swiveled his or her chair around so that it faced the television monitor, and stared intently at the staree's monitor image throughout the 30-sec recording periods. During non-staring periods, the experimenter/starer kept the chair turned away from the monitor, so that the monitor's screen could not be seen, and thought about matters unrelated to the experiment. All reflective surfaces had been carefully covered and inadvertent glimpses of the monitor screen were not possible. In consulting a session's random-sequence sheet, the experimenter/starer used a method of occluding all epoch instructions other than the present one, so that he/she could devote full attention to the assignment for that epoch without being distracted by instructions for subsequent or previous epochs.

Throughout the session, the experimenter/starer was provided with no information about the staree's ongoing electrodermal activity; the latter was continuously and automatically assessed by the computer system. The equipment sampled the staree's rectified (by means of a diode) spontaneous phasic skin resistance responses (SRR) 10 times each second for the 30 seconds of a recording epoch and averaged these measures, providing what is virtually a measure of the area under the curve described by the fluctuation of electrodermal activity over time (i.e., the mathematically integrated activity). Because of the slowly changing nature of these autonomic reactions, this relatively slow sampling rate is quite adequate. At the end of the experimental session, the computer printed the electrodermal results for each of the 20 recording periods. The experimenter filed away the printout, taking special precautions not to look at the electrodermal measures, then went to the staree's room and discussed the experiment in general terms with the staree. Neither experimenter/starer nor staree had any knowledge of the numerical results for the session. Only after all sessions had been completed did W. B. analyze the results and give the experimenter feedback. The experimenter later provided feedback to those starees who requested it.

The general procedure described above applied to both replications. The specifics, however, were changed slightly for Replication 2. For the latter, instead of 20 recording epochs (10 staring and 10 nonstaring, randomly interspersed), a session consisted of 32 recording epochs. One half of a session was an "experimental" half and included 8 staring and 8 nonstaring periods (randomly interspersed). The other half of a session served as a new empirical sham control and included 8 sham or pseudostaring periods and 8 nonstaring periods (also randomly interspersed). The experimental half provided a comparison of true staring versus the absence of staring. The sham control half provided data that were treated in the same manner as the experimental data, but actual staring did not take place during the sham/pseudostaring periods. This provided an empirical control that yielded information about the likelihood of artifactual evidence for autonomic discrimination in arbitrary subdivisions of a session. For half of the Replication 2 sessions, the experimental half preceded the sham half; for the remaining sessions, the sham half preceded the experimental half. The television monitor was turned off throughout the sham half of a session, and the experimenter/starer occupied her mind with matters unrelated to the experiment. The staree had no way of knowing, conventionally, when the real periods and when the sham periods were in effect; and he/she was monitored in an identical fashion throughout both types of period.

Personality Assessments

For Replication 1, one personality assessment was administered. This was the Social Avoidance and Distress (SAD) scale (Watson and Friend, 1969) which measures social-evaluative anxiety (the experience of distress, discomfort, fear, and anxiety in social situations) and deliberate avoidance of social situations. This self-report scale emphasizes subjective experience, and it excludes physiological signs as well as items related to impaired performance. The scale is constructed so that the opposite instance of a trait simply indicates the absence of that trait, not the presence of some other trait. For example, the opposite instance of social avoidance is simply lack of an avoidance motive, not desire to affiliate. Similarly, the opposite instance of distress is lack of unhappiness, not the presence of some positive emotion. Others have found that scoring patterns of the SAD scale were indeed predictive of behaviors in social situations. We sought to learn whether SAD scoring might also be predictive of reactions to the remote or "psi mediated" social conditions involved in remote attention (remote staring or watching).

For Replication 2, the SAD scale was used along with the Myers-Briggs Type Indicator (MBTI, Form F: see Briggs and Myers, 1957). For this study, we were especially interested in the MBTI extraversion-introversion scale because of its possible relationships with SAD scoring and with remote-staring detection effects in this psi-mediated social (staring) context.

For Replication 1, the psychological assessments were completed by the starees after their experimental sessions. For Replication 2, the psychological assessments were completed by starees during their experimental sessions.

Experimental Hypotheses

Our experimental hypotheses were that, in Replications 1 and 2, the starees would discriminate the true staring from the nonstaring periods autonomically (electrodermally)—that their levels of spontaneous electrodermal activity during the staring periods would differ from those during the nonstaring periods. Therefore, two-tailed tests were used in the analyses, with alpha set at $\leq .05$. We also predicted that, in Replication 2, no such discrimination would occur in the empirical (sham) control segments of the sessions.

Exploratory analyses examined the correlations among the magnitude of the autonomic remote staring detection effect, SAD scoring, and MBTI extraversion-introversion scoring. Since these analyses were exploratory, two-tailed tests were used in their evaluation, with alpha set at $\leq .05$.

Results

Primary Analyses

For each volunteer participant (staree), electrodermal activity was measured during 10 staring and 10 nonstaring periods (for Replication 1) or during 8 staring and 8 nonstaring periods (for Replication 2). Rather than compare these multiple scores within a given participant, we reduced the activities for an entire session to a single score for each participant and performed statistical tests using participants, instead of multiple period scores, as the units of analysis. We used the more conservative session score (a kind of single majority-vote score) in order to bypass criticisms based on possible noninde-pendence of multiple electrodermal measures taken within a given session. Although it would be possible to analyze individual epoch scores using, for example, a repeated measures analysis of variance procedure, such an analysis assumes that the autocorrelations among the measures within each session (i.e., within each participant) are constant across epochs, and that the same auto-correlation applies to all sessions (participants) (J. Utts, personal communica-tion, July 13, 1991). Because these assumptions may not be met in these experiments, we preferred to use the more conservative session-based (rather than epoch-based) analyses, even though the former are more wasteful of data and result in tests with reduced statistical power.

For each of the 30 sessions for Replication 1, a total score was calculated for all 20 recording periods (10 staring and 10 nonstaring). This total score was divided into the sum of the electrodermal activity scores for the 10 staring *(S)* periods; the process was repeated for the 10 nonstaring *(N)* periods. In the absence of a remote-staring effect, these two ratios *[S/(S+N), N/(S+N)]* should approximate 50%. A remote staring effect would be indicated by a significant departure of the scores from the 50% mean chance expectation (MCE). A single mean *t* test was used to assess the departure of the ratios from MCE (50%). This is approximately equivalent to calculating a dependent (matched) *t* test to com-pare the raw scores for each subject for staring versus nonstaring periods. We have consistently used such ratio scores in our various projects as a method of "stan-dardizing" scoring so that scoring magnitude could be compared more meaning-fully for the different dependent measures (response systems) with which we work. A similar analysis was performed for the 16 sessions of Replication 2, with 8 staring and 8 nonstaring periods contributing to each session score.

For Replication 1, mean electrodermal activity percentages were 45.15% (for the staring periods) and 54.85% (for the nonstaring periods), rather than the 50%/50% expected by chance. This scoring rate yielded a single mean t = 1.85 (29), p = .06, two-tailed, and an effect size (r) = .34. For Replication 2, mean electrodermal activity percentages were 45.66% (for the staring periods) and 54.34% (for the nonstaring periods), rather than the 50%/50% expected by chance. This scoring rate yielded a single mean t (15) = 2.08; p = .05, two-tailed, and an effect size (r) = .47. Thus, for both replications, the autonomic discrimination took the form of reduced spontaneous electrodermal activity during staring periods, compared with nonstaring periods. For the 16 sham control session scores of Replication 2 (each based on 8 pseudostaring and 8 nonstaring periods), the mean electrodermal activity percentages were, as expected, virtually identical to the 50%/50% values expected on the basis of chance. Here the scoring rates were 49.16% (for the pseudostaring periods) and 50.84% (for the nonstaring periods). This scoring rate yielded a single mean t (15) = 0.50; p = .76, two-tailed; and an effect size (r) = .08. Expanded summary statistics for Replications 1 and 2 and for the Sham Control series are presented in Table 1. For comparative purposes, the results for our previous two series with untrained and trained starees (see Braud, Shafer and Andrews, 1993) are also included in this table. Electrodermal activity rates during the staring and nonstaring periods of all four experiments, as well as for the sham control sessions, are presented graphically in Figure 1.

Table 1. Statistical Summary of Autonomic Staring Detection Results for Four Experiments and for the Sham Control Series

Series	Scoring rate $\bar{X}$	Scoring rate SD	t	df	Single mean p^a	z^b	Effect size^c r	95% Confidence interval
Untrained Ss	59.38%	14.11	-2.66	15	.02	-2.37	-.57	51.86–66.90
Trained Ss	45.45%	8.46	2.15	15	.05	1.98	.48	40.94–49.95
Replication 1	45.15%	13.85	1.92	29	.06	1.85	.34	39.97–50.32
Replication 2	45.66%	8.37	2.08	15	.05	1.91	.47	41.19–50.12
Sham control	49.16%	11.34	0.30	15	.76	0.31	.08	43.11–55.20

^{a}All ps are two-tailed. ^{b}zs are given for Stouffer z computations. ^{c}The effect size is derived from

$$r = \sqrt{\frac{t^2}{t^2 + df}}\,.$$

Secondary Analyses

Linear correlation coefficients (Pearson rs) were calculated in order to determine the interrelationships among the magnitude of the remote-staring detection effect, SAD scoring, and MBTI extraversion-introversion (E/I) scoring. To study the relationship between remote-staring detection and SAD, Pearson *r*s were computed for the percent electrodermal activity occurring during the staring periods (as in the primary analyses) versus the SAD scores (expressed as a percentage of the highest possible SAD score) for Replication 1, for Replication 2, and for the sham control sessions. Summary statistics are provided in Table 2. For Replication 1, the magnitude of the remote-staring detection effect (the degree of "calming" during the staring periods) was significantly and positively correlated with degree of social avoidance and distress; a similar trend was found for Replication 2. For the sham control sessions, on the other hand, this same correlation was small, negative, and nonsignificant.

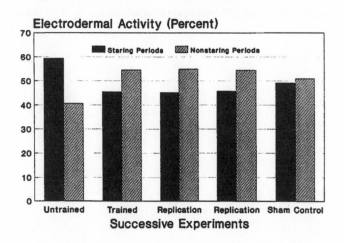

Figure 1. Percent spontaneous electrodermal activity during staring and non-staring periods for the four experimental series and for the sham control series.

Table 2. Linear Correlations Between Staring Period EDA (Percent) and Social Avoidance and Distress (SAD) Score

Series	r	df	p[a]
Replication 1	.36	28	.05
Replication 2	.43	14	.09
Sham control	−.12	14	.66

[a]All *p*s are two-tailed.

To study the remote staring detection as related to introversion, Pearson *r*s were computed for the percent electrodermal activity during the staring periods versus the continuous score for the MBTI introversion scale, for Replication 2 and for the sham control sessions. (The MBTI was not administered for Replication 1.) Summary statistics appear in Table 3. For comparative purposes, similar analyses are presented for our previous two series with untrained and trained starees (in which the MBTI, but not the SAD, had been administered). For Replication 2, there was a strong, positive, and highly significant correlation between the magnitude of the remote staring detection effect and the staree's degree of MBTI introversion. No such correlation occurred for the sham control segment of the experiment.

Table 3. Linear Correlations Between Staring Period EDA (Percent) and MBTI Extraversion/Introversion (E/I) Score

Series	r	df	p^a
Replication 2	.68	14	.0037
Sham control	.16	14	.55
Untrained	.12	14	.66
Trained	.07	14	.80

^{a}All ps are two-tailed.

The relationships between remote staring detection and SAD scoring are presented graphically in Figure 2, and the relationships between remote staring detection and MBTI introversion scoring are presented graphically in Figure 3. In these figures, the ordinate indicates the percentage of total spontaneous electrodermal activity that occurred during the staring periods; increasing departures below the 50% chance level indicate increasing remote calming effects. The abscissas indicate, respectively, increasing degrees of social avoidance/distress/anxiety and increasing tendencies toward MBTI introversion.

For Replication 2, scores were available for both the SAD assessment and the MBTI introversion assessment, and these two instruments could be intercorrelated. The Pearson r for SAD versus introversion was .53 which, with 14 df, was associated with a two-tailed p = .035. The direction of the correlation was, of course, for high social avoidance and distress to be positively correlated with introversion.

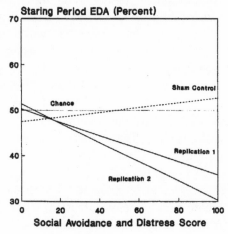

Figure 2. Linear regressions between remote staring detection (calming) and SAD scoring.

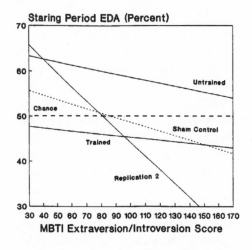

Figure 3. Linear regressions between remote staring detection (calming) and MBTI introversion scoring.

Discussion

Four separate experiments have now been carried out to determine whether persons are able to discriminate periods in which they are watched remotely by someone beyond the range of their conventional senses from periods in which such remote watching is not taking place. We reasoned that measurements of relatively unconscious autonomic nervous system activity might provide unusually sensitive indications of successful discrimination. Evidence for autonomic discrimination of remote watching or staring was indeed obtained in all four studies. This evidence reached statistical significance (as adjudged by conservative, two-tailed p values) in two earlier studies (Braud, Shafer, and Andrews, 1993) and in Replication 2 of the present paper, and very closely approached significance (p = .06, two-tailed) in Replication 1 of the present paper. The effect sizes (see Rosenthal, 1984, 1985) were all relatively large, ranging from .34 to .57. Inspection of Table 1 provides convincing evidence that autonomic staring detection occurred and was replicated in these studies. The absence of a similar effect in the special sham control trials provides another indication that the effect obtained in the real trials was not artifactual.

In the very first experiment (conducted with an untrained starer and untrained starees), remote staring (remote attention) was associated with autonomic activation. In the remaining three experiments, a remote autonomic calming effect was observed. We suggest the following interpretation for these different effects. Although we attempted to equate staring conditions as closely as possible in all experiments, different psychological conditions did nonetheless occur. In the very first experiment, the starer was uneasy and somewhat nervous about the prospect of staring at another person (via the closed-circuit television system) and felt she was "intruding" upon the starees. It is likely that the starer's anxiety (and therefore her heightened sympathetic nervous system activation) may have flavored her attempts to "purely attend" to the starees, and this increased activation may have been communicated to the starees. Prior to the second experiment, both starer and starees had undergone intensive connectedness training and everyone felt very comfortable and relaxed about staring and about "merging" with one another. The starer, in fact, reported feeling much more relaxed, positive, and nonanxious about her sessions in the second experiment, and it is likely that the starer's attention was flavored by these relaxed and comfortable feelings (and their associated sympathetic nervous system deactivation); and these feelings could have been communicated to the starees in the second experiment. Additionally, the starees themselves were comfortable and relaxed about staring and merging, as a result of their own connectedness training. The starer's relaxed and comfortable state could have carried over into Replication 2 (in which she was again the starer) and could have added a relaxed character to her remote attention, even though she was now working with new starees who had not been trained. It is also likely that the starer communicated some of her relaxed attitude, characteristics, and expectations to the three starers of Replication 1 during her training of the latter and her discussions of earlier results with these starers, and that these starer characteristics were then communicated to the Replication 1 starees. Such interpretations could be tested in analytical studies in which the attitudes and conditions of starers are deliberately manipulated.

Although the apparently discrepant findings of the first versus the remaining three studies make sense in terms of the foregoing interpretation, a conservative strategy can be used in pooling the four results: the sign of the t, z, and r scores can be reversed for the result that is inconsistent with the bulk of the results, according to a recommendation by Rosenthal (1984, p. 95). This is the reason for the minus signs in the first row of Table 1.

One of the rationales for conducting these studies in the first place was not only to study staring detection using a new (and hopefully more sensitive) methodology, but also to study a pure attention component that may have been present in all of our prior biological psychokinesis experiments, along with the specific, directional, intentional aims of those experiments. The lesson of the present series of studies is that it is difficult to isolate pure attention, that the latter is easily adulterated by other starer feelings, and that the quality of the starer's attention is important in determining the nature of the experimental outcome.

The significant correlations that obtained between the remote staring detection effect and the two personality variables of SAD and MBTI introversion in the real experiments but not in the sham control segment provide additional evidence for the reality of the remote staring detection effect and also relate the magnitude of the effect to certain psychological variables. The interpretation of these relationships is still unclear; however, certain preliminary suggestions can be offered. Because of the high, significant correlation between SAD and introversion ($r = +.53$), it may be the case that one of these is functioning as a moderator variable in the interaction of the other with the magnitude of the remote-staring detection effect.

Inspection of Figure 2 reveals that as social avoidance, distress, and anxiety increase from zero to high values, the remote calming effect increases from zero (i.e., 50% MCE) to high values. We can offer the following speculative interpretations of the greater susceptibility of those starees with high social avoidance, distress, or anxiety to the remote calming effect.

1. Persons with greater social avoidance/distress/anxiety may be more sensitive to social interactions (in a vigilant way), even when those interactions are psi-mediated. Therefore, high SAD starees may have been more likely to detect the experiment's remote-staring procedure and to have responded appropriately. (Such a suggestion is consistent with an earlier empirical finding by Watson and Friend [1969] that persons with high SAD scores tended to score more highly on an "audience sensitivity index" [Paivio, 1965] than did persons with low SAD scores.)

2. Persons with greater social avoidance/distress/anxiety are ordinarily more isolated and fearful and therefore more "needy" of social interactions. Their normal need to "connect" socially with others is ordinarily denied. Perhaps their greater need for social interaction provided greater motivation for the

efficacy of the remote-staring detection effect. This finding would parallel an earlier finding (Braud and Schlitz, 1983) of a greater remote mental influence effect in persons with greater "need" to be influenced. Stated somewhat differently, perhaps the remote-staring procedure of the present experiment provided a less threatening opportunity for social responsivity than is normally the case, and those persons with greater need took greater advantage of such an opportunity.

3. Persons with greater social avoidance/distress/anxiety may simply be more comfortable (than those with lower SAD scores) and at ease working alone and could, therefore, have been more at home in the isolated staree room and less distracted than persons with lower SAD scores (who might have felt unnaturally isolated and therefore in a less than optimal state of mind).

4. Persons with greater social avoidance/distress/anxiety may be more persuadable or more conforming (having developed such a coping mode as a means of anxiety reduction) than persons with lower SAD scores and this persuasibility or conformity may extend beyond the social realm to the psi realm. (Watson and Friend, 1969, discuss this correlation of SAD with persuasibility and conformity; they do not, of course, mention the possible extension to conformity with psi influences.)

Inspection of the Replication 2 regression line in Figure 3 reveals that MBTI introverts tend to exhibit remote calming, whereas MBTI extraverts tend to show a reversal of this effect (i.e., they evidenced remote activation during staring periods). As in the case of SAD scoring, we can offer some speculative interpretations of this relationship.

1. If becoming calmer is the appropriate response to remote staring under the conditions of the experiment, the more appropriate reaction of introverts may simply be due to their greater ease and comfort under the specific test conditions of the experiment (sitting alone in a room, essentially doing nothing other than "being with themselves" for about 30 min), compared with extraverts (who might be less comfortable, more restless, more distracted, and so forth, and whose less than optimal psychological state may reverse the direction of the psi effect).

2. There are empirical indications that introverts evidence greater sympathetic autonomic arousal than extraverts (Coles, Gale, and Kline, 1971; Geen, 1984; Sadler, Mefferd, and Houck, 1971), that introverts are more excitable or arousable than extraverts in response to given levels of stimulation (Geen, 1984), and that the optimal level of stimulation needed to produce a preferred level of physiological arousal may be lower for introverts than for extraverts (Eysenck,

1967). If it is hypothesized that directing remote attention or psi attention toward a person has a balancing or homeostasis-enhancing influence, then perhaps such a balancing influence would be in the direction of calming for introverts (who are naturally more "excitatory" and may be overaroused ordinarily), but in the direction of activation for extraverts (who are naturally more "inhibitory" and may be underaroused ordinarily). This hypothesis is not unrelated to LeShan's (1966) suggestion that a single moment of special attention in which one feels "at one" with another may be sufficient to trigger optimal self-healing or self-balancing events within that other.

3. To the degree that introversion is correlated with social avoidance/distress/anxiety, the various interpretations offered above in connection with SAD scoring would also be applicable to introversion.

Several of these postulated processes, along with still others, could have interacted to yield the obtained experimental outcomes. Further research would, hopefully, clarify these interrelationships.

In Replication 2, the staree's conscious attention was directed to a personally engaging task (completing personality assessments) during the experimental session. Nonetheless, the staree's more unconscious autonomic nervous system continued to maintain a connection with, and respond appropriately to, the attention and mental processes of another, distant person (the starer). This indicates a dissociation between the two levels of knowing/reacting, as well as the possibility of going about one's individualized activities while still remaining interconnected in an important manner with others. In this experiment, neither of these complementary processes or ways of being or knowing seemed to interfere with the other. If these sorts of physiological "coherences" can be demonstrated in the laboratory, it follows that they may also be present continuously throughout life, and may indicate that while we are all, indeed, isolated individuals, we are simultaneously interconnected members of a much more inclusive, interacting, and interdependent "long body" (see Roll, 1989).

We hope other investigators will attempt to replicate these studies. We recommend the design as one that is straightforward, has already yielded consistent positive results, and addresses a very familiar psi manifestation in a manner that is readily communicable and understandable to the experimental participants and to the public at large.

References

Braud, W., and Schlitz, M. (1983). Psychokinetic influence on electrodermal activity. *Journal of Parapsychology,* 47, 95–119.

Braud, W., and Schlitz, M. (1989). A methodology for the objective study of transpersonal imagery. *Journal of Scientific Exploration,* 3, 43–63.

Braud, W., Shafer, D., and Andrews, S. (1993). Reactions to an unseen gaze (remote attention): A review, with new data on autonomic staring detection. *Journal of Parapsychology,* 57, 373–390.

Briggs, K. C., and Myers, I. B. (1957). *Myers-Briggs Type Indicator Form F.* Palo Alto, CA: Consulting Psychologists Press.

Coles, M. G., Gale, A., and Kline, P. (1971). Personality and habituation of the orienting reaction: Tonic and response measures of electrodermal activity. *Psychophysiology,* 8, 54–63.

Eysenck, H. J. (1967). *The biological basis of personality.* Springfield, IL: Charles C. Thomas.

Geen, R. G. (1984). Preferred stimulation levels in introverts and extraverts: Effects on arousal and performance. *Journal of Personality and Social Psychology,* 46, 1303–1312.

LeShan, L. (1966). *The medium, the mystic, and the physicist.* New York: Viking Press.

Paivio, A. (1965). Personality and audience influence. In B. Maher (Ed.), *Progress in experimental personality research, Vol. 2.* New York: Academic Press.

Peterson, D. M. (1978). *Through the looking glass: An investigation of the faculty of extrasensory detection of being stared at.* Unpublished thesis, University of Edinburgh, Scotland.

Roll, W. G. (1989). Memory and the long body. In L. A. Henkel and R. E. Berger (Eds.), *Research in parapsychology 1988* (pp. 67–72). Metuchen, NJ: Scarecrow Press.

Rosenthal, R. (1984). *Meta-analytic procedures for social research.* Beverly Hills, CA: Sage Publications.

Rosenthal, R. (1965). Designing, analyzing, interpreting, and summarizing placebo studies. In L. White, B. Tursky, and G. Schwartz (Eds.), *Placebo: Theory, research and mechanisms* (pp. 110–136). New York: Guilford Press.

Sadler, T. G., Mefferd, R. G., and Houck, R. L. (1971). The interaction of extraversion and neuroticism in orienting response habituation. *Psychophysiology,* 8, 312–318.

Watson, D., and Friend, R. (1969). Measurement of social-evaluative anxiety. *Journal of Consulting and Clinical Psychology,* 33, 448–457.

Williams, L. (February, 1983). Minimal cue perception of the regard of others: The feeling of being stared at. Paper presented at the 10th Annual Conference of the Southeastern Regional Parapsychological Association, West Georgia College, Carrollton, GA.

Institute of Transpersonal Psychology
744 San Antonio Road
Palo Alto, CA 94303

9

Empirical Studies of Prayer, Distant Healing, and Remote Mental Influence

William G. Braud

This chapter provides a review of empirical studies of the influences of intercessory prayer, mental healing, and experimental analogs of direct healing. The chapter also addresses theoretical explanations for these phenomena and discusses their implications for the concept of human interconnectedness.

This material, under the title, "Healing Analog Research and Human Connectedness," originally was presented at the Annual Meeting of the Society for the Scientific Study of Religion and Religious Research Association, Virginia Beach, VA, November 9–11, 1990, and later published in Braud, W. G. (1994). Empirical explorations of prayer, distant healing, and remote mental influence. Journal of Religion and Psychical Research, *17(2), 62–73. The contents of this article are Copyright © 1994 by, and reprinted through the permission and courtesy of, the Academy of Religion and Psychical Research.* —William Braud

Introduction

Two concepts of importance to many spiritual and religious systems are (a) belief in the possibility of spiritual healing or in the efficacy of prayer in healing and (b) belief in a profound interconnectedness among human beings. While the literature discussing these topics is considerable, scientific studies that directly address these issues are extremely rare. The paucity of well-controlled investigations of these topics is understandable. Conventional scientists are reluctant to explore claims of anomalous processes that have strong spiritual or nonmaterialistic implications. There are practical difficulties in conducting projects with the requisite degree of methodological rigor. Other difficulties are engendered by attitudes of the public at large and of sophisticated spiritual teachers that mundane scientific approaches to religion or spiritual issues are, by their very nature, either inappropriate or profane. Nonetheless, a number of relevant studies have been conducted, with provocative results.

Prayer

Perhaps the first objective study of the possible efficacy of prayer was reported by the British anthropologist, Sir Francis Galton. In one of the first applications of statistics to scientific research, Galton (1872, 1883) sought to determine whether life expectancies of prayerful people (e.g., clergy) were greater than those of materialistic people (e.g., doctors and lawyers), and whether persons frequently prayed for (e.g., sovereigns) tended to live longer than others. His results led him to conclude that prayer did not influence longevity. There are, of course, many difficulties with the assumptions, designs, and conclusions of Galton's retrospective studies. However, Galton deserves credit for his view that the efficacy of prayer was amenable to empirical study and for his application of the statistical method to the study of this issue.

It was only after a very long hiatus that other researchers sought to address prayer's efficacy using more sophisticated methods in well-controlled, prospective studies. We now know that various forms of meditation and of contemplative prayer, developed in the contexts of Western and Eastern meditative, mystical, and spiritual traditions, may beneficially influence the health and well-being of those who practice such disciplines (see Benson, 1975; Benson, Greenwood, and Klemchuk, 1977). This finding does not cause dismay, since it is becoming increasingly acceptable to expect one's own mental activities, including prayer, to affect one's own bodily condition. Our understanding of these mind-body interactions has been aided by investigations of placebo effects, stress reduction, suggestion, expectancy, hypnosis, biofeedback, self-regulation, and psychoneuroimmunological principles (Achterberg, 1985; Ader, 1981; Frank, 1961; Green and Green, 1977; Justice, 1987; Locke and Colligan, 1986; Ornstein and Sobel, 1987; White, Tursky, and Schwartz, 1985; Wickramasekera, 1988).

But what about effects of prayer *on other people?* Can prayer influence the health of someone other than the person who prays, and can such effects occur even when the person prayed for is *unaware* of the fact that he or she is being prayed for? It was to answer such questions that Joyce and Welldon (1965) designed a double-blind clinical trial to test the efficacy of prayer in London Hospital out-patients suffering from chronic stationary or progressively deteriorating psychological or rheumatic disease. Matched pairs of patients with serious disorders were randomly assigned to receive or not receive prayers from distant prayer groups. Conventional medical treatments were continued for all 38 patients (19 pairs). Careful double blinds were used to assure that both the patients and their assessing physicians were unaware of which patients were receiving the prayers. Patients were evaluated at the beginning of the study, then again after 8–18 months. Only six of these patients with poor prognoses improved; five of the improved patients were in the prayer group, while only one was in the control group. During the first half of the study, the prayer group improved more than did the control group. Overall results, while encouraging, did not reach statistical significance for these small samples.

Collipp (1969) conducted a similar study with 18 leukemic children patients. Ten children were assigned randomly to a prayer condition and eight to a non-prayer, control condition. This study was triple-blind. The patients, parents, and physicians were unaware that some of the children were being prayed for, and the prayer groups (ten families in another state) were not aware that this was a study of the efficacy of prayer. After 15 months of prayer, results were tabulated and summarized. Of the ten leukemic children in the prayer condition, seven were still alive; of the eight leukemic children in the control condition, only two were alive. The superior survival rate of the prayer patients approached but did not reach statistical significance ($p = .069$, Fisher's exact test). Nonetheless, with such small sample sizes, the results are encouraging.

The most satisfactory study of prayer efficacy was conducted by cardiologist Randolph Byrd, M.D. and was published in 1988 in the well-respected, peer-reviewed *Southern Medical Journal.* Byrd utilized a prospective, randomized, double-blind protocol to study possible effects of intercessory prayer in a sample of coronary care unit (CCU) patients. Over ten months, 393 patients admitted to the CCU were, with informed consent, randomized to a prayer group (192 patients) or to a control group (201 patients). Prayer was provided by participating Christians outside the hospital. Neither patients nor their evaluating physicians were aware of which patients were receiving prayer. It was found that, although the patients were well matched at entry, the prayer patients showed significantly superior recovery compared to controls ($p < .0001$). The prayed-for patients were five times less likely than control patients to require antibiotics and three times less likely to develop pulmonary edema. None of the prayed-for patients required endotracheal intubation, whereas 12 controls required such mechanical ventilatory support. Fewer prayed-for than control patients died, but the difference in this area was not statistically significant. The design and the results of the Byrd study are impressive, and even skeptical commentators seem to agree on the significance of the findings.

Distant Healing

In these two different reports of remote healing an interesting contrast appears.

The apparent success of healing methods based on all sorts of ideologies
and methods compels the conclusion that the healing power of faith resides
in the patient's state of mind, not in the validity of its object. At the risk of
laboring this point, an experimental demonstration of it with three severely
ill, bedridden women may be reported. One had chronic inflammation of the
gall bladder with stones, the second had failed to recuperate from a major
abdominal operation and was practically a skeleton, and the third was dying
of widespread cancer. The physician first permitted a prominent local faith
healer to try to cure them by absent treatment without the patients' knowledge. Nothing happened. Then he told the patients about the faith healer,
built up their expectations over several days, and finally assured them that he
would be treating them from a distance at a certain time the next day. This
was a time in which he was sure that the healer did *not* work. At the suggested
time all three patients improved quickly and dramatically. The second was
permanently cured. The other two were not, but showed striking temporary
responses. The cancer patient, who was severely anemic and whose tissues
had become waterlogged, promptly excreted all the accumulated fluid, recov-
ered from her anemia, and regained sufficient strength to go home and
resume her household duties. She remained virtually symptom free until her
death. The gall bladder patient lost her symptoms, went home, and had no
recurrence for several years. These three patients were greatly helped by a
belief that was false—that the faith healer was treating them from a distance—suggesting that "expectant trust" in itself can be a powerful healing
force. (Frank, 1961, pp. 60–61)

However, all results must be evaluated cautiously. The most dramatic
single result I had occurred when a man I knew asked me to do a distant healing for an extremely painful condition requiring immediate and intensive
surgery. I promised to do the healing that night, and the next morning when
he awoke a "miraculous cure" had occurred. The medical specialist was
astounded, and offered to send me pre and post healing X-rays and to spon-
sor publication in a scientific journal. It would have been the psychic healing
case of the century except for one small detail. In the press of overwork, I had
forgotten to do the healing! If I had only remembered, it would have been a
famous demonstration of what can be accomplished by this method.
(LeShan, 1974, p. 125)

These two anecdotes illustrate the difficulties of studying distant healing
in everyday life settings and also indicate the powerful somatic influences of
such psychological factors as suggestion and expectancy. If one wishes to eliminate such factors completely, and study distant healing in its pure or uncont-
aminated form, it is essential that the person being healed (the "healee") be
unaware of the healing attempt (i.e., "blind" to the healing manipulation). The
easiest way to assure blindness on the part of the healee is to station the healer
and the healee at separate, distant locations and to schedule the healing
attempts at randomly selected times that are unknown to the healee. The dis-

tant isolation eliminates the possibility of subtle, unintentional cues from the healer (e.g., subtle changes in voice, breathing, or body language) that might allow the healee to know when healing attempts are in progress. The random scheduling eliminates the possibility of the healee using rational inference to determine the likely times of healing attempts. In addition, an adequate experimental design presupposes that the healee's somatic condition can be objectively and reliably measured so that changes can be properly assessed.

Therapeutic Touch

"Therapeutic touch" is a modern variation on the ancient healing procedure of laying-on of hands. It has been elaborated and used within a non-religious context by a New York University professor of nursing, Dolores Krieger, R.N., Ph.D. Since its development in the early 1970s, the therapeutic touch technique has been taught to thousands of nurses and other health care professionals who use the procedure in their practices.

In doing therapeutic touch, the practitioner first "centers" herself by shifting her awareness from an external to an internal focus, becoming relaxed and calm. She makes a mental intention to assist the healee therapeutically, i.e., to help and heal the patient. She then moves her hands over the healee's body in a prescribed manner in order to detect and correct areas of imbalance and disease. Early studies demonstrated the effectiveness of the technique in increasing patients' blood hemoglobin values, decreasing pain and anxiety, lowering blood pressure, decreasing edema, easing abdominal cramps and nausea, resolving fevers, stimulating growth in premature infants, accelerating the healing process in cases of fractures, wounds, the common cold and other infections, and increasing relaxation and well-being (Boguslawski, 1979; Borelli and Heidt, 1981; Heidt, 1981; Keller, 1984; Krieger, 1975, 1976, 1979; Krieger, Peper, and Ancoli, 1979; Kunz, 1985; Macrae, 1979). Less formal studies of healing and pain-reducing effects of similar hand manipulations and mental intention had been reported previously by Knowles (1954, 1956) and by Hubacher, Gray, Moss, and Saba (1975). Most of these early studies did not control adequately for possible suggestion and expectancy effects and have been criticized on other grounds as well (Clark and Clark, 1984).

More recent therapeutic touch studies have been conducted with improved methodologies that obviate criticism. For example, Quinn (1984) obtained significant results while using "noncontact therapeutic touch" (in which the hands are moved near but not touching the patient) and employing an excellent treatment simulation procedure to control for artifactual psychological effects. Even more impressively, Wirth (1989) was able to observe dramatic effects of noncontact therapeutic touch upon objectively measured full-thickness dermal wound healing in a carefully designed double-blind study in which the healer and healee were stationed in separate adjoining rooms. The healees extended their arms and shoulders (on which precise biopsy wounds had been made) through a special opening in the wall, designed so that they were unaware of whether or not a treatment was in progress. The healees and the assessing physician were even unaware of the nature of the treatment. Healing rate for treated subjects was compared statistically with that of untreated control subjects; highly significant differences were observed.

Remote Healing

Wirth's (1989) study of noncontact therapeutic touch by a healer in another room seems to have adequately eliminated possible confounding effects of suggestion and expectation. However, the proximity of the healer's hand passes (a few inches from the healee's skin) may still allow the concomitant influence of subtle conventional energies such as heat, electrostatic fields, and electromagnetic fields. Such conventional influences were eliminated in other studies in which great distances intervened between healer and healee. LeShan (1974), Goodrich (1974, 1976), and Winston (1975) were able to observe significant influences upon healees at distant locations when healers used a form of remote healing ("Type 1" in LeShan's terminology) in which they used meditation techniques in order to produce feelings of merging with or being "at one" with the healees. LeShan, a psychologist who developed the method, believes that the healer's alteration of consciousness to achieve "oneness" with the healee, along with a strong healing intention, evokes a self-healing process in the healee that facilitates the healee's balancing and recovery from illness and disease.

Physiological measurements have been used in some remote healing experiments. Miller (1982) measured systolic and diastolic blood pressure, heart rate, and body weight in 96 hypertensive patients who participated in a distant healing study. He reported a statistically significant reduction in systolic blood pressure for patients who received distant healing from eight healers, compared to control patients not receiving healing; diastolic pressure, heart rate, and body weight showed no such effects. A more recent study, along similar lines, conducted in the Netherlands by Beutler, Attevelt, Schouten, Faber, Mees, and Geijskes (1988) failed to find significant evidence for a remote healing influence upon blood pressure in hypertensive patients.

Remote Healing Analog Studies

The scarcity of well-designed remote healing studies attests to the difficulty of conducting such studies with actual patients in clinical contexts. It is not always easy to establish accurate pre- and post-healing diagnoses and to properly control or assess the myriad environmental, pharmacological, dietary, exercise, and other physical and psychological factors or treatments that could confound one's observations and lead to ambiguous conclusions. There are also ethical issues involved in properly "blinded" studies and in providing unvalidated remote healing treatments to some needy patients but not to the equally needy patients in a control group. There may be opposition to remote healing treatments by the patients, their families, or their attending physicians. For these and other reasons, researchers have developed an alternative strategy for studying remote healing influences—they have designed and conducted "healing analog" studies. Rather than investigate healers and actual patients in everyday life contexts, researchers have abstracted and simplified the healing interaction in order to create laboratory models or analogs of those interactions. The experiments involve the remote mental influence of living systems (biological "targets") under well-controlled conditions. One arbitrarily selects a target organism, then selects a readily measured aspect of that organism's behavior or physiological activity. An actual healer, or an unselected person playing the role of a healer, then attempts to exert a remote mental influence upon the targeted activity in a prescribed manner. Appropriate non-healing controls are used, with treated and nontreated (control) organisms or activities assigned in a random fashion. The experiment is conducted repeatedly so that results may be analyzed statistically. Precautions are taken to eliminate any conventional influences that could bias the experiment, and the experimental protocol is designed so that any uncontrolled factors would be expected to "randomize out," i.e., to influence both treatment and control conditions equally.

Many healing analog experiments have been conducted within the last 40 years and have been reviewed by Benor (1984, 1986), Braud (1990b), and Solfvin (1984). Statistically significant remote mental influences have been observed in experiments with bacteria, fungus colonies, yeast, plants, protozoa, larvae, woodlice, ants, fish, chicks, mice, rats, gerbils, cats, dogs, dolphins, and humans. Additional experiments have demonstrated effects upon *in vitro* cellular preparations (blood cells, neurons, cancer cells) and enzyme activity. The great range of these experiments and their positive results suggest that the ability to mentally influence living systems at a distance may be a widespread, latent, natural ability possessed, to some degree, by all of humankind. The wide variety of obtained effects also suggests that, in principle, one could direct or focus one's remote mental influence so that it could have medically, psychologically, and socially beneficial effects.

Experiments with Electrodermal Activity

In an ongoing research program at the Mind Science Foundation we have been systematically exploring a variety of remote mental influence designs. In these experiments, we have found that individuals are able to remotely influence a wide range of biological target systems including the spatial orientation of fish, the locomotor activity of small mammals, the rate of hemolysis of human red blood cells *in vitro,* and the muscular movements and mental imagery of other persons. The details of these studies may be found in other publications (Braud, 1990a; Braud and Jackson, 1982, 1983; Braud, Davis, and Wood, 1979; Braud, Schlitz, and Schmidt, 1989).

Our most extensive research, however, has focused on the remote mental influence of another person's electrodermal activity (Braud and Schlitz, 1983; Braud and Schlitz, 1989; Schlitz and Braud, 1985). In these experiments, a "target" subject plays the role of a healee and sits in a comfortable room while his or her spontaneous skin resistance responses (SRRs) are monitored continuously by means of electronic equipment interfaced with a microcomputer. These SRRs reflect the degree of activation of the subject's sympathetic autonomic nervous system and, hence, the subject's degree of emotional, cognitive, or physical activation or arousal. Higher SRR activity is associated with physiological activation, whereas lower SRR activity reflects relaxation and calmness. In a separate, distant room (typically 20 meters away), the experimenter is stationed with another person, the "influencer," who plays the role of a healer. The ongoing SRR activity of the distant subject is displayed to the influencer by means of a polygraph (chart recorder) and also is objectively and automatically assessed by the computer system. The influencer watches the polygraph as she or he attempts to exert a remote mental influence upon the distant subject. Influence attempts are made during ten 30-second periods; these are randomly interspersed among ten 30-second control or baseline periods during which no influence is attempted. The subject, of course, is unaware of the nature, timing, and scheduling of these periods, and is physically isolated from any conventional energetic or informational signals from the influencer. Thus, the protocol completely eliminates suggestion and expectancy effects.

The aim of the influencer is to either calm, activate, or not influence the distant subject according to a prearranged random schedule. During calming attempts, the influencer relaxes and calms himself or herself, intends and gently wishes for the subject to become calm, and visualizes or imagines the subject in a relaxing, calming setting. During activation attempts, the influencer tenses his or her own body, intends and wishes for the subject to become more active, and images the subject in activating, energizing or arousing settings and situations. During the noninfluence control periods, the influencer attempts to keep his or her mind off of the subject and to think about matters unrelated to the experiment. The influencers may use the polygraph tracings as feedback to indicate how well their influence attempts are succeeding. Alternatively, they may proceed without such feedback and simply close their eyes and intend and visualize the desired outcomes. We have found that both feedback and nonfeedback strategies are effective.

Thus far, we have completed 15 electrodermal remote influence experiments, with the number of subjects in each experiment ranging from 10 to 40.

In all, there have been 323 sessions conducted with 271 different subjects, 62 influencers, and 4 experimenters. The experiments have been quite successful. Thirteen of the 15 studies yielded overall results in the expected direction. Six of the 15 experiments (40 percent) were independently significant statistically (i.e., had ps less than .05); this is to be compared with the 5 percent experimental success rate expected on the basis of chance. Fifty-seven percent of the individual sessions were successful (i.e., yielded results in the expected direction); this is to be compared with the 50 percent session success rate to be expected on the basis of chance. The series as a whole yields a combined (Stouffer) z score of 4.08, with an associated $p = .000023$ (i.e., odds against chance of approximately 50,000 to 1); the average effect size for all experiments is 0.29. Recently, we have completed additional experiments with 32 new subjects (Braud, Shafer, and Andrews, 1990); these experiments also yielded significant outcomes. We conclude that an individual is indeed able to directly, remotely, and mentally influence the physiological activity of another person through means other than the usual sensorimotor channels.

LeShan (1990) has discussed four interpretations or explanations of the means through which distant healing is accomplished. These are classes of explanations that have been offered by various healers. In these four interpretations, healing effects are attributed, respectively, to: (a) divine intervention, (b) spirit intervention, (c) some type of "energy" mediator, and (d) increased "self-repair" on the part of the healee that is induced by the healer and the healee sharing in a unitive experience. Interpretations (a) and (b) are well known and require no additional elaboration. Interpretation (c) will be treated below. Interpretation (d) is the one preferred by LeShan himself and was referred to under "Remote Healing" in this paper.

Explanations

Similar interpretations could be offered for the effects observed in our remote influence experiments. We ourselves, however, believe that there are two general classes of explanation for our obtained effects. The first is that the effects are mediated by an unusual form of "energy"—either an unfamiliar form of physical energy or a novel, quasiphysical energy that has unusual features. The "operating characteristics" of the remote influence effect are unusual. It does not vary in a familiar manner as a function of spatial distance or time, and it is not influenced importantly by physical barriers, shields, or the nature of the particular system that is "targeted." Perhaps the only conventional energy that may qualify as a potential mediator is extremely low frequency (ELF) electromagnetic radiation. The latter has excellent "penetrating" properties and can travel great distances. Some investigators, notably Persinger (1979), have seriously advanced ELF fields as mediators for the anomalous effects we have been discussing. The major problems with possible ELF mediators, however, are that (a) they would have to behave in highly unusual ways with respect to *time* in order to explain the time-displaced mental effects that have been observed in certain experiments, (b) they would have to carry more information than they would appear capable of carrying, and (c) they would have to be encoded by the influencer's brain (or other bodily process) and decoded by the subject's brain (or other bodily process) in ways that we do not understand and for which we have no known mechanisms. Therefore, an ELF-mediated carrier remains, while not entirely impossible, a highly implausible hypothesis.

The second explanation of our obtained effects is that *mind is nonlocal* and that under special conditions its nonlocal nature is manifested. According to this view, energy or information does not "travel" from one place to another or from one mind to another, but is already "everywhere." The influencer's mind and the subject's mind may not really be as distinct, separate, and isolated as they appear to be but, rather, may be profoundly interconnected, unified, omnipresent, and omniscient. What is available to one mind may be available to all minds and may already be part of all minds in what is analogous to a "holographic" form.

Connectedness

The prayer and remote healing findings reviewed in this paper strongly suggest a profound interconnectedness among people. Indeed, it is difficult to understand how such anomalous interactions could occur if an underlying matrix of subtle yet important connections among people did not exist. The existence of such connections is compatible with the ontological, epistemological, and ethical teachings of many of the world's religious, spiritual, meditative, and mystical traditions. Interestingly, the findings, models, and theories of leading thinkers in the physical, biological, and psychological sciences are becoming increasingly consistent with the common worldview and perennial wisdom of these traditions. Extensive and lucid expositions of these consistencies, along with other relevant discussions, may be found in works by Bohm (1980), Dossey (1982, 1989), Eccles (1980), Huxley (1945), Jung (1973), LeShan (1974), Pribram (1976), Sheldrake (1981, 1988), and Wilber (1982, 1984).

William James (1902), Walter Houston Clark (1958), and other important figures in the psychology of religion have expressed the view that all expressions of religion grow out of the mystical experiences of individuals— that religion (or, better, *spirituality*) is the inner experience of someone who senses a "Beyond" and is reflected in the behaviors of individuals as they attempt to harmonize their lives with that Beyond (Clark, 1979). The processes that underlie effective prayer and effective remote healing provide a glimpse of such a Beyond. The success of these and similar studies indicate that there are ways of empirically exploring the interactions of individuals with that Beyond, that matrix of interconnections that makes these interactions possible.

There is an increasing realization that these and other connections contribute importantly to health and well-being. Physical, mental, and spiritual balance and wholeness are facilitated when one recognizes and experiences the connections between different parts of the mind, between the mind and the body, between people, and between people and all of Nature. The exploration of the nature of these connections is an excellent focus for cooperative scientific, clinical, and practical studies.

The findings examined in this paper provide a glimpse into exceptional experiences and more extended potentials that are available, under certain conditions, to all of us. They also provide an empirical foundation that is consistent with the existence of the healing possibilities which are described in the world's major spiritual traditions.

References

Achterberg, J. (1985). *Imagery and Healing*. Boston, MA: New Science Library.

Ader, R. (1981). *Psychoneuroimmunology*. New York: Academic Press.

Benor, D. J. (1984). Fields and energies related to healing: A review of Soviet and Western studies. *Psi Research*, 3(1), 21–35.

Benor, D. J. (1986). Research in psychic healing. In B. Shapin and L. Coly (Eds.), *Current Trends in Psi Research* (pp. 96–112). New York: Parapsychology Foundation.

Benson, H. (1975). *The Relaxation Response*. New York: William Morrow and Company.

Benson, H., Greenwood, M., and Klemchuk, H. (1977). The relaxation response: Psychophysiologic aspects and clinical applications. In Z. Lipowski, D. Lipsitt, and P. Whybrow (Eds.), *Psychosomatic Medicine* (pp. 377–388). New York: Oxford University Press.

Beutler, J., Attevelt, J., Schouten, S., Faber, J., Mees, E., and Geijskes, G. (1988). Paranormal healing and hypertension. *British Medical Journal*, 296, 1491–1494.

Boguslawski, M. (1979). The use of Therapeutic Touch in nursing. *Journal of Continuing Education in Nursing*, 10(4), 9–15.

Bohm, D. (1980). *Wholeness and the Implicate Order*. London: Routledge and Kegan Paul.

Borelli, M. D., and Heidt, P. (1981). *Therapeutic Touch*. New York: Springer.

Braud, W. G. (1990a) Distant mental influence of rate of hemolysis of human red blood cells. *Journal of the American Society for Psychical Research*, 84(1), 1–24.

Braud, W. G. (1990b). On the use of living target systems in distant mental influence research. In B. Shapin and L. Coly (Eds.), *Psi Research Methodology: A Re-examination* (in press). New York: Parapsychology Foundation.

Braud, W. G., Davis, G., and Wood, R. (1979). Experiments with Matthew Manning. *Journal of the Society for Psychical Research*, 50, 199–223.

Braud, W. G., and Jackson, J. (1982). Ideomotor reactions as psi indicators. *Parapsychology Review*, 13, 10–11.

Braud, W. G., and Jackson, J. (1983). Psi influence upon mental imagery. *Parapsychology Review*, 14, 13–15.

Braud, W. G., and Schlitz, M. (1983). A methodology for the objective study of transpersonal imagery. *Journal of Scientific Exploration*, 3, 43–63.

Braud, W. G., Schlitz, M., and Schmidt, H. (1989). Remote mental influence of animate and inanimate target systems: A method of comparison and preliminary findings.

Proceedings of Presented Papers, 32nd Annual Parapsychological Association Convention, San Diego, CA, pp. 12–25.

Braud, W. G., Shafer, D., and Andrews, S. (1990). Electrodermal correlates of remote attention: Autonomic reactions to an unseen gaze. *Proceedings of Presented Papers, 33rd Annual Parapsychological Association Convention,* Washington, D.C., pp. 14–28.

Byrd, R. C. (1988). Positive therapeutic effects of intercessory prayer in a coronary care unit population. *Southern Medical Journal,* 81(7), 826–829.

Clark, P., and Clark, M. (1984). Therapeutic touch: Is there a scientific basis for the practice? *Nursing Research,* 33(1), 37–41.

Clark, W. H. (1958). *The Psychology of Religion.* New York: Macmillan.

Clark, W. H. (1979). The religious perspective on psychic phenomena. *Journal of the Academy of Religion and Psychical Research,* 2(1), 7–13.

Collipp, P. J. (1969). The efficacy of prayer: A triple-blind study. *Medical Times,* 97(May), 201–204.

Dossey, L. (1982). *Space, Time and Medicine.* Boulder, CO: Shambhala.

Dossey, L. (1989). *Recovering the Soul.* New York: Bantam.

Eccles, J. C. (1980). *The Human Psyche.* New York: Springer International.

Frank, J. D. (1961). *Persuasion and Healing.* Baltimore, MD: The Johns Hopkins Press.

Galton, F. (1872). Statistical inquiries into the efficacy of prayer. *Fortnightly Review,* 12, 125–135.

Galton, F. (1883). *Inquiries into Human Faculty and Its Development* (pp. 277–294). London: Macmillan.

Goodrich, J. (1974). *Psychic Healing: A Pilot Study.* Unpublished doctoral dissertation, Union Graduate School.

Goodrich, J. (1976). Studies of paranormal healing. *New Horizons,* 2(2), 21–24.

Green, E., and Green, A. (1977). *Beyond Biofeedback.* New York: Dell Publishing Company.

Heidt, P. (1981). Effects of therapeutic touch on anxiety level of hospitalized patients. *Nursing Research,* 30(1), 32–37.

Hubacher, J., Gray, J., Moss, T., and Saba, F. A laboratory study of unorthodox healing. *Proceedings of the Second International Congress on Psychotronic Research,* Monte Carlo, pp. 440–443.

Huxley, A. (1945). *The Perennial Philosophy.* New York: Harper Colophon Books.

James, W. (1902). *The Varieties of Religious Experience.* New York: Longmans, Green.

Joyce, C. R., and Welldon, R. M. (1965). The objective efficacy of prayer: A double-blind clinical trial. *Journal of Chronic Diseases,* 18, 367–377.

Jung, C. G. (1973). *Synchronicity: An Acausal Connecting Principle.* Princeton, NJ: Princeton/Bollingen.

Justice, B. (1987). *Who Gets Sick: Thinking and Health.* Houston, TX: Peak Press.

Keller, E. K. (1984). Therapeutic touch: A review of the literature and implications of a holistic nursing modality. *Journal of Holistic Nursing,* 2(1), 24–29.

Knowles, F. W. (1954). Some investigations into psychic healing. *Journal of the American Society for Psychical Research,* 48(1), 21–26.

Knowles, F. W. (1956). Psychic healing in organic disease. *Journal of the American Society for Psychical Research,* 50(3), 110–117.

Krieger, D. (1975). Therapeutic touch: The imprimatur of nursing. *American Journal of Nursing,* 75(5), 784–787.

Krieger, D. (1976). Healing by the "laying-on" of hands as a facilitator of bioenergetic change: The response of in-vivo human hemoglobin. *Psychoenergetic Systems,* 1, 121–129.

Krieger, D. (1979). *Therapeutic Touch: How to Use Your Hands to Help or Heal.* Englewood Cliffs, NJ: Prentice-Hall.

Krieger, D., Peper, E., and Ancoli, S. (1979). Therapeutic touch: Searching for evidence of physiological change. *American Journal of Nursing,* 79, 660–662.

Kunz, D. (1985). *Spiritual Aspects of the Healing Arts.* Wheaton, IL: Theosophical Publishing House.

LeShan, L. (1974). *The Medium, the Mystic, and the Physicist.* New York: Viking Press.

LeShan, L. (1990). Explanations of psychic healing. *ASPR Newsletter,* 16(1), 1–3.

Locke, S., and Colligan, D. (1986). *The Healer Within.* New York: Dutton.

Macrae, J. (1979). Therapeutic touch in practice. *American Journal of Nursing,* 4, 664–665.

Miller, R. N. (1982). Study of the effectiveness of remote mental healing. *Medical Hypotheses,* 8, 481–490.

Ornstein, R., and Sobel, D. (1987). *The Healing Brain.* New York: Simon and Schuster.

Persinger, M. A. (1979). ELF field mediation in spontaneous psi events: Direct information transfer or conditioned elicitation? In C. Tart, H. Puthoff, and R. Targ (Eds.), *Mind at Large* (pp. 191–204). New York: Praeger.

Pribram, K. (1976). Problems concerning the structure of consciousness. In G. Globus (Ed.), *Consciousness and the Brain.* New York: Plenum Press.

Quinn, J. (1984). Therapeutic touch as energy exchange: Testing the theory. *Advances in Nursing Science,* 1, 42–49.

Schlitz, M., and Braud, W. G. (1985). Reiki-Plus natural healing: An ethnographic/experimental study. *Psi Research,* 4, 100–123.

Sheldrake, R. (1981). *A New Science of Life.* Los Angeles, CA: J. P. Tarcher.

Sheldrake, R. (1988). *The Presence of the Past.* New York: Times Books.

Solfvin, J. (1984). Mental healing. In S. Krippner (Ed.), *Advances in Parapsychological Research,* Vol. 4 (pp. 31–63). Jefferson, NC: McFarland.

Wickramasekera, I. E. (1988). *Clinical Behavioral Medicine.* New York: Plenum.

Wilber, K. (1982). *The Holographic Paradigm and Other Paradoxes.* Boulder, CO: Shambhala.

Wilber, K. (1984). *Quantum Questions.* Boulder, CO: Shambhala.

Winston, S. (1975). *Research in Psychic Healing: A Multivariate Experiment.* Unpublished doctoral dissertation, Union Graduate School.

Wirth, D. P. (1989). Unorthodox healing: The effect of noncontact therapeutic touch on the healing rate of full thickness dermal wounds. *Proceedings of Presented Papers, 32nd Annual Parapsychological Association Convention,* San Diego, CA, pp. 251–268.

White, L., Tursky, B., and Schwartz, G. E. (1985). *Placebo: Theory, Research, and Mechanisms.* New York: Guilford Press.

Institute of Transpersonal Psychology
744 San Antonio Road
Palo Alto, CA 94303

10

Helping Others Concentrate Using Distant Mental Influence

William Braud, Donna Shafer, Katherine McNeill,
and Valerie Guerra

This chapter describes experiments in which persons' focused atten-tion, ability to concentrate, and freedom from distractions could be facili-tated by distant "helpers" through the use of distant mental intentional and attentional influences. The effect was greatest in persons with greatest need to be helped (as indicated by measures of concentration difficulties and difficulties in attending). This work extends the distant mental influ-ence process from physiological to cognitive "targets," and it indicates an area in which this process can be practically applied.

This work originally was published in Braud, W. G., Shafer, D., McNeill, K., and Guerra, V. (1995). Attention focusing facilitated through remote mental interaction. Journal of the American Society for Psychical Research, *89(2), 103–115. The contents of this article are copyright © 1995 by, and reprinted by permission of, the* Journal of the American Society for Psychical Research. —*William Braud*

A version of the paper was presented at the 36th Annual Parapsychological Association Convention, Toronto, Canada, August 15–19, 1993.

We are indebted to Helmut Schmidt for his help with the computer program for random-izing the experimental periods and monitoring and assessing the distraction-indicating but-ton presses, and to Russell Targ for suggesting that we calculate separate effect sizes for the "more needy" and "less needy" participant subgroups. We also thank the 60 participants for their contributions to our research. Thanks are also due to Sperry Andrews and to Jessica Utts for stylistic and statistical suggestions.

Abstract: This study is part of a program that is beginning to assess direct mental influences of one person upon a variety of *nonphysiological* activities (cognitive, emotional, social, psychic) being carried out simultaneously by another, distantly isolated, person. Sixty volunteer participants, during individual 16-minute sessions, focused attention upon an object while indicating each time the mind wandered from this focus (i.e., each time the mind was distracted) by pressing a hand-held button. A computer recorded these distraction-indicating button-presses. During eight 1-minute Help periods, another person in a distant room attempted to help the participant by focusing on an identical object and intending for the participant to attend well and not be distracted. During eight 1-minute Control periods, the helper did not attempt to influence the participant but, rather, thought about irrelevant matters. The random schedule of the two types of periods was unknown to the participant. Participants evidenced significantly greater focused attention (fewer distractions) during Help than Control periods, $t(59) = 2.00$, $p = .049$, two-tailed, effect size = .25. the magnitude of the remote mental helping effect was significantly correlated ($r = .26$ and $r = .32$) with two measures of the participant's "need to be helped" (measures of concentration difficulties and difficulties in attending). The effect size for the needy participants was .56, whereas the effect size for the non-needy participants was –.03.

For several years, we have been exploring direct mental interactions with remote, spatially distant living systems. In most of these investigations, we have studied the ability of one person to influence, mentally and at a distance, the ongoing physiological activity of another person who is located in a separate room and isolated from the first person in terms of conventional energetic and informational connections. The first person (the "influencer") uses mental processes of attention, intention, and imagery in order to influence the distant second person (the "influencee") in a prespecified manner. The rationales, procedural details, and experimental findings of these investigations have been presented in two recent summary papers (Braud and Schlitz, 1989, 1991). The findings show that direct mental interactions can occur under well-controlled laboratory conditions, that the effect is replicable and relatively robust, and that the overall effect size for these studies is not trivial ($r = .33$).

A general implication of this work is that in any dyadic situation, the mental activity of one member of the dyad could directly influence the course of some physical, physiological, emotional, or cognitive process in the other member of the dyad. In various educational, counseling, therapeutic, or healing contexts, the actual progress of the student, counselee, client, or patient might be directly facilitated if appropriate and powerful attention, intentions, and images are held concurrently in the mind of the teacher, counselor, therapist, or healer. Direct mental influence, therefore, may be a useful adjunctive tool in these and other dyadic practices. The present experimental report represents an initial test of this general implication or application in a new context (i.e., a focused attention task) and uses a cognitive instead of a physiological measure (i.e., frequency of self-reported mental distractions). It also explores the interaction of the direct mental influence effect with several relevant psychological characteristics of the influencees (e.g., assessments of their ability to focus attention and of their concentration difficulties in every-day life).

Method

Experimental Design

The experimental design consisted of a participant seated in a quiet room attempting to focus attention upon a particular object. Whenever the participant's mind wandered from this attentional focus (i.e., was distracted), this lapse was registered by pressing a button that was recorded by a computer. A helper was stationed in a distant room, isolated from all conventional energetic or informational interactions with the participant. During Control (baseline) periods, the helper occupied her mind with everyday matters and did not think about the participant or the experiment. During Help periods, the helper focused her own attention on a similar object and concurrently maintained an intention for the distant participant to focus well on his or her object and remain free from mental distractions and thus be better equipped to succeed in the attentional task. Control and Help periods were randomly interspersed; the participant was not told when the control or helping periods were in effect. Incidence of mind-wandering (frequency of registered distractions) was compared for the control versus the helping periods. At the end of the session, the participant completed several psychological assessments that measure attention, concentration, distractability, and absorption. A measure of the efficacy of "remote helping of attention" was correlated with the various psychological measures in order to assess psychological interactions.

Hypotheses

The primary experimental hypothesis was that the participants' distraction scores would differ for Control (baseline) and remote Help periods. Of secondary interest in this experiment was an examination of possible interrelationships among the psychological assessments and between the psi scores and the psychological assessment scores. Finally, and for descriptive purposes only, we planned to characterize the nature of the participants' experienced distractions in terms of (a) how quickly the participant became aware of a distraction, and (b) the typical time frame (past, present, future) of the distractions.

Participants

Sixty unpaid volunteers, ranging in age from 16 to 65 years, participated (44 females, 16 males). Most of the volunteers were college-aged friends, acquaintances, fellow students, or co-workers of the three "helpers."

Approximately 75% of the participants were female, and 25% were male. Participants were solicited by the three helpers and were asked whether they would be interested in participating, for about an hour, in a laboratory experiment investigating attention and remote mental influence. We were not interested in working with a so-called "random sample" of participants. Instead, we used "purposive sampling" to pick persons with the requisite interest in the processes we were studying. We have no desire to generalize our results (nomothetically) to the population at large, but only to the population of similar, self-selected individuals.

Three of the co-authors functioned as "remote helpers" in this study, each working with 20 participants. These sample sizes were planned in advance. D. S. has an undergraduate psychology degree and has extensive experience in conducting parapsychological and psychological experiments as a research assistant at the Mind Science Foundation. K. M. has an undergraduate psychology degree and had recently been trained in parapsychological research in the Summer Studies Program at the Foundation for Research on the Nature of Man (Durham, NC). V. G. conducted her portion of the study as part of an independent studies program at a local college through which interested students may participate in research practica (internships) at the Mind Science Foundation. The three helpers served dual roles as helper and experimenter during their own portions of the study.

The study was conceived, designed, analyzed, and written by senior author W. B. in collaboration with the three co-authors. W. B. also trained the helpers for their experimental roles, set up the necessary equipment, and selected the psychological assessments used in the study.

Procedure

Physical layout. During the experimental sessions, it was essential to guarantee that there be no sensory cues that could inadvertently let the participant know which condition was in effect at any given time. This was accomplished by situating the participant and the helper in separate closed rooms. The floor plan of the laboratory is given in Figure 1. The helper and participant were isolated from each other during the actual session by means of two closed doors and an intervening corridor. Verbalizations or other distinctive sounds by the helper were disallowed.

Participant's and helper's activities. After being greeted and engaged in a few minutes of rapport-building general discussion, the participant was given instructions for the attention task. This task was a variation on one introduced by Van Nuys (1971). The participant was to attend as fully as possible to the attentional focus object, a lighted votive candle in a pale blue, transparent, glass candle holder placed on a small table approximately two meters away. The participant sat in a comfortable armchair and was instructed to press a hand-held button whenever he or she observed the mind wandering away from the focusing object (the candle holder and candle). Button-presses thus served to register frequency of mental distractions away from the object of concentration. Participants were instructed that after pressing the button, they should gently return attention to the candle holder and candle and to attend fully to it once again. Participants were told that the entire session would last approximately 20 minutes, and that at random times during the session a helper would concentrate upon a similar candle while attempting mentally and at a distance to help the participant pay attention to the object. Participants were asked not to attempt to figure out when the helping periods were occurring but, rather, to be open to such help throughout the session. To minimize the "task" and "success" aspects of maintaining attention on the object, the participants were told that there was really no success or failure, and that our interest was in learning how people actually respond in such a situation and whether that response could be remotely influenced.

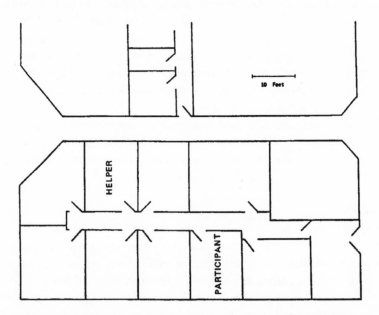

Fig. 1. Laboratory floor plan showing locations of Helper and Participant; the two persons are in independent rooms separated by closed doors and a corridor.

Each of three "helpers" worked with 20 participants. After instructing the participant and returning to her own room, the helper activated a computer program that controlled the experiment and monitored results. The participant had been instructed to take a few minutes to settle down and prepare for the beginning of the session. When ready, the participant pressed the hand-held button one time. This first press served as a signal to the computer to begin the session; it was not counted as a distraction. When it detected this start signal, the computer program began a sequence of 16 one-minute periods. These 16 periods were arranged in 8 pairs. For each pair, the computer determined (by means of a random algorithm) whether the pair sequence would be Control/Help or Help/Control. The eight independent random orderings of the two types of periods within each pair were determined by the random algorithm operating upon a seed number that was based upon the value of the computer's internal clock at the time of the experimenter's initiation of a session. The number of the current period, and the type (whether Help or Control) was indicated to the helper by means of a monitor display. In addition, Help periods were signaled auditorily to the helper by means of a low-volume, low-pitched tone through the helper's headphones. This signal could not be heard by the distant participant. During the 8 randomly scheduled Control (baseline) periods, the helper attempted not to think of the participant or of the experiment, but to think instead of everyday matters. During the 8 randomly scheduled Help periods, the helper focused her own attention fully upon her own candle and holder and concurrently maintained an intention for the distant participant to sustain attention upon the focusing object and to be free of distracting thoughts. The helper did not receive any real-time feedback of the participant's button presses. At the conclusion of the session, the computer provided a paper printout of the participant's distraction scores during each of the 16 periods. In order to rule out possible disruptive emotional reactions to early data returns, the helper did not observe the scores for the sessions. The helper carefully removed the data printout without looking at the scores, folded the printout, and deposited it into a special file folder. The helper then returned to the participant's room and administered the psychological assessments. Upon the completion of the assessments, the helper discussed the experiment, in general terms, with the participant. The participant did not receive any numerical feedback, since the helper herself was unaware of the scores.

We chose to use the computer simply as a device to randomize and control the order and timings of various experimental events and to objectively record the distraction (button-press) responses. We could have stored the button-press results in computer files, as well as or instead of the printout we actually used, but chose not to do so. We have just as much confidence and trust in paper printouts as in electronic events stored in computers. In fact, it could be argued that computer stored information is *more* liable to destruction, loss, or tampering than is information on paper printouts. We think this is something that should be borne in mind in these days of computer-fetishism. Many experimental tasks can be done just as validly, reliably, and objectively without computers as with them. The printout results were checked carefully and redundantly by several persons during stages of data reduction and data analysis.

Psychological assessments. Immediately upon returning to the participant's room, the helper asked the participant to complete four psychological assessments. The first assessment was a one-item visual analog scale on which the participant marked a 160-mm long line to indicate how well attention had been maintained on the candle holder and candle for the overall session. The two extremes of the line were labeled "not successful at all" and "extremely success-ful." The second assessment was a questionnaire on which the participant: (a) indicated how quickly he or she realized that the mind was wandering (immedi-ately, after some time, or long after the distraction had occurred); (b) indicated whether the distracting thoughts tended to be of past, present, or future events; and (c) provided a general description of the types of distracting thoughts that had occurred during the session. The third assessment was a 15-item measure of the degree to which the participant experienced difficulties in focusing attention or concentrating in everyday life. The fourth assessment was a 34-item Absorption scale of the tendency to become totally absorbed in everyday events (Tellegen and Atkinson, 1974). The completed questionnaires and assessments were coded with the participant's number and were stored with the computer printout in the helper's special file until the conclusion of the experiment.

Assessment scoring and data reduction. When all 60 experimental sessions had been completed, the distraction score printouts were examined and ana-lyzed by W. B.; scoring was double-checked by D. S. The psychological assess-ments were scored by D. S.

Results

Psi Scoring (Remote Helping Effect)

The presence of a remote mental influence upon participants' ability to sustain a focused state of attention was examined by comparing the number of button presses as a measure of distraction during Control periods with those occurring during remote mental Help periods. It was planned in advance that we would pool the data from the three experimenter/helpers. In fact, a one-way analysis of variance (ANOVA) comparing the 3 sets of 20 difference scores (20 scores for each of the 3 experimenter/helpers) indicated no significant dif-ferences among the three data sets, yielding $F(2,57) = 0.79$, $p = .46$. This ANOVA result indicated that it was appropriate to pool the three sets of scores.

For each of the 60 participants, distraction scores were summed across the 8 one-minute Control (baseline) periods and the 8 one-minute Help periods. The mean numbers of total distractions (button presses) during the Control and Help periods were 13.60 and 12.43, respectively. These numbers corre-spond to distraction rates of 1.70 and 1.55 distractions per minute, respec-tively. A matched t test calculated for these measures indicates a significant difference between the Control and Help distraction scores, yielding $t(59) = 2.0023$, $p = .049$, two-tailed. The effect size (r) associated with this t is .25. An appealing presentation of effect size is the binomial effect size display (BESD) which converts an effect size to the change in success rate (e.g., survival rate, improvement rate) that would be expected if a treatment or procedure having that effect size were to be instituted (Rosenthal, 1984). According to a BESD, a baseline treatment which ordinarily produces, for example, a 37.5% average survival rate in some population can be augmented by another treatment with an effect size of .25 (the effect size of the remote mental interaction in this experiment) to a 62.5% average survival rate. This is hardly a trivial effect.

Not all statisticians agree with Rosenthal about the appropriateness of the BESD for summarizing data. For those who question such a measure, we can summarize our results even more conservatively by simply noting that in the present study there was a 9% decrease in distractions in the Help periods compared to the Control periods. In other words, out of 100 possible periods of distraction under normal circumstances, these results suggest that there would only be about 91 such episodes during Help periods.

Interrelationships Among Measures

For purposes of examining the interrelationships among the five major measures in this experiment (total distractions, psi influence, estimated attentional success, concentration difficulties in everyday life, and absorption), a Pearson r correlation matrix was produced. This correlation matrix is presented in Table 1.

Table 1. Correlation Matrix of Pearson r Correlates for Five Measures

	Total Distractions	Psi Influence	Attention Success	Concentration Difficulties	Absorption Score
Total Distractions	1.00	.09	–.36**	.18	.26*
Psi Influence		1.00	–.32*	.26*	.13
Attention Success			1.00	–.32*	–.09
Concentration Difficulties				1.00	–.10
Absorption					1.00

*$p < .05$, two-tailed
**$p < .01$, two-tailed

Of the 10 meaningful correlations in Table 1, five are statistically significant. Two of these are "reasonable" correlations between psychological variables. The significant negative correlation between self-estimated success in the attentional task and total number of distractions indicates a congruence between subjective and behavioral assessments of distractions to focused attention within the experimental setting. The significant negative correlation between self-reported difficulties of concentration in everyday life and self-estimated success in the attentional focusing task in the experimental setting is an expected one, given validity of the two measures, and it indicates the generality of the attentional measures.

The third significant correlation, which involves strictly psychological measures, is that between absorption score and total distractions. It is in an unexpected direction, and its interpretation is unclear. It may be that persons with high absorption scores are more aware of their internal processes, including distractions, and therefore are more likely to report such distractions. It is recognized that the self-report assessments for attentional success in the experiment and for concentration difficulties in everyday life were both made immediately after the laboratory attention task and that this circumstance could have been influenced by perceived performance in the task. This was done because: (a) the self-rated attentional success assessment necessarily had to follow what was being rated and, in fact, we simply wished to compare self-rated success with a more "objective" behavioral measure of success (button presses), and (b) we wished to assess both laboratory distractibility and everyday life distractibility under as identical conditions as possible, and this required that we make the assessments at the times we chose to do them.

Of much greater interest are the two significant correlations involving the psi measure. The magnitude of the psi influence score is positively correlated with degree of concentration difficulties in everyday life. This finding is consistent with a "need-related" consideration of psi: Those persons who generally have difficulties concentrating or focusing attention are most in need of attention-focusing assistance, and they indeed show a stronger remote mental attention-focusing effect; those most in need of psi assistance indeed appear to derive more of this assistance.

The second psi-related correlation is between degree of self-estimated success in the attentional focus task in the laboratory and the magnitude of the psi influence effect; this correlation is significantly negative. This again is consistent with a need-related consideration of psi. Those persons who generally had difficulty concentrating in the experimental setting are those who are most in need of remote, mental attentional assistance; the greater the need, the greater was the observed psi effect. It should be emphasized that these obtained need-reflecting relationships are not trivial ones that could be attributed to statistical artifacts such as regression to the mean. The psi influence measure is a *relative* one that is measured *within* a given participant and is not necessarily dependent upon absolute level of responding.

In order to assess the contribution of "participant need" in determining the size of the psi influence effect, a post hoc analysis was performed in the following manner. This method was chosen as the most reasonable one to use in defining participant need, it was the only method used, and it was decided upon before looking at how the data would fall as a result of its use. The 60 participants were dichotomized at the median according to their scores on the two "need" assessments (i.e., self-estimated success on the attentional focus experimental task and concentration difficulties in everyday life). The participants then were categorized as either "more needy" or "less needy" according to a method we are calling "conjoint classification." The more needy participants were those who scored below the median on the attentional success measure *and* who also scored above the median on everyday concentration difficulties. Nineteen of the 60 participants met this conjoint classification, which measured the degree of need to focus attention in both lab *and* life. The less needy participants were those who scored above the median on the attentional success measure and who also scored below the median on everyday concentration difficulties. Nineteen of the 60 participants met this conjoint classification that measured relative freedom from need to focus attention in both lab and life. For the 19 more needy participants, an analysis of their frequencies of distractions during the Control period (18.08) was significantly higher than during the Help period (14.42), yielding a matched t (18) = 2.86, p = .01, two-tailed, and an effect size r = .56. On the other hand, a similar analysis for the 19 less needy participants yielded means of 8.87 for the Control and 8.97 for the Help periods; t (18) = –0.14, p = .89, two-tailed, and an effect size r = –.03. Thus, there was a strong psi influence in the expected direction in the "more needy" participants, but an effect of essentially zero magnitude (in fact, an extremely weak effect in the *reversed* direction) in the "less needy" participants. This is consistent with a need-related theory of psi. Further, this finding conceptually replicates a similar finding observed earlier in our laboratory under very different conditions in which persons who were more needy with respect to autonomic reactivity evidenced a strong psi influence effect in the expected direction, whereas those who were less needy showed a very weak and nonsignificant reversed effect (Braud and Schlitz, 1983).

Descriptive Analysis of Distractions

The distractions were analyzed by counting the frequencies of participants who indicated that they noticed their minds wandering "right away" (33), "some time" after the distraction was already in progress (24), and "a long time" after the mind had wandered from the candle holder and candle focusing object (3). We also counted frequencies of participants who indicated that their minds wandered to "past" (12), "present" (40), or "future" (8) events or time periods.

The dominant tendency, in this particular setting, is for participants to detect distractions relatively quickly and for most of the distractions to be related to the present, rather than to past or future events or time-frames. Both of these patterns (speed of detecting distractions and time frame of distractions) differ significantly from equiprobable expected frequency distributions (yielding chi squares of 23.7 and 30.4, respectively, which, with 2 df, are both highly significant).

Discussion

In previous work in this laboratory [Mind Science Foundation], we found evidence that one person's mental processes of attention, intention, and imagery could interact with another, distant, person's physiological activity.

We view that work as providing an experimental model or analog for at least certain subcomponents or subtypes of mental healing *at a somatic level*. In the present experiment, we extend this work to the mental level. We have found that one person's mental activity, in the form of attention, intention, and focusing, can interact significantly with the mental activity (i.e., attentional processes and freedom from distractions) of another, distant, person. We suggest that this sort of experimental design holds promise as an experimental model or analog for at least some subcomponents or subtypes of mental healing of *mental difficulties*. Under special conditions, calming and quieting my mind can help calm and quiet yours, even when we are spatially separated and have no conventional means of intercommunicating.

Throughout this paper we have attempted conscientiously to use the phrase "direct mental interaction" to indicate that there is indeed an interaction or interrelationship between the mental intentions of the helper and the mental activities and behavioral reactions of the participants. Although we prefer to think of the former as influencing the latter, we recognize this is one psychic interpretation among many possible psychic interpretations. Because we know of no foolproof method, given the present state of the parapsychological art, that can be used with certainty to distinguish telepathy, clairvoyance, precognition, and psychokinesis, we have chosen the more neutral term *interaction,* rather than *influence,* to describe the direct or remote relationships we are observing. Even when we occasionally lapse and use the term "influence," we continue to be aware that this is an interpretation of an empirical relationship that can, at least in principle, be interpreted in other ways. It is important to note, however, that the empirical relationship observed continues to hold independently of the chosen theoretical interpretation or explanation. For example, our aim was to learn whether participants evidenced fewer distractions during periods when helpers were mentally and remotely assisting them (with the latter assistance operationally defined in terms of our protocol). We found indications that this was indeed the case. It may be that the psychic action is entirely within the participants, who may have telepathically, clairvoyantly, or precognitively distinguished the Help and Control periods and then responded appropriately through unconscious or conscious self-regulation of attention or of behavioral responding. The net outcome of such psychic maneuvers is the same as that expected on the basis of direct mental influence of participant by helper. For possible practical applications, it is outcome that matters, rather than inferred process or "mechanism" (this is especially relevant to the possibilities suggested in the last paragraph of this Discussion). In fact, we would argue that questions about the type or source of exhibited psi are outmoded and unproductive, given the emerging view of psi as a dynamic *process* that involves a *field* of persons and events—a field that is transspatial, transtemporal, and transpersonal. We lapse into terminology such as "X influencing Y" for convenience of expression and because of old linguistic habits. Recognizing such outmoded expressions and questions is, at least, the first step toward aligning our language and concepts with the new lessons that psi is teaching us.

It is of interest to note that remote mental interactions occurred in the present study without the provision of immediate sensory feedback to the helper. Feedback was deliberately excluded from the design in order to make it resemble more closely various everyday life situations in which such feedback may not be present or possible.

In both the somatic case (Braud and Schlitz, 1983) and in the mental case, the magnitude of the obtained psi-mediated helping effect was positively and significantly related to the helpee's experienced need to be helped. This latter need may be defined in terms of departure from balance or departure from homeostatis in some particular dimension or aspect of functioning. It appears that psi-mediated helping provides a balancing or normalizing function, helping to return the helpee to a less extreme state or condition. Similar findings have been observed in the experimental work of others (e.g., Grad's 1965 work with saline-stressed or dryness-stressed seeds as opposed to seeds growing under normal, optimal conditions). Similar observations have been made in clinical practice and in theoretical conceptualizations of mental healing (e.g., LeShan's [1974] view that a momentary "union" or "merging" experience of healer and healee may activate the healee's self-healing capabilities in the direction of balance and away from previously distorting or interfering influences on the healee's health and well-being).

If the attention-focusing or concentration exercises of the present study are viewed as protomeditational in nature, then the present findings suggest that one person's meditation process may be directly influenced by the concurrent meditation of another person. This is consistent with anecdotal reports of meditation being easier or more profound in group, as opposed to individual, settings. It is also consistent with reports of meditation in a disciple or trainee being facilitated by the presence of a master or teacher. The present findings are also consistent with the controversial claims within the Transcendental Meditation tradition that meditation by a critical number of meditators can exert unconventional influences upon the social activities of persons in the local geographical vicinity (see Orme-Johnson, et al., 1988; Schrodt, 1990).

We suggest that this simple experiment and its encouraging results point to the feasibility of exploring a wide range of cognitive, emotional, social and spiritual processes that could be facilitated in dyadic situations through the practice of specific mental activities on the part of one member of the dyad. Interesting experiments could be designed to explore possible practical applications of direct mental influence as they might occur in meaningful everyday life processes such as education, counseling, therapy, healing, and spiritual development. As a complement to such an experimental approach, we recommend that equal attention be directed to the study of similar processes in more natural, nonlaboratory settings using alternative research methodologies (see, e.g., Lincoln and Guba, 1985).

References

Braud, W., and Schlitz, M. (1983). Psychokinetic influence on electrodermal activity. *Journal of Parapsychology*, 47, 95–119.

Braud, W., and Schlitz, M. (1989). A methodology for the objective study of transpersonal imagery. *Journal of Scientific Exploration*, 3, 43–63.

Braud, W. G., and Schlitz, M. J. (1991). Consciousness interactions with remote biological systems: Anomalous intentionality effects. *Subtle Energies*, 2, 1–46.

Grad, B. (1965). Some biological effects of the "laying on of hands": A review of experiments with animals and plants. *Journal of the American Society for Psychical Research*, 59, 95–127.

LeShan, L. (1974). *The medium, the mystic and the physicist.* New York: Viking.

Lincoln, Y. S., and Guba, E. G. (1985). *Naturalistic inquiry.* Beverly Hills, CA: Sage Publications.

Orme-Johnson, D., Alexander, C. N., Davies, J. L., Chandler, H. M., and Latimore, W. E. (1988). International peace project in the Middle East: The effects of the Maharishi Technology of the Unified Field. *Journal of Conflict Resolution*, 32, 776–812.

Rosenthal, R. (1984). *Meta-analytic procedures for social research.* Beverly Hills, CA: Sage Publications.

Schrodt, P. (1990). A methodological critique of a test of the effects of the Maharishi technology of the unified field. *Journal of Conflict Resolution*, 34, 745–755.

Tellegen, A., and Atkinson, G. (1974). Openness to absorbing and self-altering experiences ("absorption"), a trait related to hypnotic susceptibility. *Journal of Abnormal Psychology*, 83, 268–277.

Van Nuys, D. (1971). A novel technique for studying attention during meditation. *Journal of Transpersonal Psychology*, 3, 125–133.

Institute of Transpersonal Psychology (Braud)
744 San Antonio Road
Palo Alto, California 94303

11
Distant Mental Influence and Healing: Assessing the Evidence
Marilyn Schlitz, Ph.D., and William Braud, Ph.D.

This chapter provides a brief overview of experimental analogs of distant healing and a summary and meta-analysis of 30 formal experiments in which self-reported healers, psychics, and other self-selected volunteers were successful in influencing the autonomic nervous system activity of distant persons. These distant mental influence effects have important implications for our understanding of the possible mechanisms of distant healing, the nature of the mind-body relationship, and the role of consciousness in the physical world.

This work originally was published in Schlitz, M., and Braud, W. G. (1997). Distant intentionality and healing: Assessing the evidence. Alternative Therapies in Health and Medicine, 3(6), 62–73. *This article is reprinted with permission. Copyright © 1997* Alternative Therapies in Health and Medicine. —*William Braud*

Since the 1950s, researchers have attempted to understand reports of distant or "psychic" healing, developing experimental protocols that test the distant healing hypothesis by measuring biological changes in a target system while ruling out suggestion or self-regulation as counterexplanations. This article provides a brief overview of these "healing analog" experiments. It also provides a summary and meta-analysis of 30

Marilyn Schlitz is director of research at the Institute for Noetic Sciences in Sausalito, Calif. William Braud is professor and research director at the Institute of Transpersonal Psychology in Palo Alto, Calif.

formal experiments in which self-reported healers, "psychics," and other self-selected volunteers attempted to influence autonomic nervous system activity in a distant person. Results across the experiments showed a significant and characteristic variation during distant intentionality periods, compared with randomly interspersed control periods. Possible alternative explanations for the reported effects are considered. Finally, the implications of distant intentionality are discussed for an understanding of the possible mechanisms of distant healing, the nature of the mind-body relationship, and the role of consciousness in the physical world. (*Alternative Therapies in Health and Medicine.* 1997;3(6):62–73)

In Siberia, a middle-aged woman contacts a shaman, whom she trusts to cure her ailing daughter, even though there are hundreds of kilometers between the healer and his patient. Healing, the woman believes, is possible through the power of the shaman's thoughts.

In a jungle in Papua New Guinea, members of the Kaluli tribe gather around a man whom they believe was killed by a distant sorcerer. It is a tragic circumstance, but not surprising for members of a culture in which it is believed that thoughts can create action at a distance.

In an urban setting in Northern California, a woman faces the serious illness of her elderly mother. With the aid of a prayer group from her church, she asks for divine help to restore her mother's health. Although the elderly woman has had no knowledge of her daughter's efforts, she reportedly recovers from her life-threatening illness within hours of the prayer.

These are isolated stories—but they are also connected. Indeed, from *botanicas* in Mexico to street markets in Senegal to the desert of the Kalahari to healing shrines in Japan to suburban neighborhoods in the United States, we find people who believe that the intentionality[1] of one person can influence another person's health or state of being, even at a distance. Some believe that such influences—though typically associated with healing—may also be used for harm, depending on the intentionality of the practitioner or the actions of the patient.[2,3]

Are such beliefs misguided? It is well established in psychology that popular beliefs and attitudes are poor arbiters of "objective" truth, and anecdotes do not carry the same level of reliability as does the scientific method. Anthropological reports provide rich descriptions of what appear to be cases of intentionality at a distance. However, few attempts have been made to account for the observed effects (from seemingly miraculous healings to hex deaths) beyond psychological or psychosomatic explanations.[3-5] Is it possible that the ostensible efficacy of distant healing may be more than a psychological or self-regulatory effect? If so, how would we know?

To answer these provocative questions, we may turn to the area of psi research, where, for more than a century, small numbers of researchers have been applying strict scientific standards to the study of distant intentionality phenomena. Psi research involves the scientific study of anomalous phenomena including telepathy, clairvoyance, precognition, and psychokinesis.[6–8] Since the 1950s, researchers have attempted to study distant or "psychic" healing by developing experimental protocols that provide "healing analogs." Here, the distant healing hypothesis has been put to the test by measuring biological changes in a range of target systems while ruling out suggestion or self-regulation as counterexplanations.[9,10] It is the goal of this article to briefly overview these healing analog experiments, evaluating the strength of the database within one specific program of distant intentionality research and exploring the implications for our understanding of the possible mechanisms of distant healing, the nature of the mind-body relationship, and the role of consciousness in the physical world.

Evaluation Issues

Before we can evaluate the evidence for distant intentionality on living systems in the context of healing analog studies, it is helpful to remind ourselves of the criteria that must be satisfied to indicate the existence or nonexistence of the phenomenon under question. Belief versus evidence is not a straightforward issue. For example, we have all heard that "exceptional claims require exceptional proof." Of course, claims are only "exceptional" if they fail to fit within a particular frame of reference. We need not look far into our past, however, to see how scientific beliefs about the nature of reality have shifted from one view to another. The discovery of meteorites comes to mind. For centuries, peasants reported stones falling from the sky.[11] The French Academy of Science dismissed the peasants' stories as incredible at the time, but today scientists have no problem accommodating meteorites. Contemporary evidence that calls into question the dividing line between mind and matter raises provocative empirical challenges—in the same way that meteorites, radioactivity, atomic fission, and radio waves once did.

The field of psi research has been controversial throughout its history. Strong views frequently resist change, even in the face of data. Many people—including scientists—make up their mind about whether distant intentionality is fact or fiction without examining any data at all.[12–14] As one critic and skeptic of the field noted in a recent review: "The level of the debate during the past 130 years has been an embarrassment for anyone who would like to believe that scholars and scientists adhere to standards of rationality and fair play."[15] Although much of the skeptic/proponent debate has been useful, leading to stronger research designs and more sophisticated analyses, it has limited the ability to conduct a clear and unbiased evaluation.

Fortunately, we need not rely on arbitrary criteria to conduct a credible evaluation of the evidence for distant intentionality on living systems. Over the past half century, researchers have developed techniques for measuring possible distant intentionality effects on living systems, and for assessing probabilities so that chance expectation can be determined and criteria can be established for rejecting the null hypothesis.[9,10,16,17] Typically, the goal of these experiments has been to influence some objectively measured process in another living system. The best experiments use careful, controlled designs that rule out conventional sources of apparent effect, including physical manipulations, suggestion, and expectancy.

Recently, researchers have used meta-analysis as a tool for assessing large bodies of data. A meta-analysis is a critical and technical review of a body of published literature.[18] Going beyond the typical narrative literature review, a meta-analysis applies a variety of statistical inference techniques to reported data and attempts to draw general conclusions. The emphasis is on determining the level of replication across experiments of a specific type.[19] A good description of the issues pertinent to evaluating statistical replication can be found in Utts,[20] but they include the following concepts:

• *File Drawer.* Often in social and psychological sciences the measure of success is a P value equal to or less than .05. That is, assuming the truth of the null hypothesis, there is a 5% chance of observing a deviation as large in an independent test. Although the trend is changing, many researchers and technical journals have treated this value as a sharp threshold—studies quoting a P value of .055 are not published; those quoting .045 are. The "file drawer" represents those studies that were conducted but failed to meet the .05 threshold and were not published. Obviously, if researchers were only publishing 1 in 20 studies in the literature, a nonexistent effect would look "real" according to these standards. Any review of evidence must estimate the number of studies in the file drawer.

• *Statistical Power.* P values strongly depend on the number of trials in a study. An experiment may "fail to replicate," not because the phenomenon in question is not real, but because there were not enough trials in the study and therefore not enough statistical power to detect an effect. Rosenthal[18] and others have addressed this question by proposing a trial-independent measure of success called the "effect size." Reviewers must be cognizant of such "threshold" problems.

• *A Replication Issue.* The evidence for the existence of a phenomenon cannot rest on a single investigator or a single laboratory. How other laboratories attempt to conduct a replication of an earlier experiment is a general problem for social sciences.[12] For example, there is such a thing as an exact replication in which independent investigators try to duplicate the original protocol as closely as possible. These studies contribute toward the overall evidence; however, it is possible that some undetected artifact that is subtly embedded in the protocol might cause misinterpretation of the result. A conceptual replication, in which experiments address the broad concept but contain appreciably different methodological details, protects against such misinterpretations, lessens the chances for fraud, and guards against the possibility of inappropriate techniques. In a review of evidence in social sciences, conceptual or heterogeneous replications carry considerably more weight than do exact or homogeneous replications.

Following are (1) an overview of the field of distant intentionality on living systems research and (2) an evaluation of the robustness of the database within one specific research program through the use of a preliminary meta-analysis.

The Scope of Distant Intentionality Research

A range of so-called target systems has been used to study the possible effect of distant intentionality on living systems, with a range of possible studies that is nearly as diverse as are the processes within an organism that might be influenced.[21] Research participants have included healers, psychics, and unselected laboratory volunteers. The existing literature shows the typical stages of a research paradigm, moving from less to more systematic research over a period of 40 years. Despite vast differences in the database of more than 150 studies, the experiments generally fall into two major categories.

The first category is a direct analog of actual healing practices. It consists of studies in which a healer seeks to influence and mitigate a deleterious process or condition in a target organism. The aim is to improve the organism's vitality or decrease its morbidity. For example, biologist Bernard Grad, a pioneer in this field of study, watered seeds with saline solution that had been treated by a healer or solution that had not. In a careful, double-blind design, Grad found that the seeds watered with healer-treated saline were more likely to sprout and grow successfully.[22]

Another biologist, Carroll Nash, reported that the growth rate of bacteria could be influenced by conscious intention in controlled, double-blind studies.[23] Likewise, psychological researcher William Braud found a highly significant reduction, attributable to the effect of intention, in hemolysis rates of the participant's own blood cells held in a saline solution in test tubes in a distant room.[24]

Some studies in this category involved an attempt to influence the course of a naturally occurring disease or condition. For example, healers have successfully reduced the growth of cancerous tumors in laboratory animals, compared with growth rates for unhealed control animals.[25] In another example, volunteers successfully minimized complications related to heart disease in hospitalized patients, compared with untreated control patients.[26] It is in this latter case that we find research that bears the closest resemblance to healing per se.

Closely related to these experiments is a subset involving attempts to influence the course of an artificially induced disease or condition. For example, in a series of studies using mice—controlling for possible artifacts such as extra warmth from the hands—Grad and colleagues found that dermal wounds healed more rapidly when treated by healers.[27] Apparently, healers also have been able to increase the recovery rate of experimentally imposed wounds on the skin of human volunteers.[28] Statistical analysis typically shows that the rate of wound healing in the treatment group is significantly faster compared with a control group that is otherwise similar but receives no intentional healing treatment.

A second major category of distant intentionality on living systems involves the measurement of ongoing normal processes or behaviors in target organisms. The typical experiments are designed to have either neutral or beneficial effects. The research includes effects on long-term factors such as growth of plants or cell cultures[29] and short-term changes in motor behavior or physiological activity.[30] For practical reasons, the study of ongoing normal processes has received the most experimental attention.

In particular, numerous studies have addressed the question of whether physiological measures—specifically autonomic nervous system activity in humans—might be susceptible to distant intentionality. In one series of experiments, electrodermal activity (EDA) fluctuations were chosen as the physiological measure. Such measurements are readily made, are sensitive indicators, are known to be useful peripheral measures of the activity of the sympathetic branch of the autonomic nervous system, and have relevance to the area of healing research. Studies using these measures represent a coherent and methodologically consistent subset of the overall database of studies of the influence of distant intentionality on living systems. What follows is an overview and analysis of this subset of experiments; for specific details on each study, please refer to the original study report. We have chosen to focus on these experiments because they represent a well controlled and systematic program of study, because there have been heterogeneous replications in numerous laboratories by independent investigators, and because they are an area in which the authors have extensive experience.

Beginning in the 1970s, a series of experiments was conducted in which skin resistance was measured in the target person while an influencer in a separate room attempted to interact with the distant person by means of calming or activating thoughts, images, and intentions.[31] Based on relatively standard protocols across the 30 studies in this database, simple physiological measures showed a highly significant and characteristic variation during the distant intentionality periods, compared with randomly interspersed control periods.

The experiments involving autonomic nervous system activity as the dependent measure will be divided into two major sets of studies. In the direct intentionality set, 19 experiments involved an attempt by one person to influence the physiological activity of a distantly situated other person, without any direct perception of—but with some type of physiological feedback about—the latter. A total of 434 different people participated in these experiments: 317 as influencees, 105 as influencers, and 12 as experimenters.

In the remote observation set, 11 experiments were conducted that allowed the influencer to observe the distant person during random periods via closed-circuit television. A total of 230 individual sessions were conducted in these 11 experiments.

Direct Intentionality Experiments

Whereas the specific details of the experiments differed slightly, the general method across studies involved the instructed generation of specific intentions by one person, and the concurrent measurement of autonomic nervous system activity in another person. Throughout the experiment the two persons occupied separate, isolated rooms, and all conventional sensorimotor communication between them was eliminated to ensure that any obtained effects were truly attributable to distant intentionality.

In a typical experiment, person A was instructed to try to induce a specific physiological change in person B. The expected psychophysiological effect was assessed by measuring the spontaneous EDA (skin resistance responses, or SRR) of person B during randomly selected recording epochs. During half of these epochs, interspersed randomly throughout the session, person A generated imagery designed to produce a specific somatic effect (decreased sympathetic nervous system activity in some cases, increased sympathetic activation in other cases). The remaining half of the epochs served as control periods during which person A did not generate the relevant intention. Person B, of course, was unaware of the sequence of the two types of epochs and was also "blind" to the exact starting time of the experiment, the number and timing of the various periods, and so on. In the majority of experiments, the influencee was instructed to make no deliberate effort to relax or to become more active, but rather to remain in as ordinary a condition as possible and to be open to and accepting of a possible influence from the distant influencer whom he or she had already met. The influencee was asked to allow his or her thought processes to be as variable or random as possible and to simply observe the various thoughts, images, sensations, and feelings that came to mind without attempting to control, force, or cling to any of them.

The influencer sat in a comfortable chair in front of a polygraph in another closed room. During control periods, the influencer attempted not to think about the influencee or about the experiment, but to think of other matters. During influence periods the influencer used the following strategies (either alone or in combination) in an attempt to influence the somatic activity of the distant influencee:

1. The influencer used imagery and self-regulation techniques to induce the intended condition (either relaxation or activation, as demanded by the experimental protocol) in himself or herself, and imagined (and intended for) a corresponding change in the distant subject.

2. The influencer imagined the other person in appropriate relaxing or activating settings.

3. The influencer imagined the desired outcomes of the polygraph pen tracings (i.e., imagined few and small pen deflections for calming periods and many and large pen deflections for activation periods).

Rest periods in the various experiments ranged in duration from 15 seconds to 2 minutes between recording epochs. During those periods, the influencer was able to rest and prepare for the upcoming epoch.

To eliminate the possible influence of common internal rhythms and to remove the possibility that the influencer and the influencee just happened to respond at whim in the same manner and at the same time, it was necessary to formally assign to the influencer specific times for engaging in imagery. Such assignments had to be truly random, counterbalanced, and, of course, could not be known to the influencee (lest the influencee self-regulate his or her own physiology on the basis of such knowledge to confirm the expectations of the experimenter). The influencee's blindness with respect to the imagery/nonimagery sequence was maintained by keeping all participants (including the experimenter) blind regarding the sequence until preparatory interactions with the influencee had been completed and the session was about to begin. Only then, when the influencee and the influencer/experimenter team were stationed in their separate rooms, did the experimenter become aware of the proper epoch sequence for that session.

To evaluate the outcome of the protocol just described, the amount of EDA during the intentionality epochs is compared with that of the control epochs using conventional parametric statistical techniques. If the experimental protocol just described is not violated, and yet it is found that significantly greater somatic activity of an appropriate type occurs during the intentionality periods than during the control periods, we can conclude with confidence that a distant intentionality effect has occurred, and that the results cannot be attributed to (1) conventional communication channels or cues (because the two parties are isolated from contact with each other through the use of distant, isolated rooms); (2) common external signals, common internal rhythms, or rational inference of the imagery/nonimagery schedule and resultant appropriate self-regulation (because the imagery/nonimagery schedule is truly randomly determined and is unknown to person B); or (3) "chance coincidence" (because the level of responding to be expected on the basis of chance alone may actually be determined and compared statistically with the obtained response levels). What follows is an overview of the 17 experiments in the direct intentionality set.

Experiments 1 through 4—Mind Science Foundation

These experiments, designed by William Braud and colleagues at the Mind Science Foundation, were considered demonstration-of-effect or proof-of-principle studies.[30,32,33] The experiments involved male and female volunteers as "receivers" of the distant intentionality effects. These participants were not selected on the basis of any special physical, physiological, or psychological characteristics; they could best be described as "self-selected" on the basis of their interest in the topics being researched. In experiment 1, the distant intentionality influencer was the experimenter; in experiment 2, it was a well known psychic healer; in experiments 3 and 4, this role was played by unselected volunteers. Overall significance in the distant intentionality effected was reported in three of the four experiments. There appeared to be no important differences in the effect due to the type of influencer involved.

Experiments 5a and 5b—Mind Science Foundation

In this experiment investigators[31] were interested in whether those with a greater "need" for a possible calming influence would evidence stronger results than those without such need. Therefore, for that experiment, individuals who self-reported symptoms of greater than usual sympathetic autonomic activation (e.g., stress-related complaints, excessive emotionality, excessive anxiety, tension headaches, high blood pressure, ulcers, or mental or physical hyperactivity) were selected as influencees. The latter were also screened in an initial EDA recording session to guarantee that they did, in fact, exhibit greater than average sympathetic autonomic activity. For this study, the influencers were the experimenters. A significant calming effect was observed for the group who had greater sympathetic autonomic activation (i.e., greater need to be helped remotely).

Experiment 6—Mind Science Foundation

This experiment explored the role of feedback to the influencer.[34] Three experimenters tested the abilities of 24 unselected volunteer influencers to decrease the spontaneous EDA of 24 distant volunteer influencees. For half of each session, trial-by-trial polygraph feedback of the EDA of the influencees was provided to the influencer and experimenter. For the other half of each session, feedback was not provided, and the influencer simply closed his or her eyes and imagined the desired outcome. Significant differences were found in the nonfeedback condition, but not in the feedback condition.

Experiment 7—Mind Science Foundation

This experiment[34] investigated the target person's ability to block an unwanted distant intentionality influence upon his or her own physiological activity. Two experimenter/influencers attempted to increase the EDA of the distant persons. Sixteen influencees were instructed to "shield" themselves—using psychological attention, imagery, and intentional strategies—from the distant intentionality influence. Sixteen persons were instructed to cooperate with the distant intentionality effect, without knowing, of course, when such efforts were being attempted. The influencers were unaware of whether a particular influencee was "blocking" or not.

Experiment 8—Mind Science Foundation

This experiment[34] explored the specificity or generality of the effect by means of simultaneous measurements of several physiological systems (EDA, pulse rate, peripheral skin temperature, frontalis muscle tension, and breathing rate). There were two conditions in the experiment. In the standard condition, three experimenter/influencers attempted to calm the distant person's physiology. In the instructed specificity condition, the influencers were asked to attempt to make their distant intentionality influence as specific as possible. Feedback was provided for only the EDA measure.

Experiment 9—Mind Science Foundation

Experimenters served as influencers for 30 sessions in which the aim of the study was to determine whether increments or decrements in EDA might be more readily produced via distant intentionality influence.[30]

Experiment 10—Mind Science Foundation

Experimenters served as influencers for 30 sessions in which a within-subjects design was used to learn whether the magnitude of the remote intentional influence could be voluntarily self-modulated by the influencers;[30] there were attempts to produce large or small changes on different occasions.

Experiment 11—Mind Science Foundation

This experiment involved unselected volunteers and three trained Reiki healing practitioners.[35]

Experiments 12, 13a, and 13b—Mind Science Foundation

This involved a pilot study and two formal experiments designed to test a particular theoretical interpretation of remote intentionality effects, an "intuitive data sorting" or "decision augmentation" model (a description of which goes beyond the scope of this article).[36] The test of the model involved two methods of randomly selecting the sequences of remote calming and remote activation intentions. The pilot involved a total of 40 sessions in which selected influencers worked with unselected influencees. No overall significance was found. In the formal studies, 32 sessions were conducted in which there were fewer opportunities for a decision augmentation to occur. In these sessions, a significant remote intentionality effect was reported (experiment 13a). In the 32 sessions in which there were greater opportunities for decision augmentation, however, there was not a significant remote intentionality effect. Although consistent with the existence of a remote intentionality effect, these outcomes were not supportive of the decision augmentation model that was being tested. Additional details may be found in Braud and Schlitz.[37]

Experiment 14—University of Edinburgh

This was a first attempt at closely replicating the Mind Science Foundation remote intentionality work by researchers at another laboratory. The study was conducted at the University of Edinburgh, Scotland, by three experimenters.[38] This replication study consisted of 16 sessions involving a total of two new experimenters, six new influencers, and nine new influencees. The obtained t score, though not independently significant, yielded an effect size virtually identical to the average effect size obtained in prior studies of this type ($r = .27$).

Experiment 15—University of Edinburgh

This was a conceptual replication of the EDA remote intentionality work in which positive versus neutral emotions were experienced by the influencers and intended to affect the remote influencees.[39] An identical response measure of the effect yielded a P of .08 and an effect size identical to the mean effect size observed in prior studies ($r = .25$). Other response measures, not reported here, also yielded significant outcomes.

Experiment 16—University of Nevada, Las Vegas

This conceptual or systematic replication by Wezelman et al.[40] involved 11 sessions with three new experimenters, three new influencers, and three new influencees. It did not yield a significant outcome, and results went in a direction not predicted.

Experiment 17—University of Nevada, Las Vegas

An additional systematic or conceptual replication[41] involved 16 sessions, two new experimenters, two new influencers, and two new influencees. The experiment yielded a highly significant remote intentionality outcome and large effect size.

Remote Observation Experiments

Many people have had the experience of being stared at from a distance, only to turn around and discover a pair of gazing eyes focused on them. Indeed, survey data support the widespread distribution of these experiences. As early as 1913, JE Coover reported that 68% to 86% of respondents in California had

had this type of experience on at least one occasion. A survey of the Australian population reported that 74% of the respondents had had such an experience,[42] 85% within a student population at Washington University in St. Louis,[43] 94% of those surveyed in San Antonio, Texas,[44] and 80% of those informally surveyed in Europe and America.[45] Several attempts have been made to explore these claims within a laboratory setting. A review of this literature was reported by Braud et al.,[44] who identified four studies prior to the ones reported here that made use of conscious guessing as the dependent measure.

Based on these four studies, Braud et al.[44] concluded that there is suggestive evidence to support the hypothesis that people can consciously discriminate periods of covert observation from nonobservation under conditions that controlled for subtle sensory cues. The effect size in these studies was not particularly strong, however. According to Braud and colleagues,[44(p376)] this could have been due to the fact that "the testing method used in these studies was not the most appropriate one." In particular, the authors argued that the use of conscious guessing might be less relevant to everyday life experiences, in which detection of an unseen gaze takes the form of bodily sensations and spontaneous behavioral changes. For example, people frequently report the prickling of neck hairs or the tingling of the skin. What follows is a summary of the 11 experiments in this set.

Experiments 1 through 4—Mind Science Foundation

Braud and colleagues[44] designed an experimental procedure based on the hypothesis that remote observation may be detected at the level of sympathetic autonomic nervous system activity. In a series of four experiments,[46] a person stared at a distant participant through the use of a closed-circuit television system while the autonomic nervous system (electrodermal) activity of the latter person was being monitored via chart recorder and computer. The experimental design, as in previous studies involving remote mental influences on human physiology,[30,37] allowed a within-subjects evaluation of covert observation compared with nonobservation (control) periods. The researchers reported that the EDAs of "starees" correlated significantly with the intense attention of the isolated and remote observers in each of the four experiments; effect sizes ranged from .25 to .72. Results were bidirectional, depending on the attitude of the observer and the psychological conditions in effect at the time of the session.

Experiments 5 and 6—Cognitive Sciences Laboratory, Science Applications International Corporation

Designed by Marilyn Schlitz and Stephen LaBerge, these experiments examined intentionality influences of a remote observer, using within-subjects evaluations of experimental sessions that compared mean level of skin conductance response during the covert-observation and control conditions.[47] The goal was to replicate the effect in an independent laboratory, and to focus the effect in the direction of increased EDA. As predicted, skin conductance activity during the covert observation periods was significantly elevated compared with control periods in each of two studies. The effect sizes were .36 and .44.

Experiments 7 through 9—University of Hertfordshire

Three attempted replications of the Braud and Schlitz work on autonomic nervous system detection of remote staring were carried out by researchers at the University of Hertfordshire.[48,49] None of these three studies was independently significant; effect sizes ranged from .26 to .14.

Experiment 10 and 11—University of Hertfordshire

Because differences in results have been correlated with experimenters, this study was designed to test for a distant intentionality experimenter effect. Using the same laboratory, procedure, equipment, and participant population, two researchers (Richard Wiseman and Marilyn Schlitz) replicated his or her initial results—Schlitz's data producing significant deviations from chance and Wiseman's data producing a chance result.[50] This experiment suggests that the intentionality of the experimenter may be an important variable in the outcome of distant intentionality studies.

Results Across Experiments

Thirty experiments using the methods described above have been published as of this writing. In most experiments, the primary method of analysis involved a comparison of the proportion of EDA, which occurred during the distant intentionality epochs of a session, with the proportion expected on the basis of chance alone (i.e., .50). Chi-square goodness-of-fit tests indicated that the distribution of obtained session scores did not differ significantly from a normal distribution; therefore, parametric statistical tests were used for their evaluation. Single-mean t tests were used to compare the obtained session scores with an expected mean of 0.50.

Summary statistics for the 19 direct intentionality experiments are presented in Table 1. For experiments (such as direct intentionality experiments 5 and 13) in which significant differences obtained between different subconditions and/or in cases in which a priori decisions had been made to evaluate certain groups separately, scores are presented for each subcondition; otherwise, scores of subconditions are combined and presented for the experiment as a whole. The number of sessions contributing to each experiment varied from 10 to 40. The single-mean t tests produced independently significant evidence for the distant intentionality effect (i.e., an associated P of .05 or less) in 7 of the possible 19 cases, yielding an experiment-wise success rate of 37%. The experiment-wise success rate expected on the basis of chance alone is 5%.

Results for the 19 direct intentionality experiments are presented in other forms in Table 1 and Figure 1. For these presentations, we calculated z scores and effect size scores for the overall results of each experiment. The z scores were calculated according to the Stouffer method,[19] which involves converting the studies' obtained P values into z scores, summing these z scores, and dividing by the square root of the number of studies being combined; the result is itself a z score that can be evaluated by means of an associated P value. These 19 experiments yield an overall Stouffer z of 4.82, which has an associated P value of .0000007. The effect sizes shown in Table 1 and in Figure 1 are r values, which are particular forms of the effect size measures recommended for meta-analyses of scientific experiments.[19,51,52] The r's were calculated according to the formula $r=\sqrt{[t^2/(t^2 + df)]}$. These effect sizes varied from –0.25 to +0.72, with a mean r = +.25, and compare favorably with effect sizes typically found in behavioral research projects.

Table 1. Studies of Direct Intentionality Influences on Electrodermal Activity: Statistical Summary of 19 Successive Experimental Series

Experimental series	Single mean t	df	P*	z†	Effect size r
Braud and Schlitz[29]					
Experiment 1	3.07	9	.0065	2.73	.72
Experiment 2	2.04	9	.035	1.81	.56
Experiment 3	2.96	9	.0077	2.42	.70
Experiment 4	–0.76	9	.736	–0.63	–.25
Experiment 5a	2.40	15	.014	2.20	.53
Experiment 5b	–0.09	15	.537	–0.09	–.02
Experiment 6	1.77	23	.043	1.72	.35
Experiment 7	1.15	31	.13	1.13	.20
Experiment 8	0.45	29	.33	0.44	.08
Experiment 9	0.44	29	.33	0.43	.08
Experiment 10	1.31	15	.10	1.28	.32
Experiment 11	0.62	14	.28	0.58	.16
Experiment 12	0.21	39	.41	0.23	.03
Experiment 13a	2.41	31	.02	2.08	.40
Experiment 13b	–0.53	31	.70	–0.52	–.09
Radin et al.[38]					
Experiment 14	1.07	15	.15	1.04	.27
Delanoy and Sah[39]					
Experiment 15	1.41	31	.08	1.41	.25
Wezelman et al.[40]					
Experiment 16	–0.58	10	.71	–0.56	–.18
Rebman et al.[41]					
Experiment 17	4.07	15	.0005	3.30	.72
Overall results for					
19 experiments		398	.0000007	4.82	.25

*All P values are one-tailed
†z values are presented for Stouffer z purposes

Summary statistics for the 11 remote observation experiments are presented in Table 2 and Figure 2. The number of sessions contributing to each experiment varied from 16 to 30. The single-mean t tests produced independently significant evidence for the remote observation effect (i.e., an associated P of .05 or less) in 7 of the possible 11 cases, yielding an experiment-wise success rate of 64%, compared with a success rate, expected on the basis of chance alone, of 5%. These 11 experiments yielded an overall Stouffer z of 3.87, which has an associated P value of .000054. The effect sizes ranged from −.57 to +.50, with a mean of +.25, which is identical to the mean effect size obtained in the 19 direct intentionality experiments. Shown also in Table 2 and Figure 2 are results of a "sham control" test conducted in connection with experiments 3 and 4. In this sham control, data were treated as they were in a true remote observation study, but remote observation did not, in fact, occur. Chance results were found, as expected, in this special control condition.

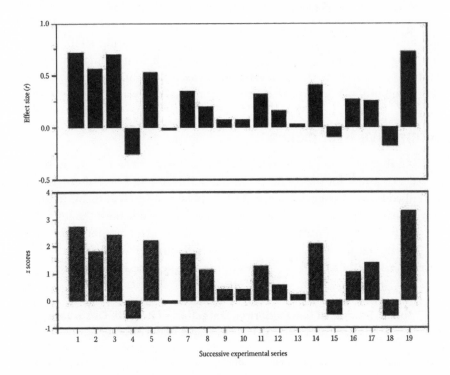

Figure 1. Effect sizes *(r)* and z scores for 19 successive experiments in which one person attempted to influence another person's electrodermal activity through remote intentionality. Negative signs are used to indicate results inconsistent with the direction of the overall findings.

If the results of all 30 EDA experiments are combined for the purpose of a global evaluation, the overall summary statistics are as follows: the single-mean

t tests produced independently significant evidence for the remote intentionality or remote observation effect (i.e., an associated *P* of .05 or less) in 14 of
the possible 30 cases, yielding an experiment-wise success rate of 47%, compared with a success rate, expected on the basis of chance alone, of 5%. These
30 experiments yielded an overall Stouffer *z* of 6.17, which has an associated
P value of 4.58 x 10^{-10}.The average effect size *(r)* is +.25.

Inspection of Tables 1 and 2 and Figures 1 and 2 suggests that a distant
intentionality effect did not occur in all experiments, but that across experiments the data show a relatively consistent effect size that appears replicable
and robust. In terms of its magnitude, the effect is not a negligible one. Under
certain conditions, the distant intentionality effect can compare favorably with
an imagery effect upon one's own physiological activity. Although it is not
reviewed in this article, an autonomic self-control experiment was conducted
immediately following direct intentionality experiment 5.[31] In this psychophysiological self-regulation study, volunteers attempted to calm themselves using relaxing imagery during ten 30-second periods, and their EDA
during those periods was compared with activity levels during 10 interspersed
nonimagery control periods. The strength of the self-control imagery effect in
that study (an 18.67% deviation) did not differ significantly from the strongest
distant intentionality effect of experiment 5 (a 10% deviation).

Alternative Hypotheses

It is important to address various alternative hypotheses that might be proposed to account for the results presented in this review. These alternatives are
described below, along with rationales for discounting each of them. In the following paragraphs, the term "observer" is used to include both the influencers
of the direct intentionality studies and the starers of the remote observation
studies; the term "observee" is used to include both influencees in the direct
intentionality studies and the starees in the remote observation studies.

**Table 2. Studies of Electrodermal Detection of Remote Observation:
Statistical Summary of 11 Successive Experimental Series**

Experimental series	Single mean *t*	*df*	*P*	*z*	Effect size *r*
Braud et al.[44,46]					
Experiment 1: Untrained participants	−2.66	15	.02*	−2.37	−.57
Experiment 2: Trained participants	2.15	15	.025	1.98	.48

Experimental series	Single mean *t*	*df*	*P*	*z*	Effect size *r*
Experiment 3:					
Replication 1	1.92	29	.03	1.85	.34
Experiment 4:					
Replication 2	2.08	15	.025	1.91	.47
Schlitz and LaBerge[47]					
Experiment 5:					
First experiment	1.88	23	.036	1.80	.36
Experiment 6:					
Second experiment	2.36	23	.014	2.20	.44
Wiseman and Smith[48]					
Experiment 7:					
Electrodermal					
activity					
experiment	1.45	29	.08	1.41	.26
Wiseman et al.[49]					
Experiment 8:					
First experiment	0.66	21	.26	0.64	.14
Experiment 9:					
Second experiment	0.91	19	.19	0.88	.20
Wiseman and Schlitz[50]					
Experiment 10:					
Wiseman experiment	0.48	15	.32	0.46	.12
Experiment 11:					
Schlitz experiment	2.25	15	.02	2.07	.50
Braud et al.[46]					
Sham control	0.30	15	.38	0.31	.08
Overall results for					
11 experiments		230	.000054	3.87	.25

*This *P* value is two-tailed; all others are one-tailed

1. *The results are due to sensory cues or other uncontrolled external stimuli.* Based on the experimental design, this alternative hypothesis can be rejected. There were no known or obvious factors that could have influenced the observee based on the random schedule of experimental and control periods.

2. *The results are due to internal rhythms that may have influenced the observee's autonomic nervous system activity.* This potential artifact has been ruled out with the use of a random and counterbalanced schedule of experimental and control periods.

3. *The results are due to chance correspondences between the observer's observations and the observee's physiological responses.* The use of conventional statistical

techniques, as well as the existence of nontrivial effect sizes in the predicted direction, minimize the likelihood of coincidence.

4. *The results are due to recording errors or motivated misreadings of the data.* The data were recorded through the use of an automated procedure that eliminated human error in data recording.

5. *Observees knew the target sequence and so manipulated their physiology to conform to the experimenter's expectations.* The use of a random sequence that was accessed after all preexperimental interactions with the observee ruled out this potential artifact.

6. *The results are due to arbitrary selection of data.* The number of trials and subjects was specified in advance and the reported analyses include all recorded data that fell within the experimental protocol.

Discussion

This article began by asking whether beliefs in distant intentionality and healing may simply be misguided. Although reported instances of healing in everyday life may turn out to be effects of ordinary treatments, physical influences from the healer (such as heat or electromagnetic or static electrical fields), or other conventional sources including the simple passage of time (spontaneous recovery),[53] they may also include conscious or unconscious distant intentionality effects.[54] A primary research goal has been the development of methods through which distant intentionality effects might be isolated and distinguished from other influences.

Based on a review of this experimental literature, the statistical results are beyond what is expected by mean chance expectation. With relatively consistent findings from different laboratories, it is unlikely that the results are due to some systematic methodological flaws. It is with confidence, therefore, that researchers reject the null hypothesis. Whereas the distant intentionality effect sizes are small, they are comparable to—or, in some cases, eight times larger than—those reported in some recent medical studies that have been heralded as medical breakthroughs.[20,55] In short, based on the standards applied to other areas of science, distant intentionality effects on biological systems, like other areas of psi research,[56,57] appear promising for future inquiry.

Distant intentionality is a challenging, important, and potentially useful area for scientific research. There is increasing interest in and use of alternatives to conventional medicine. In service of this interest, there is a need for carefully controlled experimental research to assess the viability and the proper application of intentional healing and related practices.

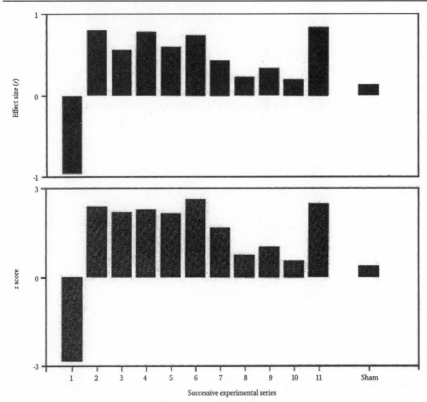

Figure 2. Effect sizes *(r)* and *z* scores for 11 successive experiments in which electrodermal activity was used as an indicator of autonomic detection of remote staring. The "sham" indications are for control sessions in which remote staring did not occur. Negative signs are used to indicate results inconsistent with the direction of the overall findings.

A particularly intriguing possibility is that the various remote intentionality influences reviewed in the report may occur not only nonlocally with respect to space (as these studies already have indicated) but also nonlocally with respect to time. Such a possibility could allow direct attentional and intentional influences to be directed "backward in time" to influence probabilistic events involved in seed moments or initial formative conditions harmful or helpful to health and well-being. These processes could provide adjunct modalities for preventive healthcare. We explored such retroactive intentionality effects empirically as early as 1979,[33] and discuss such possibilities more thoroughly in a separate article.[29] The authors are engaged in ongoing research projects addressing this issue.

As with any progressive research program, the results of this work present new problems for future research. The scientific community seems to believe that explanations for the claimed results of the distant intentionality work will be forthcoming through modifications in our current scientific models.[58,59] Although this may be the case, it is equally possible that the data will help us to reflect on and potentially revise the epistemological and ontological assumptions that are used to guide modern science itself. In this way, distant intentionality research may lead us to a new way of knowing about the world and our place within it.

Following the relativistic views of science recently advanced by works in the history, sociology, and philosophy of science, we may recall that science deals with models and metaphors representing certain aspects of experienced reality.[60] Any model or metaphor may be permissible if it is useful in helping to order knowledge, even though it may seem to conflict with another model that is also useful. (The classic example is the history of wave and particle models in physics.) It is a peculiarity of modern science that it allows some kinds of metaphors and disallows others. It is perfectly acceptable, for example, to use metaphors that derive directly from our experience of the physical world (such as "fundamental particles" or "acoustic waves"), as well as metaphors representing what can be measured only in terms of effects (such as gravitational, electromagnetic, or quantum fields). It has also become acceptable in science to use more holistic and nonquantifiable metaphors such as organism, personality, ecological community, or universe. It is taboo, however, to use nonsensory "metaphors of mind"—metaphors that tap into images and experiences familiar from our own inner awareness.[61] We are not allowed to say, scientifically, that some aspects of our experience of reality are reminiscent of our experience of our own minds—to observe, for example, that distant intentionality phenomena might indicate some supra-individual, non-physical mind.

Social philosopher Willis Harman[62] speaks of the need for a new epistemology that employs broader metaphors and recognizes the partial nature of all scientific concepts of causality. (For example, the "upward causation" of physiomotor action resulting from a brain state does not necessarily invalidate the "downward causation" implied in the subjective feeling of volition.) A new epistemology would implicitly question the assumption that a nomothetic science—one characterized by inviolable "scientific laws"—can in the end adequately deal with causality. In our search for a better understanding of the mechanisms underlying distant healing, the most fundamental issue is whether consciousness is real in some nontrivial sense. Can it be "causal"? Results reported here and elsewhere[63,64] suggest that consciousness may be causal, or that, in some ultimate sense, there may be no causality—only a whole system evolving.[65] In the latter case, distant intentionality might not be an anomaly . . . but part of another order of reality.

Acknowledgments

This work represents almost two decades of research. The authors wish to thank the following organizations for their support over the course of the authors' research: Mind Science Foundation; Parapsychology Foundation;

Cognitive Sciences Laboratory, Science Applications International Corporation; Thomas Welton Stanford, Psychical Research Chair, Department of Psychology, Stanford University; Institute of Transpersonal Psychology; Esalen Institute; Hodgson Fund, Department of Psychology, Harvard University; Department of Psychology, University of Hertfordshire; and the Institute of Noetic Sciences.

Various colleagues helped the authors at different stages of this work. They include Jeanne Achterberg, Daryl Bem, Gary Davis, Larry Dossey, Willis Harman, Gary Heseltine, Wayne Jonas, Stephen LaBerge, Ed May, Michael Murphy, Dean Radin, Donna Shafer, Matt Smith, Jerry Solfvin, Leanna Standish, Elisabeth Targ, Jessica Utts, Richard Wiseman, and Robert Wood.

References

1. Schlitz M. Intentionality in healing: mapping the integration of body, mind, and spirit. *Altern Ther Health Med.* 1995;1(5):119–120.
2. Halifax-Grof J. Hex death. In: Angoff A, Barth D, eds. *Proceedings of an International Conference on Parapsychology and Anthropology, 1973.* New York, NY: Parapsychology Foundation; 1974:59–79.
3. Cannon WB. 'Voodoo' death. *Psychosom Med.* 1957;19(3):182–190.
4. Benson H. *Relaxation Response.* New York, NY: Random House; 1992.
5. Kleinman A. Why do indigenous practitioners successfully heal? *Soc Sci Med.* 1979;13b:7–26.
6. Mitchell ED. *Psychic Exploration: A Challenge for Science.* New York, NY: GP Putnam's Sons; 1974.
7. Wolman B. *Handbook of Parapsychology.* Jefferson, NC: McFarland and Co; 1986.
8. Edge H, Morris R, Palmer J, Rush J. *Foundations of Parapsychology: Exploring the Boundaries of Human Capability.* New York, NY: Routledge; 1986.
9. Benor DJ. *Healing Research: Holistic Medicine and Spiritual Healing.* Munich, Germany: Helix Verlag; 1993.
10. Solfvin J. Mental healing. In: Krippner S, ed. *Advances in Parapsychological Research.* Vol 4. Jefferson, NC: McFarland and Co; 1984:31–63.
11. Westrum R. Science and social intelligence about anomalies: the case of meteorites. In: Collins H, ed. *Sociology of Scientific Knowledge: A Sourcebook.* Bath, England: Bath University Press; 1982:185–217.
12. Collins HM, Pinch T. *The Golem: What Everyone Should Know About Science.* Cambridge, England: Cambridge University Press; 1994.
13. Hess DJ. *Science in the New Age: The Paranormal, Its Defenders and Debunkers, and American Culture.* Madison, Wis: University of Wisconsin Press; 1993.
14. Schlitz MJ. A discourse-centered approach to scientific controversy: the case of parapsychology. Paper presented to the American Anthropological Association, Smithsonian Institute, Washington DC; 1993.
15. Hyman R. A critical overview of parapsychology. In: Kurtz P, ed. *A Skeptic's Handbook of Parapsychology.* Buffalo, NY: Prometheus Books; 1985:1–96.
16. Dossey L. *Healing Words: The Power of Prayer and the Practice of Medicine.* San Francisco, Calif: HarperSanFrancisco; 1993.
17. May E, Vilenskaya L. Some aspects of parapsychological research in the former Soviet Union. *Subtle Energies.* 1994;3:1–24.

18. Rosenthal R. *Meta-Analytic Procedures for Social Research.* Newbury Park, NJ, and London, England: Sage Publications; 1991.
19. Rosenthal R. Meta-analytic procedures and the nature of replication: the Ganzfeld debate. *J Parapsychol.* 1986;50:315–336.
20. Utts J. Replication and meta-analysis in parapsychology. *Stat Sci.* 1991;6(4):363–403.
21. Schlitz MJ. The possible application of psi to healing. In: Roll WG, Beloff J, White RA, eds. *Research in Parapsychology, 1982.* Metuchen, NJ: The Scarecrow Press; 1983:266–268.
22. Grad B. Some biological effects of the 'laying on of hands': a review of experiments with animals and plants. *J Am Soc Psychical Res.* 1965;59:95–127.
23. Nash CB. Psychokinetic control of bacterial growth. *J Am Soc Psychical Res.* 1982;51:217–221.
24. Braud WG. Distant mental influence on rate of hemolysis of human red blood cells. *J Am Soc Psychical Res.* 1990;84:1–24.
25. Grad B. Healing by the laying on of hands: review of experiments and implications. *Pastoral Psychol.* 1970;21:19–26.
26. Byrd RC. Positive therapeutic effects of intercessory prayer in a coronary care unit population. *South Med J.* 1988;81(7):826–829.
27. Grad B, Cadoret RJ, Paul GI. The influence of an unorthodox method of treatment on wound healing in mice. *Int J Parapsychol.* 1961;3:5–24.
28. Wirth DP. The effect of noncontact therapeutic touch on the healing rate of full thickness dermal wounds. *Subtle Energies.* 1990;1:1–20.
29. Braud W, Schlitz M. Consciousness interactions with remote biological systems: anomalous intentionality effects. *Subtle Energies.* 1991;2:1–46.
30. Braud W, Schlitz M. A methodology for the objective study of transpersonal imagery. *J Sci Exploration.* 1989;3(1):43–63.
31. Braud W, Schlitz M. Psychokinetic influence on electrodermal activity. *J Parapsychol.* 1983;47:95–119.
32. Braud W. Lability and inertia in conformance behavior. *J Am Soc Psychical Res.* 1980;74:297–318.
33. Braud W, Davis G, Wood R. Experiments with Matthew Manning. *J Soc Psychical Res.* 1979;50(782):199–223.
34. Braud W, Schlitz M, Collins J, Klitch H. Further studies of the bio-PK effect: feedback, blocking, generality/specificity. In: White RA, Solfvin J, eds. *Research in Parapsychology,1984.* Metuchen, NJ: The Scarecrow Press; 1984:45–48.
35. Schlitz M, Braud W. Reiki plus natural healing: an ethnographic and experimental study. *Psi Res.* 1985;4:100–123.
36. May EC, Utts JM, Spottiswoode SJP. Decision augmentation theory: toward a model for anomalous mental phenomena. *J Parapsychol.* 1995;59:195–220.
37. Braud WG, Schlitz MJ. Possible role of intuitive data sorting in electrodermal biological psychokinesis (bio-PK). *J Am Soc Psychical Res.* 1990;83:289–302.
38. Radin DI, Taylor RD, Braud WG. Remote mental influence of human electrodermal activity: a preliminary replication. Proceedings of presented papers, 36th Annual Parapsychological Association Convention; Toronto, Canada; 1993:12–23.
39. Delanoy DL, Sah S. Cognitive and physiological psi responses to remote positive and neutral emotional states. Proceedings of presented papers, 37th Annual Parapsychological Association Convention; Amsterdam, The Netherlands; 1994:128–138.
40. Wezelman R, Radin DI, Rebman JM, Stevens P. An experimental test of magic healing rituals in mental influence of remote human physiology. Proceedings of presented

papers, 39th Annual Parapsychological Association Convention; San Diego, Calif;1996:1–12.

41. Rebman JM, Radin DI, Hapke RA, Gaughan KZ. Remote influence of the autonomic nervous system by a ritual healing technique. Proceedings of presented papers, 39thAnnual Parapsychological Association Convention; San Diego, Calif; 1996:133–148.

42. Williams L. Minimal cue perception of the regard of others: the feeling of being stared at. Paper presented at the 10th Annual Conference of the Southeastern Regional Parapsychological Association, West Georgia College; Carrollton, Ga; 1983.

43. Thalbourne M, Evans L. Attitudes and beliefs about, and reactions to, staring and being stared at. *J Soc Psychical Res.* 1992;58:380–385.

44. Braud W, Shafer D, Andrews S. Reactions to an unseen gaze (remote attention): a review, with new data on autonomic staring detection. *J Parapsychol.* 1993;57:373–390.

45. Sheldrake R. *Seven Experiments That Could Change the World.* London, England: Fourth Estate; 1994.

46. Braud W, Shafer D, Andrews S. Further studies of autonomic detection of remote staring: replications, new control procedures, and personality correlates. *J Parapsychol.* 1993;57:391–409.

47. Schlitz M, LaBerge S. Autonomic detection of remote observation: two conceptual replications. Proceedings of presented papers, 37th Annual Parapsychological Association Convention; Amsterdam, The Netherlands: Parapsychological Association; 1994:352–360.

48. Wiseman R, Smith MD. A further look at the detection of unseen gaze. Proceedings of presented papers, 37th Annual Parapsychological Association Convention; Amsterdam, The Netherlands: Parapsychological Association; 1994:465–478.

49. Wiseman R, Smith MD, Freedman D, Wasserman T, Hurst C. Examining the remote staring effect: two further experiments. Proceedings of presented papers, 37th Annual Parapsychological Association Convention; Durham, NC: Parapsychological Association; 1995:480–490.

50. Wiseman R, Schlitz M. Experimenter effects and the remote detection of staring. Proceedings of presented papers, 39th Annual Parapsychological Association Convention; San Diego, Calif: Parapsychological Association; 1996:149–157.

51. Cohen J. *Statistical Power Analysis for the Behavioral Sciences.* New York, NY: Academic Press; 1969.

52. Glass G, McGaw B, Smith M. *Meta-Analysis in Social Research.* Beverly Hills, Calif: Sage Publications; 1981.

53. O'Regan B, Hirshberg C. *Spontaneous Remission: An Annotated Bibliography.* Sausalito, Calif: Institute of Noetic Sciences; 1993.

54. Schlitz M. Intentionality and intuition and their clinical implications: a challenge for science and medicine. *Advances.* 1996;12(2):58–66.

55. Honorton C, Ferrari DC. 'Future telling': a meta-analysis of forced-choice precognition experiments, 1935–1987. *J Parapsychol.* 1989;53:281–308.

56. Bem DJ, Honorton C. Does psi exist? Replicable evidence for an anomalous process of information transfer. *Psychol Bull.* 1994;115(1):4–18.

57. Utts J. An assessment of the evidence for psychic functioning. *J Parapsychol.* 1995;59:289–320.

58. Stapp HP. Theoretical model of a purported empirical violation of the predictions of quantum theory. *Am Physical Soc.* 1994;50(1):18–22.

59. Stokes DM. Theoretical parapsychology. In: Krippner S, ed. *Advances in Parapsychology Research*. Vol 5. Jefferson, NC: McFarland Press; 1987:77–189.

60. Lakoff G, Johnson M. *Metaphors We Live By*. Chicago, Ill: University of Chicago Press; 1980.

61. Schlitz M, Harman W. The implications of complementary and alternative medicine for science and the scientific process. In: Jonas W, Levin J, eds. *Textbook of Complementary and Alternative Medicine*. Baltimore, Md: Williams and Wilkins. In press.

62. Harman W. A re-examination of the metaphysical foundations of modern science: Why is it necessary? In: Harman W, Clark J, eds. *New Metaphysical Foundations of Modern Science*. Sausalito, Calif: Institute of Noetic Sciences; 1994:1–15.

63. Jahn R, Dunne B. *Margins of Reality*. San Diego, Calif: Harcourt Brace and Co; 1989.

64. Radin DI. *The Conscious Universe: The Scientific Truth of Psychic Phenomena*. New York, NY: HarperEdge; 1997.

65. Wilber K. *A Brief History of Everything*. Boston, Mass: Shambhala; 1996.

12

Health Implications of "Backward-in-Time" Direct Mental Influences

William Braud, Ph.D.

An unusual alternative healing modality is suggested, in which healing intentions—in the form of direct mental interactions with biological systems—may act in a "backward," time-displaced manner to influence probabilities of initial occurrence of earlier "seed moments" in the development of illness or health. Because seed moments are more labile, freely variable, and flexible, as well as unusually sensitive to small influences, time-displaced healing pathways may be especially efficacious. This unusual hypothesis is supported by a review of a substantial database of well-controlled laboratory experiments. Theoretical rationales and potential health applications and implications are presented.

This work originally was published in Braud, W. G. (2000). Wellness implications of retroactive intentional influence: Exploring an outrageous hypothesis. Alternative Therapies in Health and Medicine, *6(1), 37–48.* This article is reprinted with permission. Copyright © 2000 Alternative Therapies in Health and Medicine. —William Braud

Virtually all medical and psychological treatments and interventions—conventional as well as complementary and alternative—are assumed to act in present time on present, already well-established conditions. An alternative healing pathway is proposed in

---

William Braud is a professor and research director at the Institute of Transpersonal Psychology in Palo Alto, Calif., and codirector of the institute's William James Center for Consciousness Studies.

which healing intentions—in the form of direct mental interactions with biological systems—may act in a "backward," time-displaced manner to influence probabilities of initial occurrence of earlier "seed moments" in the development of illness or health. Because seed moments are more labile, freely variable, and flexible, as well as unusually sensitive to small influences, time-displaced healing pathways may be especially efficacious. This unusual hypothesis is supported by a review of a substantial database of well-controlled laboratory experiments. Theoretical rationales and potential health applications and implications are presented. (*Altern Ther Health Med.* 2000;6(1): 37–48)

> Backward, turn backward, O Time, in your flight,
> Make me a child again, just for tonight.
> —Elizabeth Akers Allen, *Rock Me to Sleep, Mother*

> What may be done at any time will be done at no time.
> —Old Scottish proverb

What if it were indeed possible to turn time backward in its flight and exert real influences on what is "past"? This logically outrageous possibility has been alluded to several times in previous issues of this journal. Dossey[1] and O'Laoire[2] considered the feasibility of time-displaced or backward-acting influences in the context of prayer, and Schlitz and Braud[3] broached this possibility in their review of studies of direct mental influence upon biological systems. Dossey[4] has explicitly raised the possibility of "time-displaced health" and "time-displaced illness," and has provided clinical examples that may be consistent with the action of consciousness outside concurrent time. The idea that mental intentions in the present could have direct, observable influences on the past may at first seem like science fiction. Upon closer examination, however, it can be discovered that there exist surprisingly strong theoretical rationales for—and a substantial body of empirical findings consistent with— this unusual idea. In this article I will review the thinking and evidence bearing on this outrageous hypothesis and discuss the implications of the evidence for physical and psychological health and well-being.

Evidence for Concurrent Direct Mental Influence ("Real-Time Psychokinesis")

There exists a substantial experimental database for concurrent direct mental influences, or what might be called "real-time psychokinesis (PK)" (mind-over-matter) effects. Careful laboratory research, conducted since the 1930s, has yielded evidence consistent with the conclusion that, under certain conditions, people are able to influence sensitive, labile physical systems by intending, willing, imagining, visualizing, or "wishing for" desired outcomes. These outcomes occur when the human influencers are at a distance from the target systems and the targets are effectively shielded from all conventional informational and energetic influences.

The inanimate target systems for these studies typically include random mechanical systems (such as bouncing dice or other small objects) or electronic random event generators (REGs) that operate on the basis of radioactive decay or thermal noise in semiconductor components. Exhaustive meta-analyses of inanimate PK studies have yielded impressive results. Radin and Ferrari[5] systematically analyzed 148 studies of direct mental influence of dice, conducted between 1935 and 1987, and concluded that real direct mental influence (PK) effects existed in this database of nearly 2.6 million trials. They also presented and successfully addressed the various criticisms that had been marshaled against such experiments. Radin and Nelson[6] reported a similar extensive meta-analysis of 597 studies of direct mental influence of electronic REGs that had been conducted between 1959 and 1987. Here, too, they found strong and consistent evidence for real PK effects on these inanimate target systems. Subsequent meta-analyses of still more REG data yielded similar, positive findings.[7–11]

In addition to inanimate targets, animate target systems have been used extensively and successfully in direct mental influence experiments. Well-designed studies demonstrating distant mental influence of living systems were carried out as early as the 1920s and 1930s by experimental physiologists working in Russia. In a series of careful experiments, investigators were able to observe direct mental influences on motor acts, visual images and sensations, sleeping and waking, and physiological reactions (changes in breathing and electrodermal activity [EDA]) in people stationed at remote locations and shielded from all conventional interactions. These Russian research teams included many investigators who are now well known for their work in more conventional areas of physiology, conditioning and learning, and higher nervous activity—investigators such as Vladimir Bekhterev, K. I. Platonov, A. G. Ivanov-Smolensky, and Leonid Vasiliev. This early work has been summarized by Vasiliev[12] and by Braud.[13,14]

During this same period, similar studies were carried out in other countries. There were French experiments on inducing hypnosis at a distance (conducted by such notable investigators as Pierre Joire, Joseph Gibert, Pierre Janet, and Charles Richet) and Dutch experiments on remote influence of motor actions (by H. Brugmans at Groningen, in the northeastern Netherlands).[12–14]

Since then, hundreds of experimental studies of distant mental influences on biological systems have been conducted; many of these have been designed as experimental models or analogs for the study of distant or mental healing. Reviews of these studies have been provided by Braud,[13–15] Braud and Schlitz,[16,17] Braud and colleagues,[18] Solfvin,[19] Benor,[20] and Targ.[21] Remote mental influence studies designed to explore the efficacy of prayer as well as healing intentions have been reviewed by Braud[22] and Dossey.[4] In the November 1997 issue of *Alternative Therapies,* Schlitz and Braud[23] presented a meta-analysis of studies demonstrating direct mental influences of the intention and attention of one person on the ongoing physiological (electrodermal) activity of another person, monitored at a distance and shielded from conventional sensorimotor influences.

The conclusion reached in these reviews and meta-analyses of concurrent direct mental interactions with biological target systems is that, in certain circumstances, the appropriate deployment of attention and intention is associated with directional shifts in objectively measured activities in distant and shielded biological systems. These anomalous influences occur under conditions that preclude chance coincidence and mediation by conventional physical processes. The biological target systems that have been influenced successfully in these studies have varied widely and have included bacteria, yeast colonies, motile algae, plants, protozoa, larvae, wood lice, ants, chicks, mice, rats, gerbils, cats, and dogs, as well as cellular preparations (e.g., blood cells, neurons, cancer cells) and enzyme activity. In human "target persons," eye movements, gross motor movements, EDA, plethysmographic activity, respiration, and brain rhythms have been influenced successfully.[4,12–23] The "psychokinetic" effects observed in these studies typically are small, yet they are reliable and consistent and may be produced not only by those selected for special talents, but also by unselected research participants who try their hands (or, better yet, their minds) at such feats for the very first time. Taken together, these studies provide a sound empirical foundation for considerations of the "mechanisms," implications, and possible practical applications of what has been called mental or spiritual healing.

The Possibility of Time-Displaced Direct Mental Influences

In the studies described in the previous section, objective changes in target system activities were measured while the influence attempts were occurring (i.e., "real time" or concurrent PK influences were being studied). In 1971, a novel twist was introduced into such studies by theoretical physicist Helmut Schmidt. Inspired by the apparently nonlocal and acausal nature of PK effects, and informed by the measurement problem, observer effects, and other paradoxical phenomena within quantum theory, Schmidt conducted preliminary studies to determine whether PK effects could be found using prerecorded targets (i.e., whether direct mental influences might occur in a time-displaced or "backward-acting" manner).

To study possible retro-PK, Schmidt[24(pp268–269)] devised an ingenious experimental design:

Consider the following experiment: A random number generator [operating on the basis of the truly random, unpredictable, and conventionally uninfluenceable, physical process of radioactive decay] is activated to produce a string of N binary numbers. These numbers are automatically recorded on magnetic tape, paper punch tape, or some other reliable recording medium. Nobody is present during this generation and recording, and nobody looks at the data until at some later time the recorded sequence of "heads" and "tails" is played back [for the very first time] to a subject in a PK test situation. During the slow playback each recorded head or tail makes a red or green lamp light up while the subject tries mentally to enforce an increased lighting rate of the red lamp.

One might think that in this situation the subject could not succeed because the decision as to how many heads and tails will appear has already been made before the test session. But one can also present arguments that PK might still operate, and that, furthermore, such PK tests with time displacement could give some interesting new insights into the physics and psychology of psi [psychic or paranormal phenomena].

Schmidt actually began carrying out such experiments in 1971 and published a formal report of his findings in 1976. He found that time-displaced PK influences of prerecorded but previously unobserved target events were indeed possible, and that the likelihood or strength of such influences did not differ appreciably from that of real-time or concurrent influences.[24]

In France, Pierre Janin[25] had conducted his own exploratory experiments on "psychokinesis into the past" in 1974, publishing a report of his findings in 1975. Janin found evidence for significant time-displaced PK influences on 2 types of random systems: a radioactivity-based electronic REG and a mechanical system involving randomly occurring right and left movements of steel marbles. He also conducted experiments on concurrent PK on the same 2 target systems and found no significant difference between concurrent and time-displaced PK. In Janin's experiments, the initial random events had been translated into treble and bass sounds on magnetic tape. These sounds were played back, for the first time, to research participants who attempted to influence the events that the sounds represented and that had initially occurred a day or so earlier.

In the Schmidt and Janin studies as well as subsequent retro-PK studies, the interval between the initial occurrence of the influenced events and the time of the later "intentional effort" was 1 or more days. The "instructed aim sequence" of which events were to be influenced in which ways was determined after the events, but before the "effort." This sequence was unknown to the experimenter at the time the initial events were recorded and was based on an intervening quasi-random outcome (such as weather information or the nature of a particular digit in a complex algorithmic calculation). The existence of a PK effect was determined by comparing segments of the prerecorded record that were "wished for" in particular ways (directions) with other segments that were "wished for" in other ways or not at all (control segments), and also by comparing obtained event frequencies with theoretically expected frequencies (based on mean chance expectations). Both empirical and theoretical baselines were therefore used for evaluating the departures that could be attributed to direct mental influence. Appropriate randomness trials and tests were used to ensure that the random processes indeed behaved randomly when they were not being subjected to direct mental influence.

Prior to this published empirical work on retroactive PK, the possibility of time-displaced PK influences had been mentioned or alluded to in theoretical papers published in 1973 by Janin,[26] in 1975 by Evan Harris Walker,[27] and in 1975 in a mathematical theory developed by Schmidt himself.[28]

Formal Time-Displaced Direct Mental Influence Experiments Conducted Under Independent Supervision

Time-displaced PK tests provide an interesting methodological feature: because an objective record of the to-be-influenced physical events already exists, before the time of the later PK influence efforts, such a record can be given to an independent supervisor for safekeeping and for later checking to ensure that mistakes or deliberate fraud on the part of the experimenter have not taken place. This special feature can allow positive results to be channeled directly to a skeptic or critic of this kind of research. Indeed, if the protocols are properly followed, a retro-PK experiment with a successful outcome is perhaps the most methodologically "safe" and potentially convincing evidence for paranormal or psychic functioning. Schmidt has conducted 5 formal, supervised experiments of this kind. These studies, which are summarized in Table 1, provide strong evidence for the existence of a time-displaced PK effect on prerecorded inanimate events.

Table 1. Statistical Summary of All Experiments on Time-Displaced Direct Mental Influence of Prerecorded Inanimate Random Events Conducted Under Independent Supervision

Study report	Obtained z score	P
Schmidt and colleagues[29] (1986)	2.71	.0034
Schmidt and Schlitz[30] (1988)	1.66	.049
Schmidt and colleagues[31] (1990)	0.62	.27
Schmidt and Braud[32] (1993)	1.98	.024
Schmidt and Stapp[33] (1993)	1.23	.11
Overall results for all 5 experiments	3.69*	.0001

*Stouffer $z = \sum z / \sqrt{N}$, where N = 5.

Additional Time-Displaced Phenomena

In addition to the formal time-displaced PK studies just described, there are other phenomena—found both in everyday life and in the laboratory—that suggest a kind of action working backward into the past. In a sense, all actions that are performed in the service of future goals exemplify the idea of a future event or outcome in some way influencing present actions. The influence of the present by the future is, of course, formally identical to the influence of the past by the present. Once—in Aristotelian thinking, for example—there was a place in philosophy for final causes or teleological action. With natural science's increasing emphasis on efficient causes—and its great success in explicating such causes—ideas of final cause, teleology, intention, and purpose were increasingly banished. Today, when we speak of future events or goals "causing" our present actions, we either use the term figuratively or smuggle the future goal into the present (where it may have legitimate efficacy) via our present anticipations, apprehensions, or expectations about the future events. It is currently politically correct to attribute causality only to such presently acting expectations, rather than to future goal events themselves.

It is more difficult to account for accurate premonitions or precognitive experiences in present-time-only terms. Many apparent premonitions can be explained away as coincidences, subtle rational inferences, or distortions of perception or memory. There remain, however, convincing anecdotal accounts of extremely detailed and accurate "foretellings" that cannot be dismissed so readily. Moreover, meta-analyses of carefully conducted precognition laboratory experiments have yielded strong evidence for the reality of precognitive effects. For example, Honorton and Ferrari[34] analyzed the results of 309 forced-choice precognition experiments conducted between 1935 and 1987. These experiments involved approximately 2 million trials, during which participants were asked to guess which of several alternative outcomes would be randomly selected to occur at some future time, ranging from milliseconds to a year. Their meta-analysis revealed strong evidence for accurate and reliable precognitive effects in this database.

Successful precognition usually is understood as someone's "mind" somehow reaching out into the future, accessing future information, and bringing this information back into the present. Even on this view, half of the explanatory process already involves something akin to backward action. It is possible, however, to reverse our usual thinking about precognition and conceptualize it as a future event somehow reaching back in time to influence a present mind. If the future-action-influencing-present-mental-activity schema of precognition is reversed to involve a future-mental-activity-influencing-present-action schema, we have a model that exactly duplicates the arrangements and outcomes found in the successful retro-PK studies described above. The entire body of existing evidence for precognition (paranormal knowledge of the future) could therefore easily be recast as evidence for backward-acting influences.

Additional laboratory findings are suggestive of processes involving influences acting backward in time. Klintman[35,36] has reported experiments in which people's reaction times in identifying color patches were faster when a name that matched that color was presented quickly afterward compared to when a name that mismatched that color was quickly presented afterward. This effect—a kind of "time-reversed interference"—occurs under conditions in which the matching or mismatching color name is randomly determined and the nature of the future name is unknown—in a conventional sense—to the person when the reaction time to the color patch is measured.

Similarly, Radin[37,39] and Bierman and Radin[38] found that people evidence differential autonomic nervous system reactions (heart rate, EDA, and plethysmographically monitored finger blood volume) to emotional versus non-emotional slides 5 seconds *before* the slides are randomly selected and exposed. The differential autonomic reaction (a kind of anticipatory orientation reaction) occurred during a time when the emotional or non-emotional nature of the upcoming slide was still unknown—in a conventional sense—to the participant. This effect has been termed a "presentiment (pre-feeling)" effect, and it is taken to reflect precognition operating at an unconscious, bodily level. It could just as well be interpreted as an objective event (the presentation of an emotional or non-emotional slide picture itself or the person's future *reaction* to the slide picture) acting backward in time to influence a person's physiological activity.

Time-Displaced Studies of Direct Mental Influence of Living Systems

With the above considerations and findings as a preface, we come now to the heart of this paper: the direct mental influence of "past" biological activities by intentions active in the "present." Is it possible to influence prerecorded but previously unobserved biological activities through a time-displaced, retroactive PK process? Inspired by Schmidt's early PK experiments with inanimate, prerecorded targets, in 1978 colleagues and I conducted a preliminary study of possible time-displaced PK using an animate, prerecorded target.[18] The target activity chosen was prerecorded EDA. Ten participants contributed fluctuating EDA tracings while sitting quietly in a room. These EDA tracings were transduced and stored on magnetic tape, but remained unobserved until they were later presented (decoded and displayed as polygraph pen tracings), for the first time, to a selected research participant who attempted to influence segments of the records in prescribed directions. The intentional "efforts" took place 1 to 7 days after the initial EDA had been emitted and recorded. For each tracing, a quasi-random sequence of 10 influence (i.e., attempt to mentally activate) and 10 non-influence control periods was determined just before each later "intentional effort" session. The influence and control epochs were each 30 seconds in duration.

The overall results for these 10 prerecorded EDA PK sessions did not differ significantly from chance. However, some interesting secondary evidence that influences may have been occurring on a session-by-session basis (based on possible correspondences of session outcomes with observations of the influencer's changing motivations from session to session) was noted and described in the original report. The results for this study[18] are given as "experiment 1" in Table 2.

Since that initial exploration of a possible time-displaced direct mental (PK) influence on prerecorded activity of an animate target system, 18 similar studies have been conducted. A statistical summary of the results of all 19 of these conceptually similar studies is presented in Table 2.

Table 2. Statistical Summary of Results of Studies of Time-Displaced ("Backward") Direct Mental Influence of Living Systems

Experimental series	No. of sessions	P	z	Effect size (r)
Braud and colleagues[18] (1979)				
Experiment 1	10	.68	−0.47	−.15
Gruber[40] (1979)				
Experiment 2	10	.01	2.33	.74
Experiment 3	10	.30	0.52	.16
Experiment 4	10	.50	0	0
Experiment 5	10	.10	1.28	.40
Experiment 6	10	.05	1.64	.52
Experiment 7	10	.50	0	0
Gruber[41] (1980)				
Experiment 8	10	.30	0.52	.16
Experiment 9	10	.05	1.64	.52
Experiment 10	10	.30	0.52	.16
Experiment 11	10	.01	2.33	.74
Experiment 12	10	.50	0	0
Experiment 13	10	.05	0.52	.16
Snel and van der Sijde[42] (1990)				
Experiment 14	19	.02	2.05	.47
Experiment 15	19	.02	2.05	.47
Experiment 16	19	.50	0	0
Braud (unpublished data, 1993)				
Experiment 17	15	.074	1.45	.37
Schmidt[43] (1997)				
Experiment 18	10	.00076	3.17	1.00
Radin and colleagues[44] (1998)				
Experiment 19	21	.016	2.16	.47
Overall results for 19 experiments	233	.00000032	4.98	.32

In 1979, Gruber[40] reported 6 experiments. In the 20 sessions of experiments 2 and 3 (10 with the investigator himself serving as influencer and 10 including 10 unselected research participants each contributing 1 session), the prerecorded, animate target activity consisted of locomotor activities of small mammals (gerbils running in activity wheels). The prerecorded living system activity in the 20 sessions of experiments 4 and 5 was a different form of mammalian locomotor activity (gerbils crossing a photobeam in a large cage). The prerecorded activity in the 20 sessions of experiments 6 and 7 was the photobeam-monitored locomotor behaviors of people who had been instructed to walk randomly in a dark room while listening to pink noise (which is white noise, or sounds of random frequencies and intensities, to which "red" sounds of lower frequency or pitch have been added to make the sound more pleasant). In each experiment, the activity of the living target system was converted into recorded click sounds that were stored in an unobserved form until they were played, for the first time, to influencers. The "influence efforts" occurred 1 to 6 days after the target activities initially occurred.

In 1980, Gruber[41] reported 6 additional experiments. In the 30 sessions of experiments 8, 9, and 10 (10 sessions with a special, selected participant and 20 sessions with 20 unselected participants, i.e., those without special talents), the prerecorded activity was the photobeam-monitored behavior of people entering a supermarket in Vienna. In the 30 sessions of experiments 11, 12, and 13, the prerecorded activity was provided by the photobeam-monitored frequency of cars passing through a small and short tunnel in the center of Vienna during rush hour. Again, target activities were converted to click sounds and played, for the first time, to the influencers 1 to 2 1/2 months after the activity initially occurred.

In 1990, Snel and van der Sijde[42] reported results of a study in which a paranormal healer attempted, through distant and retroactive mental influence, to prevent the spread and multiplication of blood parasites (rodent malaria organisms, *Babesia rodhani*) in red blood cells of athymic rats. The "healer" did not receive feedback regarding the dependent measure (the mean, absolute counts of infected red blood cells, microscopically monitored), but simply worked with photographs of the caged rats. It was not determined which animals were the "target" rats and which were the "uninfluenced controls" until *after* blood cell measures were completed (the condition assignment was randomized by someone not otherwise involved in the study). Measurements were taken 14, 28, and 42 days after the animals were inoculated with the parasites. Measurements on these 3 days are presented, respectively, as experiments 14, 15, and 16 of Table 2.

In 1993, Braud conducted 15 sessions in which participants attempted to mentally influence their own prerecorded spontaneously fluctuating EDA (W. G. B., unpublished data, 1993). Experimental procedures and measurement techniques were similar to techniques described by Schlitz and Braud[3] in

1997, with the important difference that this time the intentional influences were "distant" in time rather than space. The EDA had been converted from analog to digital form, stored as a file on a computer disk, and presented (35 to 40 minutes after the EDA had initially been generated and recorded) for the first time as a tracing on a computer monitor screen.

The person who had contributed the EDA tracing approximately one half-hour earlier now watched the tracing and used it as feedback while attempting to mentally influence his or her own prerecorded autonomic activity. Three types of 30-second measurement epochs were randomly interspersed, during which the influencer attempted to either increase (activate), decrease (calm), or not influence (rest) his or her own prerecorded EDA activity. Analyses were performed to determine whether the amounts of EDA during respective periods of the earlier, pre-recorded record corresponded to the influencer's later intentional aims for the various segments of the record. Results of comparisons of time-displaced activation versus calming periods are presented as experiment 17 in Table 2. Although the P value slightly exceeded the arbitrary .05 level, the difference was in the expected direction and yielded a substantial effect size.

In 1997, Schmidt[43] published the results of 10 experimental sessions (experiment 18 in Table 2) in which he attempted—successfully—to influence the durations of his own prerecorded breathing intervals, using recording, measurement, and time-displaced influence procedures similar to those described above. Schmidt also interspersed sessions in which he successfully influenced prerecorded electronic random events. Schmidt's influences on prerecorded animate activity (breathing rate) were somewhat stronger than were his influences on prerecorded inanimate activity (REG data), but not significantly so.

In this study and other time-displacement studies reviewed in this article, what is observed is a strong correlation between the occurrence of certain events in a (past) data stream and the occurrence of (future) intentions. Because there is no obvious connection between the prerecorded events and the random process that determines the sequence of later intentional aims (the intervention), and because changes in prerecorded events do not take place in the absence of the later intentional (intervention) aims (as confirmed through direct comparisons with non-influence, control periods), a claim for a form of "causation" or "influence" of the events by the intentions (beyond mere correlation) seems justified.

In 1998, Radin et al.[44] reported an experiment in which EDA and other autonomic measures were successfully influenced by influencers who were distant from the target activities in space (6000 miles) and time (2 months). The to-be-influenced autonomic activity records were produced in Las Vegas, Nevada, and were stored and remained unobserved until they were influenced 2 months later by healers located in Brazil. The EDA results for 21 sessions are given as experiment 19 in Table 2.

Table 2 summarizes the results of 233 experimental sessions in which participants attempted to influence a variety of living systems in a time-displaced, retroactive fashion. This table includes all animate, time-displaced studies that have been conducted to date of which the author is aware. To facilitate summary and comparisons, the differing test statistics of the various studies have all been converted to a common metric (P, z, and r scores) for the purposes of this table. The results of 10 of the 19 studies were independently significant (i.e., they yielded z scores with associated P values <.05); only 1 significant study outcome would be expected on the basis of chance alone. Using a method recommended by Rosenthal[45] for combining results of several studies, a Stouffer z score may be calculated by summing the individual z scores and dividing this sum by the square root of the number of contributing z scores (in this case, 19). The resulting Stouffer z score for the combined set of 19 studies of time-displaced direct intentional influence of living systems is 4.98, which has a highly significant associated P value of .00000032.

The effect sizes shown in Table 2 are r values, calculated according to the formula $r = z/\sqrt{N}$. These effect sizes varied from –0.15 to +1.00, with a mean r of .32. All of these statistical results compare favorably with results typically found in behavioral and biomedical research projects. Interestingly, the results are extremely similar to those of Schiltz and Braud's meta-analysis of "real time" (concurrent) intentional influences on EDA.[3] Additional compelling evidence pointing to the robustness of these findings is that—with the single exception of experiment 1—*all* of the coefficients (rs) are in the hypothesized direction.

The "Size" of These Effects

Throughout this article, statistical significance levels and effect sizes have been used to indicate the presence of direct mental influences. Historically, the arbitrary $P < .05$ criterion has been used as an indication of the reality of an effect. Many of the obtained P values in the reviewed studies reach and sometimes greatly exceed this probability criterion. More recently, however, there has been a growing movement within the behavioral and biomedical research communities to deemphasize P values (which are rarity indicators) and emphasize effect sizes (which more closely reflect the actual sizes of changes or outcomes and "correct" for differences in sample size). Typically, effect sizes in the ranges of 0.0 to 0.3, 0.3 to 0.6, and 0.6 to 1.0 are taken to represent "small," "medium," and "large" effects, respectively. On average, the effect sizes obtained in the reviewed mental influence studies are at the border between small and medium. However, they compare favorably with what is typically observed in more conventional behavioral and biomedical studies and in some cases the obtained effect sizes are quite large (Table 2). The average effect size observed in these time-displaced mental influence studies (0.32) is 10 times as great as those obtained in some representative medical study outcomes that have been heralded as medical breakthroughs (effect sizes of 0.04 and 0.03, obtained in 2 well-known studies[46,47] of the effectiveness of propranolol and aspirin, respectively, in reducing heart attacks).

To facilitate the appreciation of effect sizes, Rosenthal (a recognized research methods and meta-analysis expert) has offered a special binomial effect size display that allows us to represent a common effect size measure *(r)* in terms of the corresponding proportion of, for example, people in some sample whose health, well-being, or survival rate might be improved by an intervention or treatment with that particular effect size.[45] According to this binomial effect size display conversion, an effect size *(r)* of 0.03 would be the equivalent of 3 additional persons surviving in a sample of 100 persons. An effect size *(r)* of 0.30 (observed in the present studies) would be the equivalent of 30 additional participants surviving in a sample of 100. In life-or-death situations, especially, the outcomes associated with these effect sizes are far from trivial.

Another method for estimating the strength of these effects is to calculate the actual percentage of events or activities that change in association with the direct mental interventions. In various reported aggregations of these percent influence scores, the average influence has ranged from a fraction of a percent to a few percent (in cases of random generator influence) to 4% or 8% (in certain electrodermal influence studies) to 80%, 90%, and even 100% changes in individual sessions. In special experiments, remote, direct mental influence effects on EDA did not differ appreciably from the size of deliberate, self-regulation effects on these same activities.[16,48] Again, expressed in these percent change terms these effects are far from negligible.

Replication Considerations

Many of the trials, sessions, and studies in these and other areas of mental influence research do *not* show an effect. Such replication failures are not unexpected in areas that are being freshly explored and in which the effects and measures are relatively subtle. I offer 2 speculations regarding replication failures. The first is that the emergence of these mental influence effects may depend on the simultaneous presence of a complex and interactive set of physical, physiological, psychological, and even social and cultural factors. If all requisite ingredients of such a complex recipe are not present, or present in insufficient degrees, the effect may not occur. The nonlocal or field-like nature of these phenomena suggests that critical variables may reside not only in the immediate influencers, influencees, and target systems, but may also be present in other people or situations that are spatially and temporally removed from the test situation but meaningfully connected with the experiment.

Until we learn more about the limits and boundary conditions of these effects, such "remote" contributors will remain difficult to isolate and control.

It is crucial to begin identifying the critical independent, contextual variables that might facilitate or impede these effects. Two of the most crucial variables may be the potential for free variability in the target system and the fullness of the intentions of the influencers (the presence of a strong *need* being one guarantee of strong intentions). Experimenter effects themselves no doubt contribute to the variability in study outcome. The challenge of investigating experimenter effects has been recognized and systematically explored in these areas much more than it has in other research areas.[49] Additional important variables have been identified and discussed elsewhere.[50]

The second reason for replication failures in these areas is that, because of the extraordinary nature of the knowledge claims, many more replications are attempted in these areas than in more conventional areas of research. Even well-accepted findings do not always replicate. It would be interesting to see what would happen if conventional interventions were tested as often as anomalous claims are tested.

A final replication consideration is the "file drawer" issue: could the results of published, positive reports be canceled out by negative findings that are never published and languish in researchers' file drawers? It is highly unlikely that a huge file drawer of unreported, negative findings could cancel out the reported positive findings, for the following reasons:

1. Given the scarcity of funding for this kind of research and the small numbers of researchers who are active in these areas, it is unlikely that a large number of such studies are even conducted.

2. Unlike in other research areas, the journals devoted to studies of these types have explicit policies of publishing negative as well as positive research reports.

3. The actual extent of file drawer contributions has been formally and carefully evaluated in the relevant meta-analyses, and the analysts have concluded that the file drawer is not a major threat to the meta-analytic conclusions.

4. The overall significance levels are sufficiently rare as to exclude cancellation even by very large numbers of unreported negative or neutral findings.

Implications for Physical and Psychological Health and Well-Being

The results of the 19 experiments reviewed here suggest that it is possible for people to exert direct mental influences "into the past" to influence the preoccurring and prerecorded activities of biological systems. In these studies, the past events that were influenced in this time-displaced fashion were labile events, characterized by free variability. In addition, the records of the events were stored but never observed during the interval between their initial occurrence and the later influence attempts.

It is crucial to point out that, in the view of Schmidt and others who have conducted these studies, the present intentions, wishes, or PK influences do not *change* the past. Once an event has occurred, it remains so; it does not "unoccur" or change from its initial form. It appears, instead, that the intentions, wishes, or PK "efforts" influence what happens (or happened) in the first place. To clarify this interpretation even further, the time-displaced direct mental intervention could be said to "change" what *would have* happened, but does not change what *did* happen. If, for example, the past events consisted of holes punched in a paper tape record, the intervention does not remove holes that were already there. Instead, the intervention influenced whether certain holes were punched in the first place—if they were punched, they remain punched; if they were not punched, they remain unpunched. In this illustration, what would have happened may be inferred on the basis of a theoretical, statistical expectation (i.e., it would correspond to mean chance expectancy) or may be actually calculated on the basis of empirical contrast conditions (data segments) in which events are counted or measured in the absence of the intervention. Additional evidence that the prerecorded events "stay put"—once registered—has been provided by examining multiple records, created in various formats. All of the records correspond.

The present or future intentions seem to act on the initial probabilities of occurrence of the events and help determine which events initially come into being (i.e., which of several potential events are actualized). Psychokinesis—whether in the form of distant mental influence or time-displaced mental influence—seems to bias the probabilities of initial occurrence of random or freely variable events such that a desired, intended, or goal-serving outcome increases in likelihood. The process appears to act most efficiently on the seed moments or originations of events. Such stages would seem to be more labile, flexible, sensitive, or susceptible to influences of all kinds, including these direct mental influences.

If the findings uncovered in these laboratory experiments are indicative of general principles, these principles might be applied practically in the service of physical health and psychological well-being. Future intentions may have real influences on present seed moments or origination stages of healthful or harmful bodily events or symptoms in the present, and present intentions may have real influences on past seed moments or origination stages of healthful or harmful bodily events or symptoms in the past. Such effects would be most likely to occur in cases of seed moments that are characterized by randomness or free variability.

Consider a simple system consisting of 2 neurons and their intervening synapse. The processes at the synapse may exist in a delicate state, probabilistically balanced near a sensitive threshold that could make the difference between the firing or not firing of an adjacent neuron. Such processes could be ideal "targets" for successful direct mental influences, similar to those described in this article. Indeed, Nobel laureate Sir John Eccles has proposed that synapses may be characterized by probabilistic, random, quantum processes and may be in delicately poised conditions that might make them susceptible to mental influence.[51] Eccles suggested that such influences may be commonplace within an individual's central nervous system, providing a mechanism of action that allows ordinary volitional actions. In this view, volitional actions become instances of endogenous PK in which one's mental intentions may act on the matter of his or her labile synaptic processes to instigate the first of a series of physico-chemical-neuronal activities that eventuate—many "linkages" later—in some observable action or movement. In addition to the probabilistic synaptic activity suggested by Eccles, other substrates that might support quantum randomness effects have been proposed by others. These substrates include the presynaptic vesicular grid,[52] ion channels,[53,54] calcium ions,[55] and cytoskeletal microtubules.[56] In principle, if intention may act directly on a neural (or any anatomical or physiological) substrate in this fashion in "real time," such actions could also take place in a time-displaced fashion.

In addition, if a biological substrate possessing the requisite randomness or free variability can be susceptible to the organism's own intentions, perhaps it is susceptible to the intentions of other organisms as well. Intentional consequences may not be locally confined, but may occur nonlocally (in time and space). All of these suggestions are consistent with the theories and findings of modern PK research—some key features of which have been mentioned in this article. If these nonlocal intentions are aligned with aims or goals of health and wholeness, perhaps active intentions could be directed in the present or even into the past to promote biological and psychological seed moments favorable to physical and psychological health and well-being.

Consider another simple system—a small group of cancerous or precancerous cells at a certain location within the body and a natural killer (NK) cell that is roaming near those cells in a random or freely variable course. It is conceivable that there exists a point at which a random "choice" or "decision" occurs, and the NK cells could move, with 50-50 probability, either toward or away from those cancerous cells (seed moments of disease). In principle, PK or intentional influences could bias the probabilities of action of the NK cell sufficiently to promote movement toward and subsequent destruction of the small group of cancerous or precancerous cells, thereby terminating a seed moment that otherwise might have eventuated in illness or even death (several "linkages" down the line—through probability-pyramiding or snowballing effects).

When a patient appears in our office with a particular malady, we tend to think that the curing or healing of this condition involves using our armamentarium of conventional and unconventional treatments and interventions to slowly and progressively correct that malady in the present. We believe we should use our tools to chop away and gradually destroy an undesired condition that is already well established—working on what now exists, a system with great momentum and inertia. In addition to such real-time therapeutic influences, the findings reviewed in this article suggest an alternative healing pathway. Along with such real-time effects that are often taken for granted, it is possible that our healing intentions may be acting "backward in time" to influence the initial seed moments of the development of the malady that confronts us today. Such an alternative healing pathway or process might be a more effective and efficient one—an "easier" one—because it would be influencing a system at a more labile, flexible, sensitive, and susceptible stage in its development and progression. If such a process could act early and thoroughly enough, it might actually prevent the development of harmful physical or psychological processes. This would constitute an instance of true preventive medicine. Time-displaced healing modalities might actually have important advantages over real-time healing modalities.

If the implications just mentioned could be explored in additional, carefully designed studies, it would be possible to learn more about the ranges and limits of time-displaced, direct mental influences. As more is learned about these effects and the factors or conditions that foster or impede them, possible practical applications of these principles in health-related areas could be planned and studied.

Possible Roles of Intermediate Preobservation and Diagnostic Observations

In the time-displaced studies described above, the to-be-influenced events and record of these events are maintained in an unobserved state until the intentional influence attempts are made. There are provocative findings suggesting that preobservations of the data or records during the interval between event generation and influence attempt may influence the fate of the initial events (i.e., their susceptibility to later direct mental influence). Schmidt has found that if certain to-be-influenced events are strongly observed (with an intense and meaningful density of attention) during the intervening period, those events may no longer be susceptible to later direct mental influence. It is as though prior observation itself establishes or concretizes the reality of the observed events, "locking them in" and making them no longer susceptible to subsequent mental influence.

Such intermediate preobservation effects have been observed in the case of human, canine, and goldfish "preobservers." It should be pointed out that relatively few experiments of this kind have been performed, and their results are not always consistent.[57,58] However, the outcomes in which time-displaced PK effects may be "blocked" by prior observations are consistent with certain interpretations of the role of human and other observations or measurements in the "collapse of the state vector" in quantum systems, and there have even been empirical tests of such preobservation effects by physicists who have been uninvolved in parapsychological investigations, such as R. Smith (unpublished data, 1968) and Hall and colleagues.[59] If it could indeed be shown that conscious preobservation can block subsequent PK success, this finding could pave the way for an exciting series of studies in which the nature of the preobservation is systematically varied and its influence on PK effects is observed. Preobservations of prerecorded events by humans at various stages of their development and in various states of consciousness could be studied. Indeed, a true comparative psychology of consciousness could be developed in which organisms of various species could serve as preobservers and their outcomes could be noted. The blocking of a time-displaced PK effect could serve as a measurement method for exploring, rather directly, the phylogeny (evolutionary history) and ontogeny (individual development) of consciousness.

The influence of preobservation raises the health-related issue of the possible role of diagnostic observations in these effects. In the time-displaced studies reviewed, care was taken to ensure that the to-be-influenced events had not been consciously observed before the influence attempts were made. The very first observation of these events was in the form of a motivated intention. In fact, one rationale for choosing EDA as a target activity in our studies was that changes in the electrical activity of the sweat glands—which underlie EDA changes—are governed by the autonomic nervous system, the functioning of which we are ordinarily "unaware." A research program could be developed around studies of the nature and degree of "conscious awareness" of the to-be-influenced activities on the part of the organisms originally generating such activities. The issues that could be explored are closely related to the preobservation issues mentioned in the previous paragraphs.

If health-serving time-displaced mental influences can only occur with respect to previously unobserved activities, this might limit their application to physical or psychological conditions that have not yet been observed or noted by patients, clients, or healthcare professionals. A previously diagnosed condition would have been preobserved and therefore be less susceptible to later direct mental influence. The ambiguous empirical findings with respect to preobservation are not yet sufficiently clear to permit useful predictions about the role of diagnostic preobservations. It is likely, however, that the *nature* of the diagnostic observation could be crucial in determining its effects. Strong, clear, unambiguous diagnostic tests or measurements that are viewed by multiple observers might yield outcomes quite different from ambiguous diagnostic measures witnessed under conditions of minimal density, intensity of attention, awareness, or by only a single diagnostician.

Consider the following hypothetical case: A patient receives a pessimistic diagnosis of metastatic cancer with poor prognosis based on a radiologist's interpretation of computerized axial tomography (CAT) scan results for the liver and intestinal areas. The diagnosis consists of observations of vague spots on a CAT scan record. Subsequently, the patient engages in an intense healing program that includes self-healing components as well as the assistance of others. Strong healing intentions (of self and others) are directed toward the patient. Later, a higher-resolution CAT scan reveals a different picture of the "spots" and the initial diagnosis is now questioned. Nearly 7 years later, the patient is alive, well, and happy. Three possible interpretations of this case might be made. First, the initial diagnosis was incorrect and the patient never was ill. Second, there was indeed the beginning of a severe illness that was halted in its tracks and reversed by later psychological and life-change interventions acting conventionally in "real time." Third, there may have been a time-displaced influence of the later healing intentions on the seed moments of an illness, with the illness either not progressing or not occurring in the first place. The ambiguous nature of the first diagnostic preobservation may have allowed subsequent time-displaced mental influences to be effective. A case virtually identical to this hypothetical example actually occurred and has been reported.[60]

Usually we think of medical diagnoses as beneficial procedures that inform us about the presence of a harmful condition that is really there, allowing us to take appropriate measures to reduce or eliminate the harmful physical condition. The above considerations, however, suggest that a medical diagnosis itself could falsely indicate or *even actually produce* (through focused intentionality) a condition of illness that was not present prior to the diagnosis. Diagnoses may be both therapeutic and iatrogenically harmful. The issues raised by the nature and timing of a "diagnosis" are numerous, complex, and deep, and their adequate consideration would take us beyond the scope of this paper.

An Apparent Paradox and Its Possible Resolution

Paradox is simply the way nonduality looks to the mental level.

—Harman and DeQuincey[61(p45)]

We cannot end our discussion of possible health-serving, time-displaced intentional influences without mentioning what may be (to some) a troubling logical difficulty. If retroactive intentional effects do not *change* the past, but influence what came to be in the first place, how would one handle a case in which a patient presented with a malady that already had developed and was too obviously strongly present in this moment? If a future, intentional mental intervention were to be effective in such a case, would it not have to "undo" something that already had happened, and would this not lead to logical contradictions and paradoxical considerations?

There are several ways of dealing with such difficulties. One is to posit that it is indeed possible to change the past and not merely influence initial probabilities of occurrence. Another suggestion is that the presenting condition is complex, and that some of its synergetic, harmful components may not yet have occurred in sufficiently full form or may still be susceptible to concurrent or time-displaced mental influences. A third possibility is indicated in the figure. The nature of a symptom complex that presents itself to a physician at time T_3 may in fact be common to a family of curves (world-lines or life-lines) that describe various potential time courses of the progression of an illness.

In the figure, curve A represents a poor prognosis in which health declines progressively, eventually resulting in the death of the patient. Curve B indicates a less severe illness time course. Curve C indicates a gradual, incomplete recovery. Curve D depicts a relatively rapid and complete recovery. Note that the presenting condition at time T_3 could be on any of the 4 curves and, based only on information available at time T_3, one cannot know which curve actually may be in effect. It is possible that healing intentions generated at time T_3 might retroactively influence which of a family of possible curves is actualized at time T_2—the common seed moment for several possible progression/outcome curves.

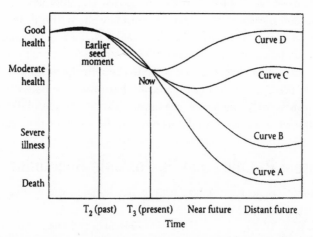

Fig. Family of world-lines or life-lines with identical past seed moment and conditions at a present observational time that yield dramatically different future outcomes.

Therefore, without violating the principle that time-displaced intentions might act only to influence but never to change the past, it is still possible to account for illness recovery curves C and D, *because one of these entire curves may have been selected, through the biasing agency of time-displaced intentionality,* from the equiprobable curves with potential origins at time T_2. Note, also, that curve D may represent an outcome that typically is termed "spontaneous remission" by the medical community. A thorough report of several hundred instances of "spontaneous remission" from metastatic cancer is available.[62] Is it possible that at least some of these spontaneous remissions are the result of time-displaced healing intentions on the part of patients, their loved ones, or health professionals?

It should be pointed out that the time-displaced possibilities presented in this section and elsewhere in the article do not preclude more conventional, "real-time" influences that could be active at time T_3 or any other time. In fact, every moment in the development of healthful or harmful conditions can be conceived as a "seed moment" for future progressions of those conditions. As such, every moment of the development or progression of a health condition may be susceptible to *both* concurrent conventional *and* complementary and alternative influences as well as possible time-displaced influences of the types proposed in this article. The latter are suggested as adjunctive and not necessarily exclusive influence pathways or possibilities.

Additional Considerations

The following areas remain to be discussed: (1) issues and research findings involving the most effective mental influence strategies; (2) whether "trial by trial" feedback to the influencers is necessary for these effects to occur (it is not); (3) possible alternative paranormal interpretations of these obtained findings (there are several, but all involve "violations" of our usual understanding of what is possible in time); (4) theoretical understandings of what might underlie these obtained effects; (5) the relevance of these findings to an intriguing suggestion made by Schopenhauer about the complex and seemingly deliberately orchestrated interrelationships of our world-lines and lives in space and time[63] and the anthropic cosmological principle;[64] and, perhaps most interestingly, (6) the import of these findings for our apprehension of the nature of time itself. These issues must await a later presentation.

Acknowledgments

All of the preceding conceptualization, analysis, and writing was done at the Institute of Transpersonal Psychology in Palo Alto, California. The already published experiment 1 (Table 2) had been performed at the Mind Science Foundation, San Antonio, Texas, and the empirical work supporting experiment 17 (Table 2) was performed at the University of Edinburgh, Scotland, and at the Institute of Transpersonal Psychology, Palo Alto, California.

References

1. Dossey L. The forces of healing: reflections on energy, consciousness, and the beef Stroganoff principle [editorial]. *Altern Ther Health Med.* 1997;3(5):8–14.
2. O'Laoire S. An experimental study of the effects of distant, intercessory prayer on self-esteem, anxiety, and depression. *Altern Ther Health Med.* 1997;3(6):38–53.
3. Schlitz M, Braud W. Distant intentionality and healing: assessing the evidence. *Altern Ther Health Med.* 1997;3(6):62–73.
4. Dossey L. *Healing Words: The Power of Prayer and the Practice of Medicine.* San Francisco, Calif: HarperSanFrancisco; 1993:127–132.
5. Radin DI, Ferrari DC. Effects of consciousness on the fall of dice: a meta-analysis. *J Sci Exploration.* 1991;5;1–24.
6. Radin DI, Nelson RD. Evidence for consciousness-related anomalies in random physical systems. *Found Physics.* 1989;19:1499–1514.
7. Dobyns YH. Selection versus influence revisited: new methods and conclusions. *J Sci Exploration.* 1996;10(2):253–268.
8. Nelson RD, Dobyns YH, Dunne BJ, Jahn RG. *Analysis of Variance of REG Experiments: Operator Intention, Secondary Parameters, Database Structure: Technical Note PEAR 91004.* Princeton, NJ: Princeton Engineering Anomalies Research Laboratory, Princeton University School of Engineering/Applied Science; 1991.
9. Nelson RD, Dunne BJ, Jahn RG. *An REG Experiment With Large Database Capability: III: Operator Related Anomalies: Technical Note PEAR 84003 (September).* Princeton, NJ: Princeton Engineering Anomalies Research Laboratory Princeton University School of Engineering/Applied Science; 1984.
10. Jahn RG, Dobyns YH, Dunne BJ. Count population profiles in engineering anomalies experiments. *J Sci Exploration.* 1991;5:205–232.
11. Dunne B, Jahn R. Experiments in remote human/machine interaction. *J Sci Exploration.* 1992;6:311–332.
12. Vasiliev, LL. *Experiments in Distant Influence.* New York, NY: Dutton; 1976.
13. Braud WG. Remote mental influence of electrodermal activity. *J Indian Psychology.* 1992;19(1):1–10.
14. Braud WG. On the use of living target systems in distant mental influence research. In: Shapin B, Coly L, eds. *Psi Research Methodology: A Re-examination.* New York, NY: Parapsychology Foundation; 1993:149–188.
15. Braud WG. Distant mental influence of rate of hemolysis of human red blood cells. *J Am Soc Psychical Res.* 1990;84:1–24.
16. Braud WG, Schlitz MJ. A methodology for the objective study of transpersonal imagery, *J Sci Exploration.* 1989;3:43–63.
17. Braud WG, Schlitz MJ. Consciousness interactions with remote biological systems: anomalous intentionality effects. *Subtle Energies.* 1991;2(1):1–46.
18. Braud W, Davis G, Wood R. Experiments with Matthew Manning. *J Soc Psychical Res.* 1979;50(782):199–223.
19. Solfvin J. Mental healing. In: Krippner S, ed. *Advances in Parapsychological Research.* Vol 4. Jefferson, NC: McFarland and Co; 1984:31–63.
20. Benor DJ. *Healing Research.* Vols 1–4. Munich, Germany: Helix Verlag; 1993.
21. Targ E. Evaluating distant healing: a research review. *Altern Ther Health Med.* 1997;3(6):74–78.
22. Braud WG. Empirical explorations of prayer, distant healing, and remote mental influence. *J Religion Psychical Res.* 1994;17(2):62–73.

23. Schlitz M, Braud W. Distant intentionality and healing: assessing the evidence. *Altern Ther Health Med.* 1997;3(6):62–73.

24. Schmidt H. PK effect on pre-recorded targets. *J Am Soc Psychical Res.* 1976;70(3):267–291.

25. Janin P. Psychocinèse dans le passé? Une expérience exploratoire. *Rev Metapsychique.* 1975;1(21–22):71–96.

26. Janin P. Nouvelles perspectives sur les relations entre la psyché et le cosmos. *Rev Metapsychique.* 1973.

27. Walker EH. Foundations of parapsychical and parapsychological phenomena. In: Oteri L, ed. *Quantum Physics and Parapsychology.* New York, NY: Parapsychology Foundation; 1975:1–53.

28. Schmidt H. Toward a mathematical theory of psi. *J Am Soc Psychical Res.* 1975;69(4):301–319.

29. Schmidt H, Morris R, Rudolph L. Channeling evidence for a psychokinetic effect to independent observers. *J. Parapsychology.* 1986;50:1–15.

30. Schmidt H, Schlitz M. *A Large Scale Pilot PK Experiment With Prerecorded Random Events: Mind Science Foundation Research Report.* San Antonio, Tex: Mind Science Foundation; 1988.

31. Schmidt H, Morris RL, Hardin CL. *Channeling Evidence for a Psychokinetic Effect to Independent Observers: An Attempted Replication: Mind Science Foundation Research Report.* San Antonio, Tex: Mind Science Foundation; 1990.

32. Schmidt H, Braud W. New PK tests with an independent observer. *J Parapsychology.* 1993;57:227–240.

33. Schmidt H, Stapp H. PK with prerecorded random events and the effects of preobservation. *J. Parapsychology.* 1993;57:331–349.

34. Honorton C, Ferrari DC. Future telling: a meta-analysis of forced-choice precognition experiments, 1935–1987. *J Parapsychology.* 1989;53:281–308.

35. Klintman H. Is there a paranormal (precognitive) influence in certain types of perceptual sequences? Part 1. *Eur J Parapsychology.* 1983;5:19–49.

36. Klintman H. Is there a paranormal (precognitive) influence in certain types of perceptual sequences? Part 2. *Eur J Parapsychology.* 1984;5:125–140.

37. Radin DI. Unconscious perception of future emotion: an experiment in presentiment. *J Sci Exploration.* 1997;11(2):163–180.

38. Bierman DJ, Radin DI. Anomalous anticipatory response on randomized future conditions. *Percept Mot Skills.* 1997;84:689–690.

39. Radin DI. Further investigation of unconscious differential anticipatory responses to future emotions. *Proceedings of Presented Papers: The 41st Annual Convention of the Parapsychological Association.* Halifax, Nova Scotia, Canada: Parapsychological Association; 1998:162–183.

40. Gruber ER. Conformance behavior involving animal and human subjects. *Eur J Parapsychology.* 1979;3(1):36–50.

41. Gruber ER. PK effects on pre-recorded group behavior of living systems. *Eur J Parapsychology.* 1980;3(2):167–175.

42. Snel FWJJ, van der Sijde PC. The effect of retro-active distance healing on Babesia rodhani (rodent malaria) in rats. *Eur J Parapsychology.* 1990;8:123–130.

43. Schmidt H. Random generators and living systems as targets in retro-PK experiments. *J Am Soc Psychical Res.* 1997;91(1):1–13.

44. Radin DI, Machado FR, Zangari W. Effects of distant healing intention through time and space: two exploratory studies. *Proceedings of Presented Papers: The 41st Annual*

Convention of the Parapsychological Association. Halifax, Nova Scotia, Canada: Parapsychological Association; 1998:143–161.

45. Rosenthal R. *Meta-analytical Procedures for Social Research.* Beverly Hills, Calif: Sage; 1984.

46. Kolata G. Drug found to help heart attack survivors. *Science.* 1981;214:774–775.

47. Steering Committee of Physicians' Health Study Research Group. Preliminary report: findings from the aspirin component of the ongoing Physicians' Health Study. *N Engl J Med.* 1988;318:262–264.

48. Braud W, Schlitz M. Psychokinetic influence on electrodermal activity. *J Parapsychology.* 1983;47:95–119.

49. Sheldrake R. Experimenter effects in scientific research: how widely are they neglected? *J Sci Exploration.* 1998;12(1):73–78.

50. Braud WG. Implications and applications of laboratory psi findings. *Eur J Parapsychology.* 1990;8:57–65.

51. Eccles JC. The human person in its two-way relationship to the brain. In: Morris JD, Roll WG, Morris RL, eds. *Research in Parapsychology 1976.* Metuchen, NJ: Scarecrow Press; 1977:251–262.

52. Beck F, Eccles JC. Quantum aspects of brain activity and the role of consciousness. *Proc Natl Acad Sci USA.* 1992;89:11357–11361.

53. Bass L. A quantum-mechanical mind-body interaction. *Found Physics.* 1975;5:159–172.

54. Donald MJ. Quantum theory and the brain. *Proc R Soc Lond.* 1991;427A:43–93.

55. Stapp HP. *Mind, Matter and Quantum Mechanics.* Berlin, Germany: Springer-Verlag; 1993.

56. Hameroff SR. Quantum coherence in microtubules: a neural basis for emergent consciousness? *J Consciousness Studies.* 1994;1(1):91–118.

57. Schmidt H. Addition effect for PK on pre-recorded targets. *J Parapsychology.* 1985;49:229–244.

58. Schmidt H. PK tests with and without preobservation by animals. In: Henkel LA, Palmer J, eds. *Research in Parapsychology 1989.* Metuchen, NJ: Scarecrow Press; 1990:15–19.

59. Hall J, Kim C, McElroy B, Shimony A. Wave-packet reduction as a medium of communication. *Found Physics.* 1977;7:759–767.

60. Targ R, Katra J. *Miracles of Mind: Exploring Nonlocal Consciousness and Spiritual Healing.* Novato, Calif: New World Library; 1998.

61. Harman WW, DeQuincey C. *The Scientific Exploration of Consciousness: Toward an Adequate Epistemology.* Sausalito, Calif: Institute of Noetic Sciences; 1994.

62. O'Regan B, Hirshberg C. *Spontaneous Remission: An Annotated Bibliography.* Sausalito, Calif: Institute of Noetic Sciences; 1993.

63. Schopenhauer A. Transcendent speculation on the apparent deliberateness in the fate of the individual. In: Payne EFK, trans. *Pererga and Paralipomena: Short Philosophical Essays.* Oxford, England: Clarendon; 1974:201–223. [Originally published 1851.]

64. Barrow JD, Tipler FJ. *The Anthropic Cosmological Principle.* Oxford, England: Oxford University Press; 1986.

Index

ABBA sequence, 9–10, 16, 31, 35, 37, 41, 43
ability to block the influence, 19
ability to influence, xxxiv
  distribution in the population, xxxii, 141
Abnormal Hypnotic Phenomena, 3
accuracy of influence, 49
Achterberg, Jeanne, 2, 122
action-at-a-distance, 119
activation aims, xxv, 8, 113
"active agent telepathy," 121
AIDS, xxxvi
Alberfan mining disaster, xxxv
Allen, Elizabeth Akers, 234
allobiofeedback, xxvi
  replications of, 44
allofeedback, xxiv–xxvi
Alrutz, Sydney, 126–127
alternative hypotheses,
  arbitrary data selection, 16, 35, 97, 226
  ceiling and floor effects, 35–36
  chance coincidence, 15, 34, 95, 225–226
  common internal rhythms, 95–96
  experimenter fraud, 17, 36–37, 97–98
  external stimuli, 15, 95, 225
  internal rhythms, 15, 225
  motivated misreading of records, 15, 34, 96, 226
  observer preknowledge, 226
  recording errors, 15, 34, 96, 226
  sensorimotor cues, 15, 95
  subject foreknowledge, 15–16, 34–35, 96–97
  subject fraud, 16–17, 36, 97
  systematic error, 16, 35, 97
alternative psi hypotheses,
  effect exerted psychically, not physiologically, 37
  effect exerted telepathically or clairvoyantly, 38
  PK influence on equipment, 37–38
Alternative Therapies in Health and Medicine, xiii, 209,
  233, 236
Alvarado, Carlos, 125
Amerindian groups, 139
amplifiers, 10, 80–81, 155
  living systems as, 129–130
analysis of variance, 61
Andrews, Sperry, 150–182
"animal magnetism," 119
A. N. Severtsev Institute of Evolutionary Morphology and
  Animal Ecology, 98
anxiety, persons with greater, 179–180
Aplysia, xxviii
apparatus, 30, 40, 155, 169
appropriateness of using living target systems, 122
  lability, 127–129
  levels of influence, 132
  living systems as detectors/amplifiers, 129–130
  multiple psi channels, 130–132
  other influences upon the target, 132–133
arbitrary data selection, 16, 35, 97, 226
Aristotelian thinking, 239
"artificial somnambulism" effects, 119
assessment instruments, standardized, xxxiv
attention alone imparting influence, xxvi, xxx, xxxiv
audience sensitivity index, 179
autobiofeedback, xxiv
autonomic discrimination, 167–168
autonomic nervous system activity level, 220
autonomic self-control, 130
  in calming other persons at a distance, 40–42
autonomic staring detection, 150–165
  methodology, 154–158
  previous staring detection experiments, 153
  results, 158–161
awareness of influencee, xxxiii
BAAB sequence, 9–10, 41

"backward-in-time" direct mental influences,
  formal time-displaced direct mental influence experi-
    ments conducted under independent supervision, 238
  health implications of, 233–256
  implications for physical and psychological health and
    well-being, 246–249
  possibility of time-displaced direct mental influences,
    236–238
  possible roles of intermediate preobservation and diag-
    nostic observations, 249–251
  replication considerations, 245–246
  size of these effects, 244–245
  time-displaced studies of direct mental influence of liv-
    ing systems, 240–244
basal skin resistance (BSR), 30, 43, 159
Bechterev, Vladimir M., 110–111, 120, 143, 235
Beloff, John, 125
Berger, Hans, 110
Bergson, Henri, xliii
BESD. *See* binomial effect size display
"Beyond," sensing existence of, 192
bidirectionality of influence, xxxiii, 142
Bierman, Dick, xxxv
binomial effect size display (BESD), 82, 84, 203
bio-PK effects, xxvi, 27–29, 34, 130–131, 133, 135 , 139–140
bioassays, 129
biofeedback, xxiv, 184
biological material, influencing, 124
blinded studies, 184, 189
blocking influences, xxxiii, 217–218, 250
blood pressure influence, studies of, 89–90
blood sample collection, 53
Bohm, David, xxxix
brain, creating consciousness, xi
Braud, William G., xii–xiii
Browning, Robert, 143
Brugmans, H., xxiii, 235
BSR. *See* basal skin resistance
Buddhist teaching, xxxviii
Byrd, Randolph, xxxv, 185
cables, shielded, 77
calming other persons at a distance, xxv, 24–45
  alternate hypotheses rejected, 34–37
  alternative psi hypotheses, 37–38
  the autonomic self-control experiment, 40–42
  experimental design overview, 29
  influence of need satisfaction factor, 28–29
  the main experiment, 29–40
  mean electrodermal activity measured, 33
  methodology, 29–32, 40–41
  present bio-PK experiment, 34
  previous bio-PK experiments, 26
  results, 32–33, 41–42
cancer, 248
Carrington, Hereward, 125–126
CAT scan results, 251
causality, 228
  final, 229
Cazzamalli, F., 110
CCU. *See* coronary care unit patients
ceiling effects, 35–36
cellular preparation (hemolysis), *in vitro* studies of, 92–93
Chalmers, David J., xi
chance, 15, 34, 95, 225–226
Chastenet, Armand-Marie-Jacques de, 119
"Chevreul pendulum," 87
chi-square goodness of fit tests, 11
Christians, participating in prayer experiments, 185
clairvoyant effect, xix, xxi, 38
Clark, Walter Houston, 192
closed-loop feedback, xxiv
"cognitive caresses" of synapses, 48
"Cohen's d" measures, 13–14, 115

coincidence, 15, 34, 95, 225–226
Commission for the Study of Mental Suggestion, 110
common internal rhythms, 95–96
community of sensation, 3
computerized axial tomography (CAT), scan results, 251
conceptual replications, 98
concurrent direct mental influence, evidence for, 233–236
conditional reflexes, 110
"conformance behavior," 137
connectedness, xliv, 192–193
   training in, 157, 163, 169
"conscious pre-observation" experiments, 102
consciousness,
   materialism dominating research in, xii
   nature and role of, xli–xlii
   "normal," xvii
   origin and destiny of, xiii
   "stream of," xxii
consistency of results, 36
"context of discovery," xviii–xxiii
control, 8, 134–135
   modalities of, 138
Control periods, 198–202, 205
conventional influences, xxxi
converging strategies approach, 136–138
Cornell University, 151
coronary care unit (CCU) patients, 185
correspondence models, xxxviii
counterbalanced design, 16
Cox, E. W., 125–126, 132
CPPC sequence, 53, 64
Crookes, William, 125
"curses," 140
Dean, Douglas, 121
death, survival of physical, xlii
"Defense Mechanism Test," 136
design considerations, 73–77, 198
   for calming other persons at a distance, 29
detectors, living systems as, 129–130
diagnostic observations, possible roles of, 249–251
Diana, Princess, funeral ceremony televised, xxxv
Dingwall, Eric, 3
direct mental influence, xxvii, xxxi
   "backward-in-time," 234
   in daily life, xli
   evidence for concurrent, 233–236
   health implications of, 233–256
   within the body, xli
direct mental interactions with living systems (DMILS),
   xxvii, 73, 77, 100, 104
discovery, "context of," xviii–xxiii
distance at which effects occur, xxxi, 19, 141
distant healing, xxxvi–xxxvii, 185–187
distant intentionality research, xxvi
   blocking influences, 217–218
   directness of, xxxi
   incrementing or decrementing EDA, 218
   "intuitive data sorting" model, 218–219
   proof-of-principle studies, 217
   replication studies, 219
   role of feedback to the influencer, 217
   scope of, 213–215
   self-modulation of effect magnitude, 218
   specificity or generality of effect, 218
   subject "need" factor, 217
   work with Reiki healing practitioners, 218
distant knowing, xxxiv
distractions, descriptive analysis of, 205
distress, persons with greater, 179–180
DMILS. See direct mental interactions with living systems
Donne, John, xliii
Dossey, Larry, xi–xiii, xxxvii
double-blind clinical trials, 184
Durov, Vladimir, 120
dyadic interactions, xlii, 100
early studies, xxiii–xxx

Eastern meditative, mystical, and spiritual traditions, 184
Eccles, John C., xxviii, 125, 248
EDA. See electrodermal activity
EEG alpha control, 137
Eisenbud, Jule, 121, 143
electrodermal activity (EDA), xxiv, 162, 214, 236,
   242–243, 245. See also skin resistance responses
   computer-monitoring of, 119
   correlates of remote attention, 85–86
   direct mental influence effect, 96
   of distant persons, xxx–xxxi
   experimental participants, 77–78
   experiments with, 189–191
   incrementing or decrementing, 21, 29–32, 218
   influence series, 77–85
   influencer's procedures, 79–80
   physiological measurements, 80–81
   precautions against conventional communication, exter-
      nal stimuli, and subtle cues, 78
   precautions against suggestion, expectancy, placebo
      effects, and confounding internal rhythms, 79
   results, 81–85
   scoring of measurements, 81
   spontaneous, 4, 81, 215
   subject's procedures, 79
electromagnetic shielding, 111
electrophysiological activity, meaning of, 10–11
ELF. See extremely low frequency radiation
Elliotson, John, 119, 143
environmental variables, compensation for, 135
equilibrium dimension, 128
equipment, PK influence on, 37–38
Esdaile, James, 119
evaluation issues, 211–213
   numbers of studies, 212
   replication, 212–213
   statistical power, 212
expectancy, 184
experimental session, representation of events of, 114
experimenters, 78
   fraud by, 17, 36–37, 97–98
Experiments in Mental Suggestion, xxiii, 111, 120
external stimuli, 15, 95, 225
extremely low frequency (ELF) radiation, xxxvii–xxxviii,
   103–104, 191–192
extraversion/introversion (E/I) scores, 72, 159, 175
eye movements, 236
eyes, staring into, 157
failures of distant mental influence, xxxiv, 20, 142
Faraday chamber screenings, xxiii
feedback to the influencer, xxxii, 217
feeling scores, 159
field-like nature of PK phenomena, 245
findings about distant mental influence, 140–142
   ability to focus effect, 141
   bidirectional effects, 142
   blocking the effect, 141
   distance at which effects occur, 141
   distribution of ability to manifest the effect, 141
   effect of focusing on desired target activity, 142
   factor of geomagnetic field (GMF) activity, 142
   failure of the effect, 142
   intention alone sufficient, 141
   nontriviality of effects, 141
   numbers of target systems susceptible to the effect, 141–142
   reliability and robustness of distant mental influence, 141
   subject need factor, 141
   subject's unawareness of influence, 141
   unconcern by subjects about influence, 141
findings about mentally protecting human red blood cells
   at a distance, 59–65
   analysis of variance, 64
   assumption of independence of measurements, 64
   experimental questions, 60
   means and standard deviations, 61

score comparison for independently significant subjects, 63
scoring rates for individual subjects, 61
findings about remote staring detection,
    association of influence with geomagnetic field (GMF) activity, xxxiii
    attention alone imparting influence, xxxiv
    awareness of the influencee, xxxiii
    bidirectionality of influence, xxxiii
    blocking unwanted influence, xxxiii
    directness of distant mental influence, xxxi
    distances spanned, xxxi
    distribution of ability to influence in the population, xxxii
    effect of focusing influence on desired target activity, xxxiii
    factors correlated with ability to influence, xxxiv
    failures of distant mental influence, xxxiv
    focusing of influence, xxxiii
    influence episodes accompanied by distant knowing, xxxiv
    influence separated in time, xxxii
    nontriviality of influence effects, xxxii
    possibility of distant influence, xxxi
    range of bodily and mental activities influenced, xxxii
    reliability and robustness of influence effects, xxxii
    replication of distant mental influence effects, xxxiv
    strategies for successful distant influence, xxxii–xxxiii
    subject concern over idea of influencing or being influenced, xxxiii
    susceptibility to influence based on need, xxxii
findings about transpersonal imagery effects,
    ability to block the influence, 19
    distance of TIE effects, 19
    distribution of TIE effect in population, 19
    effects focused on particular measures, 19
    failures of TIE, 20
    intention sufficient for effect, 19
    nontriviality of TIE phenomenon, 19
    reliability and robustness of TIE phenomenon, 19
    subject concern over being influenced, 19
    subject's unawareness of influence, 19
    susceptibility related to subject's need, 19
"First Impressions of a Psychical Researcher," xvii
flaw-determined psi, 28
floor effects, 35–36
focusing influence on desired target activity, xxxiii, 19
Fodor, Jerry A., xi
forced-choice precognition, 72, 239
foretellings, 229
formal time-displaced direct mental influence experiments, conducted under independent supervision, 238
Foundation for Research on the Nature of Man (FRNM), 30, 43, 78, 199
free variability, xxi
French Academy of Science, 211
FRNM. See Foundation for Research on the Nature of Man
Galton, Francis, 184
ganzfeld procedure, 72, 137
gaps, human need to fill, xxxix
gentle wishing, xxxiii
geomagnetic field (GMF) activity, xxxiii, 99, 142
GESP components, 38–39
Gilbert, Joseph, xxiii, 120, 235
Global Brain, The, 157
GMF. *See* geomagnetic field activity
goal-directed features, xxxix
goal-oriented imagery, xxxiii
goodness of fit tests, chi-square, 11
Grad, Bernard, 213–214
Gregory, C. C. L., 120
group consciousness, focused, xxxv–xxxvi
Gruber, Elmar, 121
Guerra, Valerie, 196–208
Gymnotus carapo, xxix, 90
Hansen, George, 64
Harman, Willis, 228
harmful influences, 139–140
Harris, W. G., xxxv

healing,
    distant, xxxvi–xxxvii
    mental, 78
    psychic, 27
    Reiki, 5, 78
    self, 47
    unorthodox, 78
"healing analogs," 211
Healing Beyond the Body, xiii
healing studies, psi-missing in, 134
health-serving mental influences, xlii, 233–256
Help periods, 198–202, 205
helper's activities, 199–201
helping others concentrate, 28, 196–208
    descriptive analysis of distractions, 205
    interrelationships among measures, 203–205
    methodology, 198–202
    psi scoring of remote helping effect, 202–203
    results, 202–205
hemolysis,
    measurements, 54–55
    rate of, xxxii, 25, 49, 52, 72, 93
    studies of an *in vitro* cellular preparation, 92–93
    trials, 53
Hermetic tradition, xxxviii
hetero-regulation, 131
"hexes," 140
holonomic models, xxxviii
Honorton, Charles, xxvii
human red blood cell effects, xxix–xxx, xxxii
    rate of hemolysis of, xxxii, 25, 49, 52, 72, 93
human welfare, relevance of Braud's research to, xii
hypnotic induction, xx, xxiv, 3, 72, 119, 121, 184
hypotheses, 172–173, 198. *See also* alternative hypotheses; alternative psi hypotheses
identity, xli
ideomotor reactions, studies of, 86–88
IDS. *See* intuitive data sorting
imagery,
    preverbal, 2
    reality of, 134
    transpersonal, 2
Imagery in Healing, 2
imagining, xxxii, 42
immediate interactions, xxxvii
independence of measurements, assumption of, 64
independently significant subjects, score comparison for, 63
individual subjects, scoring rates for, 61
influence, 8–9
    episodes accompanied by distant knowing, xxxiv
    levels of, 132
    separated in time, xxxii
    vs. interactions, 206
influencer/influencee dyad, 100
influencers, 5, 18, 39, 44, 78, 85, 119, 190
    instructions and activities, 7–9
    procedures for, 79–80
    role of feedback to, 217
influencing a distant person's bodily activity using mental imagery, 1–23
influencing factors, 98–102
    physical, 99
    physiological, 99
    psychological, 99–100
    time-displaced effects, 100–102
inhibitory influences, 28
Institut fur Grenzgebiete der Psychologie und Psychohygiene, 98
Institute for Brain Research, 110, 120
Institute of Transpersonal Psychology, 253
"instrumental" conditioning, 143
intention alone sufficient for effect, 19, 141
intentional focus, xxxiii
intentional movement, xxviii
intentionality, xxvi, 122, 210
    conventional, xl–xli

interactions. *See also* direct mental interactions with living systems
  unmediated, xxxvii
  unmitigated, xxxvii
  vs. influence, 206
intercessory prayer, xxxvi–xxxvii
  retroactive, xxxvi
interconnectedness, xviii, xliii–xliv, 154
intermediate preobservation, possible roles of, 249–251
internal rhythms, 15, 225
interrelationships among measures, 203–205
introversion, xxxiv
  scores, 159
  testing, 72
intuition scores, 159
intuitive data sorting (IDS), 21, 50, 52, 55, 57, 63, 66–68, 132, 218–219
Iroquois world view, xliii
Isaacs, Julian, 143
isolation,
  perceptual, 72
  spatial/temporal, 47–48
Ivanov-Smolensky, A. G., 111, 120, 143, 235
James, William, xvii, xxii, xxvii, xxxix–xl, 192
Janet, Pierre, xxiii, 120, 127, 235
Joire, Pierre, xxiii, 120, 235
Journal of Indian Psychology, 109
Journal of Parapsychology, 24, 150, 166
Journal of Religion and Psychical Research, 183
Journal of Scientific Exploration, 1
Journal of the American Society for Psychical Research, 46, 196
judging/perceiving (J/P) dimensions, 159
Jung, Carl, xxxviii, 128
Kaluli tribe, 210
knowing, distant, xxxiv
Kohsen, Anita, 120
Kraftwerk (musical group), 18
Krieger, Dolores, 187
kymography, 111
LaBerge, Stephen, 220
lability, 127–129
  perceived, 129
  physical, 128–129
  relative, 142
lability/inertia model of psychic functioning, xxi, 73–74
laboratory floor plans, 7–8, 56, 75–76, 112–113, 200
Lao Tsu, 118
le sommeil a distance experiments, 119, 121
Leibnitz, Gottfried Wilhelm, xxxviii
LeShan, Lawrence, 188, 191
leukemia, 185
Leyden University, 152
life forms, manipulation of, 133–134
life-lines, families of, 252
living target systems used in distant mental influence
  research, 118–149
  ability to focus effect, 141
  advantages of, 122–133
  appropriateness of, 127–133
  bidirectional effects, 142
  blocking the effect, 141
  choosing other target systems, xxvii–xxx
  converging strategies approach, 136–138
  disadvantages of, 133–136
  distance at which effects occur, 141
  distribution of ability to manifest the effect, 141
  early studies, xxiii–xxx
  effect of focusing on desired target activity, 142
  experimental control, 134–135
  factor of geomagnetic field (GMF) activity, 142
  failure of the effect, 142
  findings, 140–142
  harmful or unwanted influences, 139–140
  intention alone sufficient, 141
  logistical difficulties, 133
  manipulation of life forms, 133–134

motivation, 123
nontriviality of effects, 141
numbers of target systems susceptible to the effect, 141–142
plausibility, 124–127
psi-missing in healing studies, 134
relevance, 122–123
reliability and robustness of distant mental influence, 141
resistance, 136
statistical issues, 135–136
subject need factor, 141
subject's unawareness of effect, 141
successful outcomes, xxv–xxvi
terminology, xxvi–xxvii
time-displaced studies of, 240–244
unconcern by subjects about influence, 141
locomotor behavior of small mammals, 91–92
logistical difficulties, 133
London Hospital, 184
LSD, 129
lymphocytes, as targets, 49
Maddox, John, xi
Maimonides Medical Center, 98
Mann-Whitney *U* tests, 65, 75
Manning, Matthew, 25, 27
matched-pairs, Wilcoxon, 75
MBTI. *See* Myers-Briggs Type Indicator
MCEs. *See* mean chance expectations
McNeill, Katherine, 196–208
mean chance expectations (MCEs), 29, 32–33, 75, 82, 158–159, 173
mean electrodermal activity measured, in calming other
  persons at a distance, 33
measurements, assumed independence of, 64
meditation, 78
  Western and Eastern traditions, 184
mental healing, 78
mental interactions with remote biological systems, 71–108
  conceptual replications, 98
  electrodermal correlates of remote attention, 85–86
  electrodermal influence series, 77–85
  general design considerations, 73–77
  influencing factors, 98–102
  laboratory floor plan, 75–76
  locomotor behavior of small mammals, 91–92
  rival hypotheses, 95–98
  spatial orientation of freely swimming fish, 90–91
  statistical summary, 93–94
  studies of an *in vitro* cellular preparation (hemolysis), 92–93
  studies of blood pressure influence, 89–90
  studies of ideomotor reactions, 86–88
  studies of muscular tremor, 88–89
  uncertainties and reproducibility, 102–103
"mental or behavioral influence of an agent" (MOBIA), 121
mental processes, apprehending, xl
mental strategies of subjects, 65–66
mental suggestion, 3
mentally protecting human red blood cells at a distance, 46–70
  findings, 59–65
  intermediate phase of salinity tests, 51–52
  intuitive data sorting, 67–68
  mental strategies of subjects, 65–66
  methodology, 55–59
  pilot phase, 50–51
  protocols, 52–55
  spatial/temporal isolation, 47–48
mentation, spontaneously reported, 17
Meriones unguiculatus, xxix, 91
mescaline, 129
Mesmer, Franz Anton, 119
meta-analyses, 86, 161, 212
metaphors, in science, 228
meteorites, discovery of, 211
methodologies,
  apparatus, 30, 40, 155, 169

experimental design, 198
hypotheses, 172–173, 198
participants, 198–199
personality assessments, 172
procedure, 6–11, 30–32, 40–41, 55–59, 155–158, 169–172, 199–202
subjects, 5–6, 29–30, 40, 55, 154–155, 168–169
Mind Science Foundation (MSF), 47–48, 140, 155, 168, 199, 253
miracles, explaining, xli
misreadings of data, 15, 34, 96, 226
MOBIA. *See* "mental or behavioral influence of an agent"
mobility dimension, 128
monads, xxxviii
Monte Carlo simulation analyses, 51–52, 54
motivated misreadings of data, 15, 34, 96, 226
motivation, 122, 123
psychology of, 28
motor movements, gross, 236
Motoyama, Hiroshi, 121
MSF. *See* Mind Science Foundation
muscle relaxation, progressive, xx
muscular tremor, studies of, 88–89
Myers, Frederic W. H., 120, 126
Myers-Briggs Type Indicator (MBTI), 66, 128, 155, 159, 163, 168, 172, 175–176, 179–180
introverts, 180–181
mystical traditions, Western and Eastern, 184
Nash, Carroll, 213
National Heart, Lung, and Blood Institute, 84
natural killer (NK) cell activity, 248
need-determined psi, 28
need satisfaction factor, influence on calming other persons at a distance, 28–29
Nelson, Roger, xxxv
neural networks, 130
neurons, 247–248
"critically poised," 130
neuroscience, xi
new physics, xii
NK. *See* natural killer cell activity
noise-reduction hypothesis, xxi
noise-reorganization model, 127
"noncontact therapeutic touch," 187
nonpsi explanations. *See* alternate hypotheses
nonpsi influences, 132
nontriviality of influence effects, xxxii, 19, 141
numbers of studies, 212
observer preknowledge, 226
Ochorowicz, Julian, 120
one-way mirrors, 152
open-loop feedback, xxiv
"operant" conditioning, 143
Orwell, George, 124
"other" groups, *vs.* "own" scores, 62
overall scores and effect sizes, 12
parapsychological research, xix, 110, 163
Parapsychology Foundation, 68
Participant Information Form (PIF), 66, 155
participants, 198–201
Pavlov, Ivan P., 110, 128, 143
PCCP sequence, 53, 64
Pearson *rs*, 174–176
pendulum, "Chevreul," 87
perceived lability, 129
perceiving dimensions, 159
percent influence score, 74
perceptual isolation, 72
personality assessments, 172
phasic skin resistance responses, spontaneous, 43, 154, 157
physical factors, influencing mental interactions with remote biological systems, 99
physical health implications of direct mental influence, 246–249
physical lability, 128–129
physical layouts, 6, 199

Physician's Health Study Research Group, 84
physics,
new, xii
systems and theories of, xix
physiological factors, influencing mental interactions with remote biological systems, 99
physiological measurements, 80–81
chart tracing of spontaneous skin resistance reactions, 81
piezoelectric strain gauges, 134
PIF. *See* Participant Information Form
PK. *See* psychokinesis
placebo effects, 47, 97, 184
Platonov, K. I., 110, 235
plausibility, 122, 124–127
plethysmographic activity, 236
Poe, Edgar Allan, 143
prayer, 184–185
intercessory, xxxvi
pre-feeling studies, xxxv, 233–256
"pre-observation" experiments, 137–138
conscious, 102
precautions,
against conventional communication, external stimuli, and subtle cues, 78
against suggestion, expectancy, placebo effects, and confounding internal rhythms, 79
precognition, xix, xxi
forced-choice, 72
preliminary findings, 19–20
preparers, 10
prerecorded events, 247
presentiment studies, xxxv, 233–256
preverbal imagery, 2
procedures, 30–32, 40–41, 55–59, 155–158, 169–172
influencer's instructions and activities, 7–9
laboratory floor plans, 7–8, 56, 200
meaning of electrophysiological activity, 10–11
participant's and helper's activities, 199–201
physical layouts, 6, 199
psychological assessments, 201–202
scheduling of influence attempts, 9–10
subject's instructions and activities, 6–7
progressive errors, 16
proof-of-principle studies, 217
protection effects, 51
protocols,
blood sample collection, 53
hemolysis measurements, 54–55
hemolysis trials, 53
session sequences, 53
signaling of sessions, 53–54
spectrophotometer cooling, 53
subject instructions, 54
subjects attempting to protect their own blood cells and those of others, 52–53
pseudostaring, 171
psi, 204. *See also* alternative psi hypotheses
flaw-determined, 28
multiple channels for, 130–132
need-determined, 28
scoring of remote helping effect, 202–203
Psi Research Methodology: A Re-examination, 118
psilocybin, 129
"psychic force," 125–127
psychic healing, 27, 211
psychic "helping," xxvii
psychic optimization, xxi
psychical effect, not physiological, 37
psychical research, xix, 110
"psychoboly," 143
psychokinesis (PK), xix, xxii, xxviii
ability to focus, 141
blocking, 141
field-like nature of, 245
influence on equipment, 37–38

influencing through, 25, 27
our failure to study, xlii
our ignorance of, xlii
psychological assessments, 201–202
psychological factors,
  health implications of direct mental influence, 246–249
  influencing mental interactions with remote biological systems, 99–100
  resistance, 136, 143
Psychological Review, 136
psychology, systems and theories of, xix
psychoneuroimmunological principles, 184
Psychophysical Research Laboratories, 155
Puysegur, Marquis de, 119
quantitative summary, 12
quantum mechanical model, of PK, 127
quasiphysical energy, 191
RA predictions, 52, 55, 57, 63, 66–68
Radin, Dean, xxxv
RAND table of random numbers, 53–54, 56
random event generators (REGs),
  electronic, xxii, 100–101, 235
  sessions with, 133, 137–138
randomness, 73, 79
rapidity of influence, 49
reactions to an unseen gaze, 150–165
"real-time PK,"
  evidence for, 233–236
recalcitrant tissue, xxix
recording errors, 15, 34, 96, 226
red blood cells. *See* human red blood cell effects
reflexes, conditional, 110
REGs. *See* random event generators
Reichenbach, Hans, xviii
Reiki healing practitioners, 5, 78
  work with, 218
relaxation, xx
relevance, 122–123
reliability, 210
  of distant mental influence, 141
  of influence effects, xxxii
  of TIE phenomenon, 19
remote attention, 150–165
  electrodermal correlates of, 85–86
remote healing,
  analog studies, 188–189
remote mental influence, xxvii
remote observation experiments, 219–221
  autonomic nervous system activity level, 220
  replication studies, 221
  skin conductance response level, 220–221
remote staring detection research, xxx–xliv, 166–182
  caveats, xxxix–xl
  conventional intentionality, xl–xli
  direct mental influence in daily life, xli
  direct mental influences within the body, xli
  distant healing and intercessory prayer, xxxvi–xxxvii
  dyadic interactions, xlii
  explaining miracles, xli
  explanations and interpretations, xxxvii–xxxix
  facilitating health and healing, xlii
  focused group consciousness, xxxv–xxxvi
  frequency of occurrences, xlii
  harmful uses of, xlii
  holonomic or correspondence models, xxxviii
  human potential, xli
  implications and possible applications, xl–xlii
  intention creating spiritual experience, xli
  interconnectedness, xliii–xliv
  interpretations of, 179–181
  larger contexts, xli
  limits to distant mental influence, xli
  major findings by Braud, xxxi–xxxiv
  methodology, 168–173
  nature and role of consciousness, xli–xlii

nature of time, xlii
  our failure to study such effects, xlii
  our ignorance of these effects, xlii
presentiment (pre-feeling) studies, xxxv, 233–256
reorganization models, xxxviii
replications, extensions, and related research by others, xxxiv–xxxvii, 71–108, 209–232
results, 173–177
selfhood and identity, xli
survival of physical death, xlii
transgenerational effects, xlii
transmission models, xxxvii–xxxviii
reorganization models, xxxviii
replication of distant mental influence effects, xxxiv, 116, 212–213, 219, 221, 245–246
  conceptual, 98
  extensions and related research by others, xxxiv–xxxvii, 71–108, 209–232
reproducibility, 102–103
resistance, 136
rest periods, 216
retro-PK design, 47, 236–239
retroactive intercessory prayer, xxxvi
Rhine, J. B., 125–126
Rhine, Louisa, 125
Richet, Charles, xxiii, 120, 143, 235
robustness,
  of distant mental influence, 141
  of influence effects, xxxii
  of TIE phenomenon, 19
Rogo, D. Scott, 125
Roll's "systems theoretical" model, 133
Rumi, Jelaluddin, xvii, xxvii
Russell, Peter, 157
SAD. *See* Social Avoidance and Distress scale
salinity tests, intermediate phase of, 51–52
sample of polygraph tracing of electrodermal activity, 14
San Bernardino State College, 98
Sanskrit sayings, xliv
scheduling of influence attempts, 9–10
Schlitz, Marilyn J., 1–45, 71–108, 209–232
Schmidt, Helmut, xxii–xxiii, 100–101, 196, 236–238
Schopenhauer, Arthur, 253
Schrödinger, Erwin, xlii
score comparison, for independently significant subjects, 63
scoring of measurements, 81
scoring rates, for individual subjects, 61
Searle, John, xi
"second signaling system," 143
self-control,
  autonomic, 41, 130
  imagery effect, 15, 25
self-exploration, 78
self-healing, 47
self-modulation of effect magnitude, 218
self-regulation procedures, xxxii, 131, 184
self-reporting, 172
selfhood, xli
"semantic" conditioning, 143
sensation, community of, 3
sensing/intuition (S/N) scores, 159
sensorimotor cues, 15, 95
session, sequences, 53
sessions, signaling, 53–54
Severtsev Institute of Evolutionary Morphology and Animal Ecology, 98
Shafer, Donna, 150–182, 196–208
sham periods, 171–172, 176, 223
shaman, consulting, 210
Sheldrake, Rupert, xxxix
Sicher, F., xxxv–xxxvi
signal detection theory, 153
Simpson, O. J., murder trial, jury verdict televised, xxxv
Sinā, Ibn, 1
Sixth Psychological Congress, 127

skin conductance responses, 220–221
skin resistance responses (SRRs), 10–11, 14, 80–81,
    112–113, 159, 171, 190
  amplitude of, 162
  spontaneous phasic, 154, 157
Skinner, B. F., 143
social avoidance, persons with greater, 179–180
Social Avoidance and Distress (SAD) scale, 172, 175–176,
    179–181
somatic effects, of imagery, 3
somnambules, 119
Southern Medical Journal, 185
spatial orientation, of freely swimming fish, 90–91
spatial/temporal isolation, 47–48
specificity or generality of effect, 49, 218
spectrophotometer cooling, 53
spectrophotometry, xxx, 51, 53, 56–57, 93
spiritual experience, intention creating, xli
spiritual traditions, Western and Eastern, 184
spontaneous phasic skin resistance responses, 154, 157
SRI International, 63, 68
SRRs. See skin resistance responses
standard deviations, 61
Stanford, Rex, 137
Stanford University, 151
Stapp, Henry P., xii
staring, into eyes, 157
statistical issues, 135–136, 212
Stouffer method, 13, 82, 115, 160, 222, 244
strain gauges, piezoelectric, 134
strategies for successful distant influence, xxxii–xxxiii
"stream of consciousness," xxii
strength dimension, 128
stress reduction, 184
subjects, 5–6, 29–30, 40, 55, 154–155, 168–169. See targets
  activities of, 6–7
  attempting to protect their own blood cells and those
    of others, 52–53
  awareness of effect, 19, 141
  concern over idea of influencing or being influenced, xxxiii, 19
  foreknowledge by, 15–16, 34–35, 96–97
  fraud by, 16–17, 36, 97
  independently significant, score comparison for, 63
  instructions to, 6–7, 54
  mental strategies of, 65–66
  "need" factor, 141, 217
  procedures for, 79
  scoring rates for individual, 61
Subtle Energies and Energy Medicine: An Interdisciplinary
    Journal of Energetic and Informational Interactions, 71
successful outcomes, xxv–xxvi
suggestion, 3, 184
survival of physical death, xlii
susceptibility, 137
  related to subject's need, xxxii, 19
sympathetic nervous system activation, heightening, 178
systematic error, 16, 35, 97
"systems theoretical" model, 133
taboo-violation, 140
Targ, Russell, 192
"target persons," 121, 131, 139, 236
targets, 119
  choosing by chance and by necessity, xxvii–xxx
  phylogenetic and ontogenetic status of, 131–132, 137, 236
  special, 120
  susceptibility to the effect, 141–142
tat tvam asi, xliv
"telargy," 121
teleological action, 239
"telepathic hypnotization," 121
telepathy, xix–xxi, 38
  "at a distance," 121
terminology,
  allobiofeedback, xxvi
  bio-PK, xxvi

direct mental influence, xxvii, xxxi
direct mental interactions with living systems
    (DMILS), xxvii
distant mental influence, xxvi
remote mental influence, xxvii
transpersonal imagery effect, xxvi
therapeutic touch, 5, 78, 187–188
  "noncontact," 187
thinking/feeling (T/F) scores, 159
TIE. See transpersonal imagery effects (TIE)
time, nature of, xlii
time-based errors, 16
time-displaced effects, xxxii, 47, 234, 237–239, 249
  health implications of, 239–240
  influencing mental interactions with remote biological
    systems, 100–102
  in living systems, 240–244
  possibility of, 236–238
timing changes, 67
Titanic, sinking of, xxxv
Titchener, E. B., 151
tonic skin resistance levels, 43
Townshend, Chauncey Hare, 119
transgenerational effects, xlii
transmission models, xxxvii–xxxviii
transpersonal imagery effects (TIE), xxvi, 122
  alternate hypotheses, 15–17
  distribution in the population, 19
  implications and applications, 20
  influencing a distant person's bodily activity using men-
    tal imagery, 1–23
  methodology, 5–11
  overall scores and effect sizes, 12
  preliminary findings, 19–20
  quantitative summary, 12
  range of reactions, 17–18
  results, 11–15
  sample of polygraph tracing of electrodermal activity, 14
  summary statistics, 14
triple-blind clinical trials, 185
"true psi locus," 131
unconcern by subjects about influence, 141
unconscious, xxiv, 167
"universal fluid," 119
University of Adelaide, 153
University of Edinburgh, 253
University of Groningen, 120
University of Leningrad, 120
  Institute for Brain Research, 110, 120
unmediated interactions, xxxvii
unmitigated interactions, xxxvii
unorthodox healing, 78
unwanted influences, 139–140
Utts, Jessica, 65
van Dam, A. S., 120
Vasiliev, Leonid L., xxiii, 110–111, 115–116, 235
voluntary action, xxviii
Walker, Evan Harris, 125, 238
Watts, Alan, xxxix
web sites, xxxvi
well-being, and direct mental influence, 246–249
Western meditative, mystical, and spiritual traditions, 184
Wilcoxon matched-pairs, 75
will, action of, 125
Wiseman, Richard, 221
wishing, gentle, xxxiii
Wolfe, Thomas, xliii
"working capacity," of cerebral cells, 128
world-lines, families of, 252
World Trade Center attacks, xxxvi

Hampton Roads Publishing Company

. . . for the evolving human spirit

Hampton Roads Publishing Company
publishes books on a variety of subjects,
including metaphysics, health,
visionary fiction, and other related topics.

For a copy of our latest catalog, call toll-free
(800) 766-8009, or send your name and address to:

Hampton Roads Publishing Company, Inc.
1125 Stoney Ridge Road
Charlottesville, VA 22902

e-mail: hrpc@hrpub.com
www.hrpub.com